VOLUME SIXTY

ADVANCES IN CHEMICAL ENGINEERING

Towards Circular Economy: Closing the Loop with Chemical Recycling of Solid Plastic Waste

ADVANCES IN CHEMICAL ENGINEERING

Serial Editor

DAVIDE MOSCATELLI

Department of Chemistry,
Materials and Chemical Engineering "Giulio Natta"
Politecnico di Milano, Milano, Italy

VOLUME SIXTY

ADVANCES IN CHEMICAL ENGINEERING

Towards Circular Economy: Closing the Loop with Chemical Recycling of Solid Plastic Waste

Edited by

DAVIDE MOSCATELLI

Department of Chemistry,
Materials and Chemical Engineering "Giulio Natta"
Politecnico di Milano, Milano, Italy

MATTEO PELUCCHI

CRECK Modeling Lab, Department of Chemistry,
Materials and Chemical Engineering "Giulio Natta",
Politecnico di Milano, Milano, Italy

ELSEVIER

ACADEMIC PRESS

An imprint of Elsevier

Academic Press is an imprint of Elsevier
50 Hampshire Street, 5th Floor, Cambridge, MA 02139, United States
525 B Street, Suite 1650, San Diego, CA 92101, United States
The Boulevard, Langford Lane, Kidlington, Oxford OX5 1GB, United Kingdom
125 London Wall, London, EC2Y 5AS, United Kingdom

First edition 2022

ISBN: 978-0-323-95770-0
ISSN: 0065-2377

For information on all Academic Press publications
visit our website at https://www.elsevier.com/books-and-journals

Publisher: Zoe Kruze
Acquisitions Editor: Mariana L. Kuhl
Developmental Editor: Federico Paulo S. Mendoza
Production Project Manager: Vijayaraj Purushothaman
Cover Designer: Mark Rogers

Typeset by STRAIVE, India

Contents

Contributors

Konstantinos Alexopoulos
Department of Chemical Engineering, The Pennsylvania State University, University Park, PA, United States

Sibel Altinok
Organik Kimya, Kemerburgaz Cd. 146, Istanbul, Turkey

Pahola Thathiana Benavides
Energy Systems and Infrastructure Analysis Division, Argonne National Laboratory, Lemont, IL, United States

Ozge Deniz Bozkurt
Department of Chemical Engineering, The Pennsylvania State University, University Park, PA, United States

Lorenzo Brivio
While True s.r.l., Urgnano; Department of Chemistry, Materials, and Chemical Engineering "Giulio Natta", Politecnico di Milano, Milan, Italy

Angelo Ferrando
Basic Chemical & Plastics Research Centre, Versalis, Mantua, Italy

Matteo Ferri
Laboratory of Catalysis and Catalytic Processes, Dipartimento di Energia, Politecnico di Milano, Milano, Italy

Ulises R. Gracida-Alvarez
Energy Systems and Infrastructure Analysis Division, Argonne National Laboratory, Lemont, IL, United States

Ankush Jain
Department of Chemical Engineering, Indian Institute of Technology Madras, Chennai, India

Varaha Jayarama Krishna Jonnalagedda
Department of Energy and Mineral Engineering, The Pennsylvania State University, University Park, PA, United States

Andrea Locaspi
CRECK Modeling Lab, Department of Chemistry, Materials and Chemical Engineering "Giulio Natta", Politecnico di Milano, Milano, Italy

Matteo Maestri
Laboratory of Catalysis and Catalytic Processes, Dipartimento di Energia, Politecnico di Milano, Milano, Italy

Davide Moscatelli
Department of Chemistry, Materials and Chemical Engineering, Politecnico di Milano, Milan, Italy

Mirco Nodari
Basic Chemical & Plastics Research Centre, Versalis, Mantua, Italy

Sean Timothy Okonsky
Department of Chemical Engineering, The Pennsylvania State University, University Park, PA, United States

Matteo Pelucchi
CRECK Modeling Lab, Department of Chemistry, Materials and Chemical Engineering "Giulio Natta", Politecnico di Milano, Milano, Italy

Barbara Alejandra Perez
Department of Chemical Engineering, The Pennsylvania State University, University Park, PA, United States

Sean Riggio
Applied Materials Division, Argonne National Laboratory, Lemont, IL, United States

Francesco Serse
CRECK Modeling Lab, Department of Chemistry, Materials and Chemical Engineering "Giulio Natta", Politecnico di Milano, Milano, Italy

Mattia Sponchioni
Department of Chemistry, Materials and Chemical Engineering, Politecnico di Milano, Milano, Italy

Flavio Tollini
Department of Chemistry, Materials, and Chemical Engineering "Giulio Natta", Politecnico di Milano, Milan, Italy

Hilal Ezgi Toraman
Department of Chemical Engineering; Department of Energy and Mineral Engineering; Institutes of Energy and the Environment, The Pennsylvania State University, University Park, PA, United States

Meltem Urgun-Demirtas
Applied Materials Division, Argonne National Laboratory, Lemont, IL, United States

Ravikrishnan Vinu
Department of Chemical Engineering, Indian Institute of Technology Madras, Chennai, India

Preface

Chemical recycling of solid plastic waste (SPW) through catalytic and noncatalytic technologies is gaining an exceptional interest at both the industrial and academic levels. Chemical recycling is a process in which a polymer is converted into chemical building blocks, including monomers, which can be reused as raw materials in chemical processes (*Cefic, 2022. Chemical recycling: Enabling plastic waste to become a valuable resource. Position paper. Cefic, the European Chemical Industry Council*). Compared to landfilling, disposal, and incineration, chemical recycling enables the implementation of a virtuous circle from the perspective of sustainable waste management, circular economy, and decarbonization.

This volume is made up of contributions from academic and industrial experts ranging from molecular-scale to reactor-scale and process aspects, thus providing an exhaustive overview of the state-of-the-art and current understanding of the different chemical recycling technologies and their potentials.

In Chapter 1, Ferrando and Nodari (Versalis S.p.A., Italy) address the important topic of SPW feedstock characterization. Indeed, the strong variability and heterogeneity of SPW mixtures can strongly influence feedstock conversion, product yields, and downstream upgrading. The available experimental characterization techniques are comprehensively reviewed, touching upon infrared spectroscopy and thermal and elemental analysis techniques. Specific attention is given to the determination of ash, metal, and halogen content due to their strong detrimental impact on both chemical reactor operation and downstream upgrading.

In Chapter 2, Locaspi et al. (Politecnico di Milano, Italy) review the state-of-the-art kinetic modeling approaches ranging from the molecular-scale understanding of the chemistry of both thermal and thermocatalytic conversions to the design, assembly, and validation of comprehensive detailed and semidetailed kinetic models. The use of such models in simplified reactors allows to parametrically explore the impact of process variables such as temperature, pressure, residence time, and catalyst on feedstock conversion and product yields. Upon ad hoc model reduction, these models can also be used for simulations of real reactors, coupling chemical kinetics, heat, and mass transfer phenomena.

In Chapter 3, Vinu and coworkers (Madras Institute of Technology, India) present an overview of laboratory-scale measuring devices and diagnostics that are suitable for kinetic studies of SPW pyrolysis. Within those, pyroprobes and micropyrolyzers have extended capabilities to analyze the time evolution of pyrolysate composition in a seconds time frame and to obtain nonisothermal kinetics using the evolved gas analysis functions. The potential of pulse-heated analysis of solid reactions to obtain product evolution data in a milliseconds time frame is also discussed. This chapter discusses the perspectives for obtaining kinetic data under heat and mass transport-free conditions, paving the way for a mechanistic understanding of polymer pyrolysis.

In Chapter 4, Bozkurt et al. (Penn State University, USA) discuss the structure–performance relationships in catalytic processes that are directly or indirectly related to chemical recycling. This discussion has been based on recent experimental and computational research, assessing the proposed catalytic reaction mechanisms and overviewing from pilot to early commercial-scale catalytic chemical recycling reactors, ultimately providing projections for future directions.

In Chapter 5, Perez et al. (Penn State University, USA) provide an overview of the available experimental techniques to characterize the physicochemical properties of oil streams from thermal plastic pyrolysis. Plastic pyrolysis oil can indeed be used as fuel for many industrial applications such as turbines, boilers, and furnaces. Furthermore, upon upgrading, SPW pyrolysis oils can also serve as a drop-in alternative to commercial transportation fuels like gasoline and diesel. To these aims, together with their composition, many physicochemical properties need to be accurately determined, including heating value, viscosity, density, water content, flash point, fire point, and pour point.

In Chapter 6, Brivio and Tollini (Politecnico di Milano, Italy) focus on specific recycling technologies of an extremely diffused polymer in the food packaging industry: polyethylene terephthalate (PET). An extremely broad spectrum of the available technologies for PET recycling is presented, starting from the currently widespread mechanical process to the most promising chemical recycling solutions in the context of a circular economy. Such solutions include glycolysis, methanolysis, hydrolysis, aminolysis, ammonolysis, and enzymatic depolymerization.

In Chapter 7, Sponchioni (Politecnico di Milano, Italy) and Altinok (Organik Kimya, Turkey) review market trends and potential recycling technologies for another key polymer: polymethyl methacrylate (PMMA).

Only 10% of the currently produced PMMA is recycled, with landfilling and incineration dominating its end-of-life management. With market projections foreseeing a growth of PMMA production by ~3.4% from the current 3.9 billion metric tons per year, urgent concerns are emerging for PMMA's end-of-life management. The possibility of quantitatively depolymerizing PMMA back to methyl methacrylate is thoroughly discussed in this chapter, highlighting the clear advantages of thermal pyrolysis. To this aim, fluidized bed reactors and reactive extruders seem to be the most appealing solutions in terms of environmental footprint and efficiency.

In Chapter 8, Urgun-Demirtas (Argonne National Laboratory, USA) reports an interesting example of the application of life cycle analysis (LCA) and material flow analysis (MFA) to evaluate the degree of circularity and assess its effects on different application levels, specifically focusing on lubricant oil production from SPW. Although more diffused LCA techniques assess the environmental impact of plastic production and recycling, MFA is used to evaluate how the new circular economy approach influences the supply chain and downstream markets of new products generated from plastic wastes. The combined outcomes of LCA and MFA provide a guiding framework for a consistent and transparent evaluation of the recycling technologies, accounting for multiple driving forces such as cost, environmental benefits, availability of collection/logistics systems, market conditions, policies and regulations, social changes, and acceptance.

Davide Moscatelli and Matteo Pelucchi
Department of Chemistry, Materials and Chemical Engineering
"Giulio Natta", Politecnico di Milano, Milan, Italy

CHAPTER ONE

SPW feedstock characterization

Angelo Ferrando* **and Mirco Nodari**
Basic Chemical & Plastics Research Centre, Versalis, Mantua, Italy
*Corresponding author: e-mail address: angelo.ferrando@versalis.eni.com

Contents

Abstract

Plastic materials are of fundamental importance in the realization of manufactured articles of all kinds, and in multiple application areas. The main problem with these materials is their fate at the end of their life. Dispersion in the environment, or even solutions such as landfilling or use for energy generation should be avoided or overcome by recycling processes. Mechanical recycling, apart from the technological development of selection processes, has as its main limitation the fact that the second raw material obtained has generally inferior characteristics to virgin polymers, and therefore obliges to uses of lower value. Chemical recycling makes it possible to obtain the starting monomers—directly or through subsequent transformation processes—from plastic waste meant as raw material. Polymers obtained through the conversion of waste plastics are essentially indistinguishable in final properties from those obtained through the conversion of fossil-derived raw materials. Conversion plants of solid plastic waste (SPW) use a very heterogeneous raw material, whose characteristics must be studied taking into account both their heterogeneity and the conversion process, i.e., obtaining those compositional information necessary and useful to conduct the process in an optimal way both from the point of view of operational safety and of the final quality of the converted product. In the following paragraphs the main analytical techniques to obtain the most interesting compositional information are presented.

Advances in Chemical Engineering, Volume 60
ISSN 0065-2377
https://doi.org/10.1016/bs.ache.2022.09.001

Abbreviation

SPW solid plastic waste

1. Introduction

Plastic materials represent a fundamental share in the realization of many of the products of common use, both durable and of short use such as packaging. The increasingly widespread problem with the greatest impact on public opinion is the end-of-life solution for these materials. Unfortunately, the share of material that is destroyed by incineration or worse landfilled or dispersed in the environment is still significant. An updated view of the problem is represented by the scheme in Fig. 1, as reported in a recent study (Hundertmark et al., 2018a).

The chemical industry is now seriously considering the problem, moving from a disposable approach to a logic of circular economy, through various post-use material recycling solutions. Fig. 2, again from a study by McKinsey (Hundertmark et al., 2018b), schematically represents how the recycling of end-of-life plastics can—at various levels—be integrated into a circular logic for the production process.

With respect to the solutions generically schematized in Fig. 2, it is appropriate to make some clarifications.

Mechanical recycling: through a "positive" separation through optical sensors (usually NEAR INFRARED sensors), it is possible to separate some

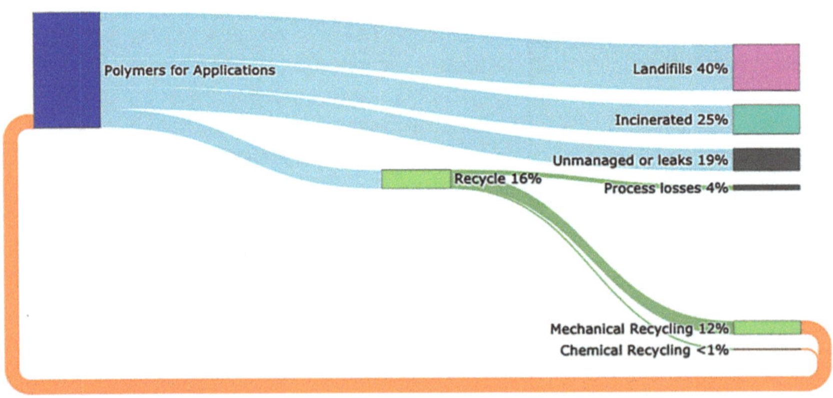

Fig. 1 The majority of plastics waste currently goes to landfills and incineration.

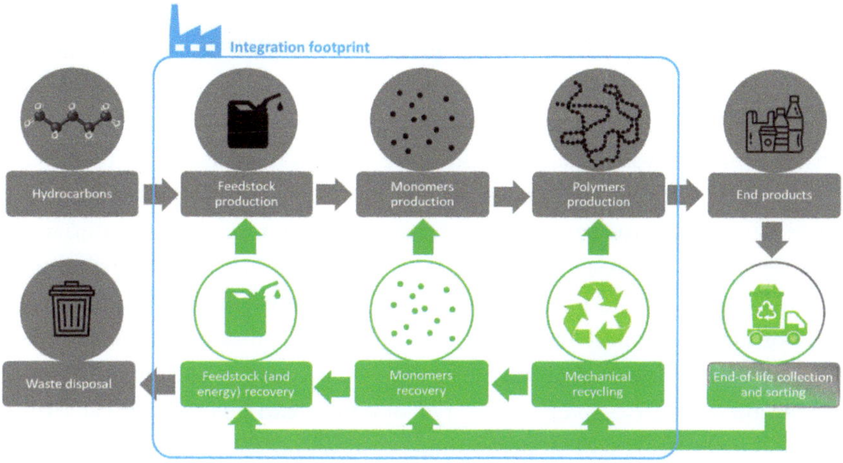

Fig. 2 How plastics waste could feed into an integrated petrochemicals complex.

types of materials for almost direct reuse. In the case of urban plastic waste, usually related to packaging, it is usually possible to separate the bottles in polyethylene terephthalate (PET)—transparent and colored—and high-density polyethylene (HDPE). A similar separation is made with regard to packaging in the form of films.

In the case of waste from other selected supply chains, such as waste from Waste Electrical and Electronic Equipment (WEEE), or from End-of-Life Vehicles (ELV), it is possible to separate selected materials such as styrenic materials (from WEEE), or styrenic and polyolefin-based materials, in the second case. In any case, materials consisting of a single polymer are destined for mechanical recycling, which can therefore be reduced to flakes or granulates and sent directly for reuse. The main limitation of this approach is that post-use polymers generally have lower performance qualities than virgin materials, therefore, they are usually destined for uses of lower value. The characterization of these materials is substantially similar to what is applied to virgin materials, due to the final use.

Monomer recycling: the approach involves, through a chemical process—usually acidic or basic hydrolysis—the reduction of the polymer into its starting monomer(s). This is a technology that is basically applicable only to condensation-type polymers, such as PET or other polyesters, or polyamides. Obviously, it is applicable only to materials already pre-selected, and which, however, are not suitable for their performance characteristics directly for reuse as they are. The characterization of these materials is mainly oriented to the determination of the polymer content and of any impurities present.

Feedstock recycling: it consists in the conversion, usually through thermal or thermo-catalytic pyrolysis processes, of the waste into reusable hydrocarbon fractions for the production of monomers and, therefore, virgin polymers. It represents the most interesting solution for the fraction of indistinct plastic waste, i.e., consisting of various materials, often not separable as they are constituents of polylaminate products, or of mixtures of materials. In this case, given the heterogeneous nature of the material, the characterization approach is aimed at evaluating the main compositional parameters, in order to guarantee processability in pyrolysis, avoiding problems both in terms of operational issues and final quality of the hydrocarbon mixtures obtainable. In the case of urban plastic waste, the mixed fraction that can be used for conversion to hydrocarbons can be estimated at around 30% of the total. Obviously, quantitative differences, even significant ones, can be found both in terms of seasonality and regionality.

2. SPW feedstock

The material destined for conversion to hydrocarbons is substantially the mixed fraction of the plastic waste collected. Its composition is very variable, but it can still be assumed that the main components present are those usually used to produce packaging, both food and other types.

Plastic Europe, the association of European plastics manufacturers, publishes the report "The Fact" (Plastics Europe, 2021) every year, in which, among other information, it reports the distribution of demand for materials, broken down by polymer and by sector of use.

The clear prevalence of PET and polyolefins is very evident with regard to packaging sector. These will then be the polymers that mainly constitute the plastic fraction present in municipal solid waste. As mentioned, a large part of PET and HDPE are recovered through selection and destined for mechanical recycling. The indistinct mixture therefore remains, still consisting mainly of polyolefins (about 70–75%), and therefore of residual PET (about 5–10%), PS (5–10%), PVC (<2%), other polymers and inert materials (about 5–10%, consisting mainly of paper, metals, organic residues, other).

As will be discussed in more detail in the following chapters, the pyrolysis process is particularly efficient when applied to polyolefins. These materials in particular make it possible to obtain mixtures of hydrocarbons consisting mainly of paraffins, isoparaffins, monounsaturated olefins, in some cases minor fractions of naphthenes. The absence of heteroatoms avoids the formation of oxygenated and/or nitrogen compounds; these substances can form potentially corrosive substances for the plant and are in any case considered impurities for the final products. Similar topics, indeed, potentially more worrying, are the presence of halogens (chlorine in particular, due to PVC): the formation of halogenhydric acids (HCl, HBr) must be managed with particular attention to guarantee the integrity of the plant assets. The presence of organic halogen compounds in the final hydrocarbon products is just problematic and to be avoided. Aromatic-based polymers such as PS lead to an increase in the fraction of aromatic compounds in the final products. Moreover, aromatic compounds can also act as precursors in the formation of solid residues—responsible for coking phenomena, which reduce the thermal efficiency of plant assets—and in any case of solid carbonaceous fractions, generally little or no exploitable at all. As previously mentioned, plastic waste destined for pyrolysis usually has a clear compositional prevalence in terms of polyolefins, usually present in mass percentages greater than 70%. The characterization of these materials will therefore be oriented to determine the content in terms of the main polymers present, i.e., first of all polyolefins (due to the presence of HDPE, LDPE, LLDPE, PP), then PET, PS, PVC and other polymers. Equally important will be the determination of the other impurities possibly present, such as humidity, metals, paper, aggregates. Obviously, the widely variable and heterogeneous nature of solid plastic waste must always be remembered; thus, also the analytical determinations need to be carried out managing an adequate compromise between the level of detail of the information and the measurement approximation.

3. Composition by infrared spectroscopy

The techniques of characterization of polymeric materials based on infrared spectroscopy represent a particularly useful solution for defining the composition of the SPW, at least in its main components. In reality, there are two similar solutions, but with specific peculiarities both in terms of approach in setting up, and of practical utility: they are in fact the mid-infrared spectroscopy, generally referred to as Fourier Transform Infrared (FTIR), and spectroscopy in the near infrared (NIR). Table 1 shows the most significant differences between the two techniques.

As is known, calibration in the case of FTIR consists first of all in identifying a spectral band that detects the material whose quantity is being evaluated; the area or height of this spectral signal, as recorded in the IR

Table 1 Comparison between the main characteristics of FTIR and NIR techniques.

	Spectral range [cm^{-1}]	Calibration	Number of calibration samples	Average weight of sample [g]
FTIR	4000–200	Lambert Beer law, monovariate	10–20	0.1–1
NIR	13,000–4000	Statistical, multivariate	100–200	10–100

absorption spectrum, will be directly proportional to the quantity of the relative polymer, through a proportionality constant "ε," defined as molar and optical path extinction coefficient.

Calibration is therefore not too demanding, as in general it does not require a large number of samples. Moreover, the technique is very fast—the analysis takes a few minutes—and usually no particular manipulations of the sample are necessary. The main limitation—in the case of the SPW analysis—of the FTIR lies in the physical form and the total amount of material to be analyzed.

The acquisition of the spectrum can take place by transmission or by reflection. In the first case, it is necessary to create a thin film (fractions of a millimeter) which represents a problem if the SPW is not extruded, but in the form of flakes: each individual flake can have its own specific composition, therefore, to use this approach in these cases it is first necessary to homogenize—as far as possible—the material under analysis, e.g., by extrusion.

A good alternative to the acquisition of the transmission spectrum is the use of the Attenuated Total Reflection (ATR) technique, which does not require you to make an adequate film first. Moreover, even with the ATR the problem of the specific and variable composition of the single flakes remains, and can be solved once again by homogenization by extrusion, or alternatively through an average obtained from many analyzes, repeated on many different flakes of material (at least about 10).

However, since the technique analyzes a very modest quantity of material (a few grams maximum, in the case of repeated analyzes), this represents a limitation considering the high heterogeneity of the SPW. So, the results obtained must be understood with an important degree of approximation (at least ±10%).

The Fig. 3 shows the FTIR spectra of the main polymers present in the SPW (see also Calero et al., 2018 and Kuptsov and Zhizhin, 1998). The most

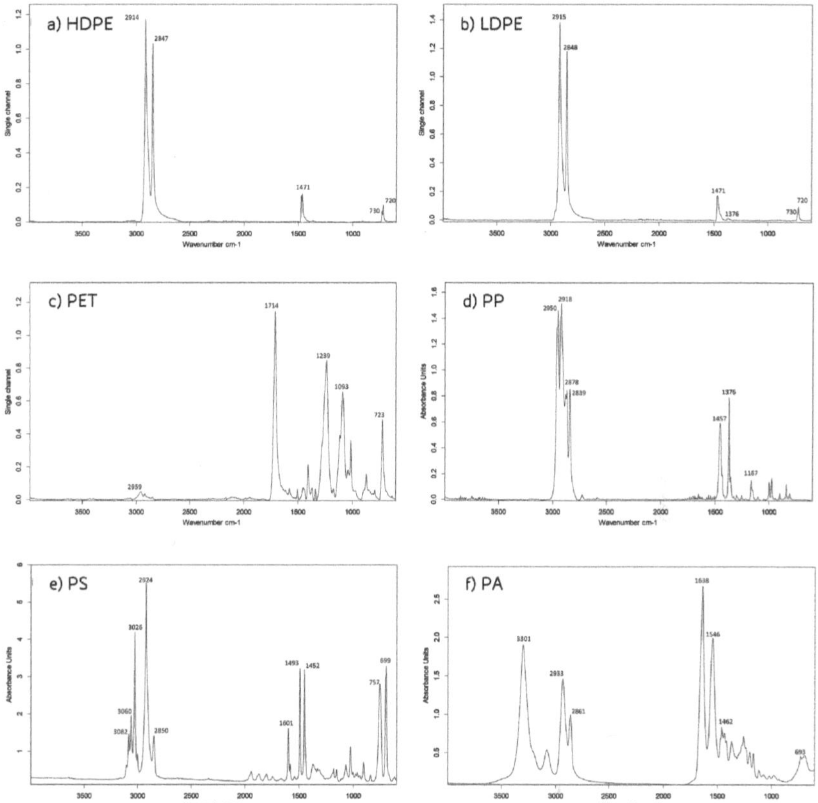

Fig. 3 FTIR spectra of pure (A) High Density Polyethylene, (B) Low Density Polyethylene, (C) PolyEthyleneTerephthalate, (D) PolyPropylene, (E) PolyStyrene and (F) PolyAmide. *Data from Versalis' spectra database.*

significant absorption bands are indicated, which can be used for the quantification of the individual polymers present in the mixture. The spectrum of the SPW consists of the superposition, due to the quantitative presence, of the spectra of pure polymers. The resulting spectrum will be as complicated as the number of polymeric species increases. In some cases, it may not be easy to recognize the presence of all the polymeric species present, since if the relative quantity is too low ($<1\%$), the specific absorption bands may not always be clearly visible and above all well measured.

The use of NIR spectroscopy involves some significant differences. First of all, the spectra obtained in the near infrared spectral region are related to transitions relative to second and third overtones, or to their combination bands. This implies that the signals are very broadened, and the sensitivity

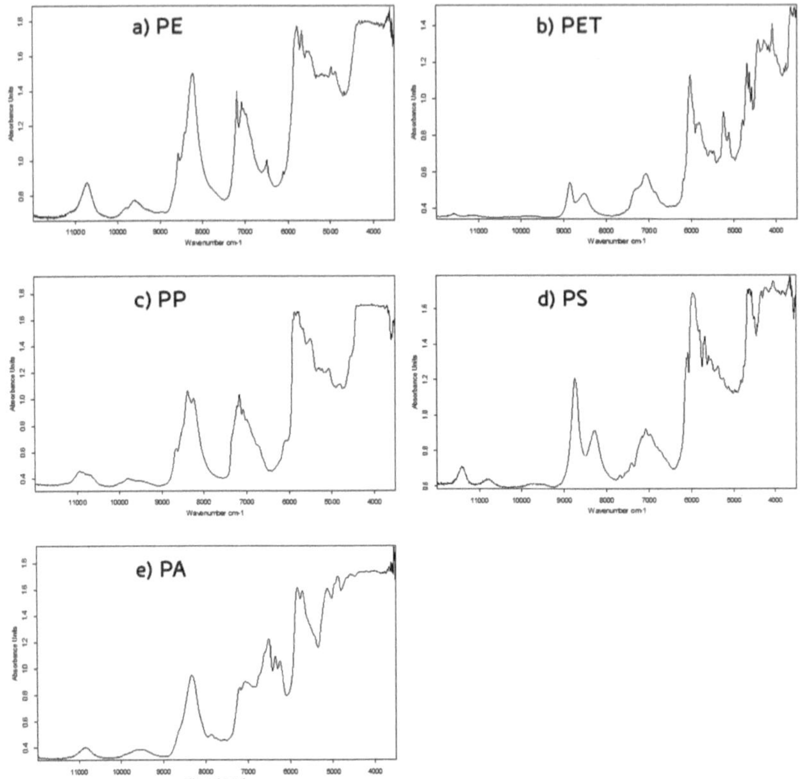

Fig. 4 NIR absorption spectra of pure (A) PE (Polyethylene), (B) PET (PolyEthyleneTerephthalate), (C) PP (PolyPropylene), (D) PS (PolyStyrene) and (E) PA (PolyAmide). *Data from Versalis' spectra database.*

of the technique is lower than the FTIR. Fig. 4 shows the NIR absorption spectra of pure polymers, which are clearly much less diagnostic than FTIR analogs.

Given the nature of the spectra of pure polymers—very broad and poorly defined bands – the calibration approach in the case of NIR spectroscopy uses chemometric analysis techniques, in particular the Partial Least Square (PLS) algorithm is generally used (Ciurczak et al., 2021). In this case, a large number of samples is required, usually about a few dozen at least, which represent as completely as possible the compositional ranges that will be studied. The approach involves the preparation of mixtures of pure polymers, with a known composition, of which the NIR spectrum must be recorded. Calibration is based on a large number of spectral bands, from several tens to hundreds, depending on the case and the spectral resolution of the instrument used. The development of the technique is therefore rather long and demanding, as regards the collection and preparation of calibration samples. On the other hand, by using the acquisition of the spectra of the samples being analyzed by reflection and subsequent collection of the signal through an integrating sphere, it is possible to analyze significant quantities of material, up to 100 g, in the most varied physical form—both flakes and granules. This aspect is certainly the most up-to-date advantage of the technique with regard to the specific application to particularly heterogeneous samples such as SPW. The analysis of important fractions of material allows to completely overcome the lower sensitivity of the technique compared to the FTIR technique; on the other hand, the resulting composition represents a significant and very realistic average of the actual quality of the material.

4. Composition by thermal analysis

Thermal analysis is widely used in the field of polymeric materials, in particular for the characterization of the latter in terms of performance for the application sector they are aimed. Consequently, single polymers' behavior in both thermogravimetric analysis (TGA) and differential scanning calorimetry (DSC) is well known in material science. Starting from this principle, lot of work has been done on characterization of polymer mixtures, as well, in order to evaluate and quantify the contribution of the various components in such mixtures. Nevertheless, this approach can be extended also to solid plastic waste with the aim of gaining as much information as possible about the actual composition of such variable and undefined mixtures.

As regards thermogravimetric analysis, both isothermal and non-isothermal methods can be performed. Kinetic data derived from TGA are conveniently used to determinate degradation profile of the mixture, inferring type and relative abundance of its constituents (Murichan and Cherntongchai, 2014). This is relatively easy to do in the case of simple mixtures and/or for mixtures in which single components have pretty different degradation behavior (see example in Fig. 5, for PVC-PET-HDPE pure components thermodegradation; see also Heikkinen and Spliethoff, 2003). On the contrary, if single components show similar TGA curves, then the behavior of their mixture will be difficult to analyze, i.e., deconvolving peaks into single constituents' contributions will be fairly inaccurate. In this case, TGA can still

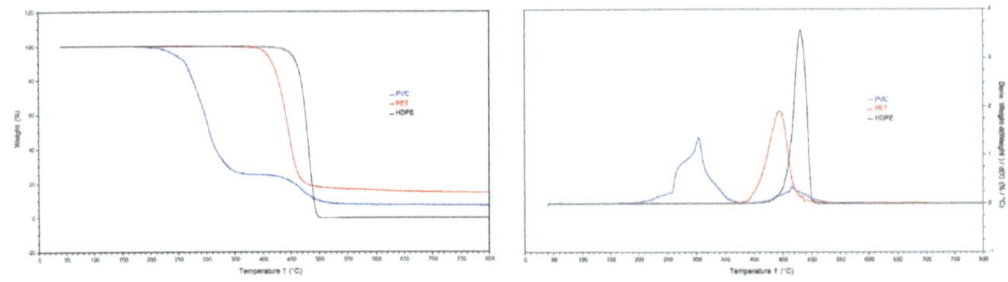

Fig. 5 TG- and DTG- (derivative thermogravimetric) curves of pure polymeric components. *Data from Versalis' internal database.*

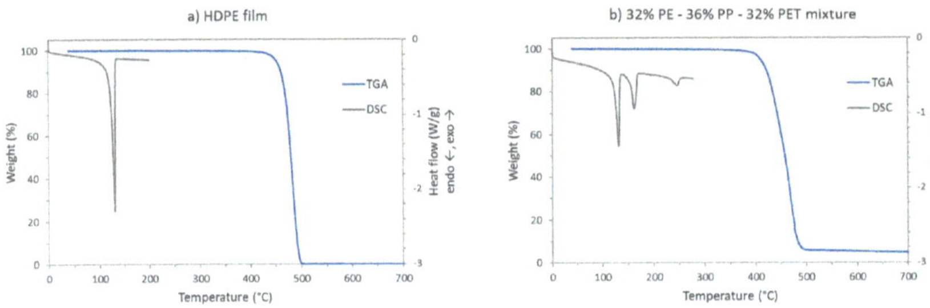

Fig. 6 TG (blue line) and DSC (gray line) curves of: (A) PE film, (B) 32% PE—36% PP—32% PET mixture. *Data from Versalis' internal database.*

provide useful figures like humidity, volatile matter, fixed carbon and ash content, that contribute to the overall characterization of the SPW.

In order to better understand the presence of single polymers within the mixture, DSC can be used as a powerful tool, both alone and in combination with TGA and/or FTIR analysis (Cafiero et al., 2015). Such hybrid approaches lead to the simultaneous detection of thermal transitions—in particular the melting of semicrystalline phases in polymers (e.g., PE, PP, PET, PBT), which are easily recognizable by their peak temperature and measurable by their enthalpy—and decomposition processes, once again attributable to different materials according both to the peaks in DSC and to the vapor species evolved during TGA, as detected by FTIR spectra (Fig. 6).

As mentioned, DSC can effectively identify and quantify polymers in plastic mixtures like SPW. Anyway, sensitivity is related to the type of polymers within such mixture, since not all the materials are semicrystalline or have distinctive elements in term of thermal transitions and decomposition patterns, when DSC is coupled with other techniques. In addition to this, the goodness of inference from experimental data may be heavily influenced by both non-plastic materials like inorganic fillers, paper, metal salts, etc., and non-canonical polymers accidentally present in real mixtures, that can act as noise for deconvolution models.

For all these reasons, thermal analysis has to be considered as only a part of a broader, multitechnical approach aimed to the complete characterization of SPW.

5. Composition by elemental analysis

The very heterogeneous nature of the SPW—as already highlighted—does not allow an easy precise determination of the chemical composition of the material, in particular as regards the distribution between the polymeric species present. This practical complication can make it useful to set up simple determinations that allow a parametric evaluation of the material, in order to define its quality in view of its conversion through chemical recycling processes. The elemental analysis (EA) allows through a quick and simple analysis the determination of the main elements present in the material: carbon, hydrogen, oxygen, nitrogen and sulfur. The analysis is carried out using dedicated tools, which allow a quick and simple calibration, which consists in the practice of analyzing standards with a known and certified formula (primary standards). Carbon and hydrogen are present in all organic materials, and therefore in all polymers; their molar ratio, and their absolute content by weight, however, are typical values for each polymer, easily obtainable from the molecular formula of the repetition unit. Table 2 shows the molar contents, for the most common polymers, of C, H, N and O.

Based on these considerations, it is possible to obtain for each polymer (known the molecular formula of the repetition unit) the indices H/C and C, according to the simple expressions:

$$H/C_{index} = \frac{100}{2} \times \frac{H_{mole}}{C_{mole}} \tag{1}$$

$$C_{index} = 100 \times (C_{mole} \times M_C)/M_{ru} \tag{2}$$

where:

H_{mole}, C_{mole} = moles of H and C in molecular formula of repetition unit (ru).

M_C = carbon standard molecular weight (12.0107 g/mol, from (IUPAC, 1997)).

Table 2 Molar content of C, H, N and O for some common polymers.

Polymer	**Molecular formula of the repetition unit**	**C**	**H**	**N**	**O**
PE	C_2H_4	2	4	–	–
PP	C_3H_6	3	6	–	–
PS	C_8H_8	8	8	–	–
PET	$C_{10}H_8O_4$	10	8	–	4
PVC	C_2H_3Cl	2	3	–	–
PC	$C_{16}H_{14}O_3$	16	14	–	3
PA	$C_6H_{11}NO$	6	11	1	1
PMMA	$C_5H_8O_2$	5	8	–	2
POM	CH_2O	1	2	–	1
Cellulose	$C_6H_{10}O_5$	6	10	–	5

Table 3 H/C index e C index of some polymers.

Polymer	**Molecular formula of the repetition unit**	**M [g/mol]**	**H/C index**	**C index**
PE	C_2H_4	28.05316	100.0	85.7
PP	C_3H_6	42.07974	100.0	85.7
PS	C_8H_8	104.14912	50.0	92.3
PET	$C_{10}H_8O_4$	192.16812	40.0	62.5
PVC	C_2H_3Cl	62.49822	75.0	38.4
PC	$C_{16}H_{14}O_3$	254.28056	43.8	75.6
PA	$C_6H_{11}NO$	113.15764	91.7	63.7
PMMA	$C_5H_8O_2$	100.11582	80.0	60.0
POM	CH_2O	30.02598	100	40.0
Cellulose	$C_6H_{10}O_5$	162.1406	83.3	44.4

M_{ru} = standard molecular weight of molecular formula of polymer repetition unit (ru).

From the above equations, the indices shown in Table 3 are obtained for the main polymers present in the SPW.

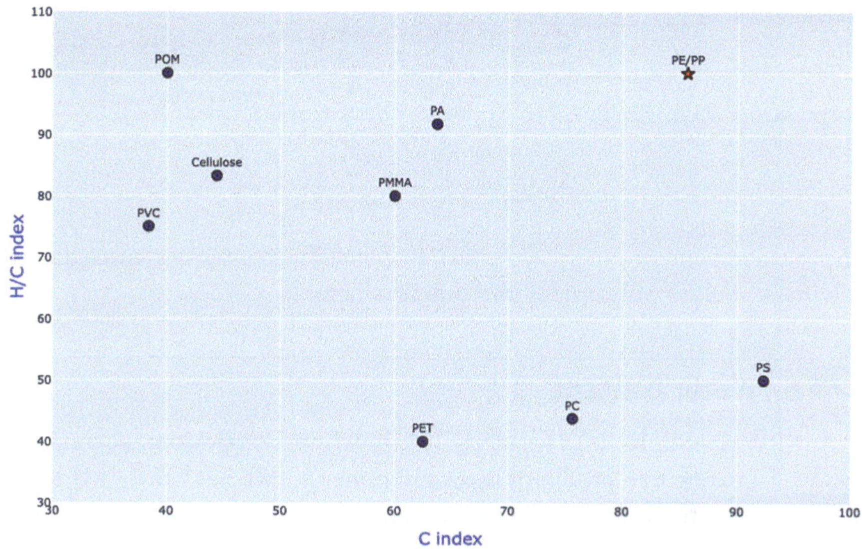

Fig. 7 H/C index versus C index, for some common polymers.

Using the combination of these two indices, it is possible to evaluate the quality of the material in semi-quantitative terms. Referring to polyolefins as optimal materials for pyrolysis processes, a synthetic indication can be obtained, which can also be easily visualized in a "topographical" way, using the graph shown in Fig. 7.

The elemental analysis allows to determine the content in percent by weight of carbon and hydrogen. From these values you can easily obtain the H/C and C indices, through a few simple steps.

$$H_{mole} = \left(\frac{\%H_w}{M_H}\right) / \left(\frac{\%H_w}{M_H} + \frac{\%C_w}{M_C}\right) \tag{3}$$

$$C_{mole} = \left(\frac{\%C_w}{M_C}\right) / \left(\frac{\%H_w}{M_H} + \frac{\%C_w}{M_C}\right) \tag{4}$$

where:

$\%H_w$ and $\%C_w$ = hydrogen and carbon content – weight percent – as obtained from EA.

H/C_{index} can then be calculated using Eq. (1).

The C_{index} is instead obtained from the Eq. (5), taking into account the content of inert, or inorganic matter, present in the SPW and obtained as ash content ($\%Ash$, as reported in par. 7).

$$C_{index} = \frac{100 \times \%C_w}{(100 - \%Ash)} \tag{5}$$

In the same way, it is possible to derive the nitrogen index N_{index}:

$$N_{index} = \frac{100 \times \%N_w}{(100 - \%Ash)} \tag{6}$$

The nitrogen index N_{index} allows to estimate the content of nitrogenous polymers, such as polyamides and polyurethane.

6. Water content

SPW feedstocks are variable materials, whose composition depends on the quali-quantitative profile of their components, but also on the type and number of treatments along the sorting process, meant in a broad sense and including many potential steps like shredding or grinding (up to micronization scale), washing or flotation, densification, extrusion and pelletization.

Due to this, water content is a relevant parameter in the characterization of such materials, since wet or dry treatment processes can bring to significatively different amounts of moisture in the samples. Order of magnitude may vary from less than 1% wt. for very hydrophobic polymer mixtures treated in dry conditions to 10% wt. or even more in the case of wet processes treating polar and hydrophilic matrixes.

For this reason, water content is generally measured through gravimetric methods instead of more sensitive analytical techniques (e.g., Karl Fischer titration) that would be more appropriate for lower amounts, or traces, of water. Many standard methods are then based on simple heating of samples in convection oven at 105 °C at least for 4 h or up to weight stability of the sample. Water content is, in the end, expressed as the percentage loss of mass during the stay in the oven, referred to the initial weight of SPW.

7. Inert content by ash determination

Several methods and procedures have been proposed to measure inert content in polymeric matrixes, according to their nature and/or composition. In the case of SPW, inert content includes all the inorganic fraction in the sample, that can be present due to both formulative reasons (inorganic fillers, catalysts, additives, reinforcement agents, etc.) and contamination along the waste management chain.

Ashes can be determined through thermoanalytical techniques such as TGA (carried out in oxidative atmosphere) or simple incineration/calcination (ISO 3451-1, 2019), with methods directly derived from the analytical experience in polymers' field. TGA is more sensitive and can give much more information about the kinetics of decomposition of the sample and the type of compounds in it (e.g., presence of carbonates in the inorganic fraction, if the correspondent decarbonation peak is detected), but lacks in robustness since it typically analyzes very small amount of sample (mg scale) and this could be risky in terms of compositional representativity of the bulk. On the contrary, incineration/calcination in muffle furnace (where temperature is generally raised up to 850°C) is a simpler method to provide ashing of the sample, even on grams or tens of grams scale, and this leads to consistent figures, in particular for inhomogeneous SPW.

8. Determination of metal residual

As for inorganic fraction, metals can be present in SPW samples rather ubiquitously, due to the widespread use of metal compounds in polymers manufacturing. Most of catalysts and additives used for the synthesis, formulation and processing of plastic materials are based on metals or metalloid species, whose detection is, therefore, crucial in terms of monitoring of the same within the SPW feedstock, also for predictive purpose on the downstream products obtained in the recycling process. Indeed, the presence of metals in the final hydrocarbon fractions may affect their own appeal as feedings for the monomers production plants, due to both safety and process issues in terms of compliance with the specs of that particular plant section, e.g., steam cracking or aromatics extraction.

Generally, the most abundant metals (and metalloids) in plastic waste samples include alkaline earth metals like Calcium and Magnesium, alkali metals like Sodium and Lithium, Aluminum, Zinc, metalloids like Silicon and Antimony, and several transition metals like Iron, Titanium, Copper, etc. Table 4 shows typical values of metal content in various packaging waste streams: this data, taken from the work of Roosen et al. (2020), serves as an example of the variety of metals and metalloids (except for Si, which wasn't investigated) and the order of magnitude of the same in SPW.

Thus, several methods can be used to detect metal residues in SPW, similarly to what is done for both polymeric materials (Alam et al., 2019) and solid waste samples (see for example (EN 15309, 2007)). Among these, the most common are the Inductively Coupled Plasma Atomic Emission

Table 4 Metal content in different packaging waste streams.

Element	PET bottles	PET trays	PE bottles	PP bottles	PP trays	EPS trays	Monolayer films	Multilayer films
	Packaging waste stream							
Li	1.1	1.1	1.6	2.0	1.5	2.6	1.1	1.3
Na	42.7	61.7	145	130	29.3	152	90.7	108
Mg	9.4	34.3	31.7	46.5	186	75.6	32.4	51.4
Al	81.1	205	228	247	145	117	236	455
Ca	334	2131	1127	1642	1348	1471	678	2770
Ti	3.5	4.8	8.1	31	46.6	21.2	35.4	39.3
Cr	0.3	0.4	0.3	0.1	1.3	0.4	0.7	1.0
Mn	0.3	1.2	0.7	0.7	1.0	1.0	2.8	1.3
Fe	11.9	29.5	13.6	15.4	270	24.3	37.5	27.5
Co	1.1	0.9	<lod	0.1	0.2	0.1	0.1	0.1
Cu	7.7	3.3	8.8	3.2	82.7	7.8	21.4	40.4
Zn	3.0	11.1	7.7	5.1	39.2	45.6	28.1	33.4
Sr	1.3	6.1	3.4	11.3	1.7	9.1	3.9	5.6
Sb	39.9	96.0	<lod	0.7	0.2	2.0	0.3	7.8

All concentrations are expressed in ppm.
Elaboration from M. Roosen, N. Mys, M. Kusenberg, P. Billen, A. Dumoulin, J. Dewulf, K.M. Van Geem, K. Ragaert, and S. De Meester, "Detailed analysis of the composition of selected plastic packaging waste products and its implications for mechanical and thermochemical recycling", Environ Sci Technol, vol. 54 (2020), 13282–13293.

Spectroscopy (ICP-AES; (Pereira et al., 2011) and the X-Ray Fluorescence spectroscopy (XRF; (Bichinho et al., 2005; Węgrzynek, 2001).

ICP-AES is a very robust and sensitive technique that can detect elemental species—after calibration—from traces (ppt) to percent. All the species of interest can be detected in short times even in small amount of sample, but in practice relatively high amount of sample has to be used to overcome inhomogeneity issues in SPW and this leads to complex preparation cycles: samples have to be mineralized, solubilized and diluted during relatively long and expensive operations to get the ready-to-analyze sample.

On the contrary, XRF requires a much easier sample preparation which is, in addition to this, non-destructive. The sample is analyzed as a pressed pad of a few grams in weight and most of elements (those with atomic number greater than Sodium) in it are detected and quantified simultaneously. Hence, XRF analysis is very simple and fast and gives good accuracy and precision. As any analytical technique does, XRF can have some limitations: in this case matrix and interelement interference issues may be encountered, in particular for elemental concentrations higher than thousands of ppm.

Other techniques can be used as well, like the ones based on atomic absorption (Flame Atomic Absorption Spectroscopy, FAAS and Graphite Furnace Atomic Absorption Spectrometry, GFAAS), and on the coupling of ICP atomization with Mass Spectroscopy (ICP-MS). All of these give excellent results, in particular in terms of low detection limits and high sensitivity, but suffer of pretty high operation costs related to preparation time, chemicals and instrument running and maintenance. Alternatively, other methods to detect specific elements can be applied, such as the hydride generation (that can be coupled with ICP-MS and AAS) for the analysis of species capable of forming volatile hydrides (As, Bi, Ge, Pb, Sb, Se, Sn and Te) and the cold-vapor technique (CV-AAS) which is specific for Hg. Due to their specificity and despite their extremely high sensitivity and effectiveness, these techniques are used only in particular cases, i.e., when the detection of such elements is critical for environmental or toxicological purposes.

9. Determination of chlorine and bromine

The presence of halogens in SPW can lead to significant problems during the pyrolysis process of the material: acid haloes can develop, and these are potentially corrosive for the plant; alkyl halides, both gaseous and liquid, can form and worsen the quality of the final products; nevertheless, in particular pyrolysis conditions the formation of highly polluting and dangerous compounds such as dioxins could be a potential risk. Halogens can arise firstly from the presence of PVC (chlorine), and secondly from usually brominated compounds used as flame retardants in thermoplastics. PVC and materials containing flame retardants are usually not present in plastic packaging waste, especially if it is urban waste, but the presence of halogens in the feed to the plant should be carefully assessed for the reasons given above.

Determination of the halogens present (Zhang et al., 2013) requires mineralization of the sample, so that halogenated compounds are brought into aqueous solution in the form of halogen acids. This procedure is usually carried out by means of a Mahler bomb, in which the sample is completely burnt in a pressurized oxygen atmosphere (20–30 bar), following an electric discharge.

The procedure ensures total mineralization of the material; the halogen acids present are absorbed in a basic aqueous solution containing carbonate. The amount of sample, accurately weighed (±0.1 mg), depends on the estimated halogen content of the sample itself; for SPW, where usually chlorine and bromine present are a few percent by weight (or fractions of a percent), 1 g or fractions of a gram is charged. The resulting quantitatively discarded solution is then analyzed by ion chromatography, using an anion separation column.

The Mahler bomb mineralization step can be avoided by using an ion chromatograph equipped with an integrated sample combustion system (Combustion Ionic Chromatograph—CIC) (ASTM D 7359, n.d.; Hemmler et al., n.d.; Huang and Rohrer, 2017).). This system has the undoubted advantage of reducing sample handling into two separate steps, thus reducing a possible source of error. However, it should be noted that the amount of sample to be subjected to analysis is considerably lower (usually only a few milligrams are used), and in the specific case of the SPW this is a disadvantage, due to the already mentioned inhomogeneity of the material.

The calibration of the ion chromatograph requires the analysis of solutions of known titre of the halogen acids to be determined. Polymeric standards of known titre, ERM-EC (EUR Report, 2022), are commercially available and allow the reliability of the entire analysis method (i.e., both the mineralization phase and subsequent chromatographic analysis) to be assessed. A typical chromatogram of a standard solution containing all the halogen acids is reported by Huang et al. (Huang and Rohrer, 2017).

Hemmler reported (Hemmler et al., n.d.) the typical chromatogram obtained for the ERM-EC680k standard, which is very similar to that obtained with a real sample.

The determination of the halogen content in the solution obtained from mineralization of the sample can be obtained by silver titration, by precipitation (Mohr's method). The $AgNO_3$ titrant solution reacts with the halides present, forming a precipitate in a quantitative manner. At the equivalence point, the color change is due to the reaction of the silver ion of the titrant with the potassium chromate indicator. The fundamental limitation of this very simple and inexpensive technique is that it does not allow a distinction to be made between chlorine and bromine.

The argentometric determination can also be carried out using a potentiometric titrator (Metrohm, 2022). In this case, in addition to making the analysis very simple and reliable, it is possible to determine chlorine and bromine simultaneously (if present in comparable quantities), thanks to the different discharge potentials of the two equivalence points. The typical potentiometric titration curve obtainable with this technique clearly shows two inflection points, related to the two halides.

10. Conclusions

The characterization of SPW requires the use of many analytical techniques in order to obtain a comprehensive description of the composition of the material in its main polymeric constituents, and to identify impurities such as halogens, metals, inerts and water, which may have important consequences on the conversion of SPW within chemical recycling processes, in terms of both operational safety and quality of final products. It is preferable that the analytical techniques used would be easily available in a normal analytical laboratory, and that they could provide reliable and robust results with analysis times as short as possible. Another aspect of fundamental importance to be taken into account is the heterogeneity of the material: this last aspect may imply the need for homogenization processes of the sample, and the replication of analytical determinations on several aliquots of material. The techniques reported above were therefore selected taking into account these general aspects, trying to represent an adequate compromise between compositional level of detail, speed and simplicity of the technique, and the representativeness of the analytical results obtained.

References

Alam O, Yang L, Yanchun X: Determination of the selected heavy metal and metalloid contents in various types of plastic bags, *J Environ Health Sci Eng* 17(1):161–170, 2019.

ASTM D 7359, n.d. "Standard test method for total fluorine, chlorine and sulphur in aromatic hydrocarbons and their mixtures by oxidative pyrohydrolytic combustion followed by ion chromatography detection (combustion ion chromatography-CIC)".

Bichinho KM, Pires GP, Stedile FC, dos Santos JHZ, Wolf CR: Determination of catalyst metal residues in polymers by X-ray fluorescence, *Spectrochim Acta Part B: At Spectrosc* 60(5):599–604, 2005.

Cafiero L, Fabbri D, Trinca E, Tuffi R, Vecchio Ciprioti S: Thermal and spectroscopic (TG/DSC–FTIR) characterization of mixed plastics for materials and energy recovery under pyrolytic conditions, *J Therm Anal Calorim* 121:1111–1119, 2015.

Calero M, Martín-Lara MÁ, Godoy V, et al.: Characterization of plastic materials present in municipal solid waste: Preliminary study for their mechanical recycling, *Detritus* 04: 104–112, 2018.

Ciurczak EW, Igne B, Workman J Jr, Burns DA: *Handbook of near-infrared analysis*, 4th edition, 2021, CRC Press.
EN 15309: *Characterization of waste and soil - Determination of elemental composition by X-ray fluorescence*, 2007.
EUR Report (2022) "The certification of the mass fractions of As, Br, Cd, Cl, Cr, Hg, Pb, S and Sb and the assignment of indicative values for Sn and Zn in two polyethylene reference materials", Certified Reference Materials ERM®-EC680k and ERM®-EC681k, European Commission, Directorate-General Joint Research Centre, Institute for Reference Materials and Measurements, EUR Report 22784 EN.
Heikkinen J, Spliethoff H: Waste mixture composition by thermogravimetric analysis, *J Therm Anal Calorim* 72:1031–1039, 2003.
D. Hemmler, C. Emmenegger, D. Schmitz, S. Kaufmann, A. Steinbach, "Determination of Halogens and Sulphur in Complex Matrices", The Column-07-07-2014, Volume 12, Issue 10.
Huang B, Rohrer J: *Thermo fisher scientific, application note n° 72573*, CA, 2017, Sunnyvale.
Hundertmark T, Mayer M, McNally C, Simons TJ, Witte C: *How plastics-waste recycling could transform the chemical industry*, 2018a, McKinsey.com.
Hundertmark T, McNally C, Simons TJ, Vanthournout H: *No time to waste: What plastics recycling could offer*, 2018b, McKinsey.com.
ISO 3451-1: *Plastics—Determination of ash—Part 1: General methods*, 2019.
IUPAC: Compendium of Chemical Terminology, In *the "Gold Book"*, 2nd ed., Oxford, 1997, Blackwell Scientific Publications., ISBN:0-9678550-9-8. Compiled by A. D. McNaught and A. Wilkinson. Online version (2019-) created by S. J. Chalk https://doi.org/10.1351/goldbook.
Kuptsov AH, Zhizhin GN: *Handbook of Fourier transform Raman and infrared spectra of polymers*, 1st edition, 1998.
Metrohm: *Potentiometric titration of chloride and bromide*, 2022. Metrohm Application Bulletin No. 27/2e.
Murichan N, Cherntongchai P: Kinetic analysis of thermal degradation of polyolefin mixtures, *Int J Chem Eng Appl* 5:169–175, 2014.
Pereira JSF, Knorr CL, Pereira LSF, et al.: Evaluation of sample preparation methods for polymer digestion and trace elements determination by ICPMS and ICPOES, *J Anal At Spectrom* 26(9):1849–1857, 2011.
Plastics Europe: *Plastics – the Facts*, 2021 https://plasticseurope.org/knowledge-hub/plastics-the-facts-2021/.
Roosen M, Mys N, Kusenberg M, et al.: Detailed analysis of the composition of selected plastic packaging waste products and its implications for mechanical and thermochemical recycling, *Environ Sci Technol* 54:13282–13293, 2020.
Węgrzynek D: Quantitative elemental mapping and characterization of the homogeneity of element distribution in polymer foils by microbeam X-ray fluorescence spectrometry, *X-Ray Spectrom* 30(1):56–61, 2001.
Zhang S, Zhao T, Wang J, Qu X, Chen W, Han Y: Determination of fluorine, chlorine and bromine in household products by means of oxygen bomb combustion and ion chromatography, *J Chromatogr Sci* 51:65–69, 2013.

CHAPTER TWO

Chemical kinetics of catalytic/non-catalytic pyrolysis and gasification of solid plastic wastes

Andrea Locaspi[a,†], Matteo Ferri[b,†], Francesco Serse[a], Matteo Maestri[b], and Matteo Pelucchi[a,*]

[a]CRECK Modeling Lab, Department of Chemistry, Materials and Chemical Engineering "Giulio Natta", Politecnico di Milano, Milano, Italy
[b]Laboratory of Catalysis and Catalytic Processes, Dipartimento di Energia, Politecnico di Milano, Milano, Italy
*Corresponding author: e-mail address: matteo.pelucchi@polimi.it

Contents

Abstract

In a circular economy perspective, solid plastic wastes (SPW) can become a valuable source of chemicals, energy vectors and fuels. Catalytic and non-catalytic pyrolysis, gasification, and partial oxidation technologies can be employed to produce hydrogen, syngas, chemicals, fuels, or energy, providing more efficient and environmentally sound options compared to landfill disposal. Product distribution optimization, experiment,

† Equal contribution statement.

Advances in Chemical Engineering, Volume 60
ISSN 0065-2377
https://doi.org/10.1016/bs.ache.2022.09.002

reactor, and process design as well as pollutant formation control can largely benefit from accurate and predictive reactor models. Chemical kinetic models constitute a core part of reactive flow modeling approaches, together with heat and mass transfer models.

Mechanistic models provide insights into the degradation process of polymers and into the chemical dynamics involved in the transformation of a polymer or a polymer mixture in valuable products. Kinetic analysis of the pyrolysis of plastics is pivotal for understanding the optimal conditions of the cracking process improving reactor design and providing a solid basis for scale-up. Beyond global models mostly aiming at representing the overall conversion of the polymer backbone, theory-based detailed and semi-detailed models allow to explore with good predictive capability a wide range of conditions also tracking the product distribution. This chapter provides an overview of state-of-the-art kinetic modeling approaches from the molecular scale understanding of the occurring chemistry in both thermal and thermo-catalytic conversion, to the design, assembly and validation of integrated semi-detailed kinetic models.

1. Introduction

Chemical recycling of plastic waste materials is of paramount importance to meet sustainability targets in both the chemical and energy industries (Al-Salem et al., 2009; Solis and Silveira, 2020). As discussed in the previous chapters, SPW is a complex mixture of different polymers, additives, inorganics, and contaminants with a time- and location- varying composition and availability. In this context, thermochemical conversion processes such as pyrolysis and gasification allow for the required flexibility in mixture handling (Arena, 2012; Chen et al., 2014; Lombardi et al., 2015). Furthermore, thermochemical and thermocatalytic recycling enable to target a wide variety of products based on the technology (e.g., gasification vs. pyrolysis) and by adapting reactor conditions (i.e., by varying temperature, residence time, catalyst). Nevertheless, the strong feedstock variability requires accurate assessment and proper tools to carefully tailor efficient process operations, minimizing costs and energy inputs and targeting the desired product distribution. To this aim, fluid dynamics and chemical kinetic modeling tools can substantially aid reactor and process design and performances optimization including pollutant formation mitigation (Dogu, Pelucchi, & Van de Vijver, 2021; Madanikashani, Vandewalle, & De Meester, 2022; Ragaert, Delva, & Van Geem, 2017).

Describing the thermochemical process requires the formulation of suitable condensed-phase models able to account for the physico-chemical transformation of a complex multicomponent and multiphase reacting mixture. Indeed, the plastic waste feedstock is a solid-phase stream but, upon heating in the reacting environment, melting, mixing phenomena, decomposition and evaporation take place. Furthermore, according to the SPW composition, formation of significant amounts of a char-inorganic solid matrix can take place. The impact of conversion of such solid residue during gasification should be also accounted for as it impacts product distribution and quality (e.g., H_2/CO ratio in syngas).

The present chapter focuses on the kinetic approaches to model the chemical evolution of SPW mixtures, addressing both catalytic and non-catalytic pyrolysis and gasification. To the authors' knowledge, few studies on the kinetics of complex mixtures degradation have been presented in the scientific literature (Dogu et al., 2021). On the other hand, thermal conversion of single polyolefins and their binary mixtures has been quite widely studied in the past decades, although further work is still required. The present chapter addresses the kinetic descriptions in a hierarchical approach. Non-catalytic conversion will be addressed first as it occurs even in the presence of a catalyst. For the same reasons, pyrolysis will be first addressed as it takes place even during gasification.

2. Non-catalytic thermochemical conversion

SPW thermochemical conversion process, even without presence of a catalyst, is a multi-component, multi-phase and multi-scale problem (Dogu et al., 2021; Madanikashani et al., 2022; Ranzi et al., 2017).

Solid Plastic Wastes are complex mixtures with a time and location-varying composition as discussed in Chapter "Chemical kinetics of catalytic/non-catalytic pyrolysis and gasification of solid plastic wastes" by Locaspi et al., but several key compounds are usually observed as shown in Fig. 1. Modeling the degradation of the SPW mixture requires characterizing its thermophysical properties and constituents. In general, the mixture behavior cannot be assumed as a linear combination of its constituents (Bhaskar et al., 2004; Knümann and Bockhorn, 2010). Indeed, interactions among the chemical compounds can lead to different final products which may require further downstream processing to treat the flue -streams. The variability of the SPW mixture can be described using few reference constituents. According to the desired detail, the mixture can be represented as a single pseudo-species (Koo and Kim, 1993) or accounting for the polymers and additives included (Sommariva et al., 2011). The complexity and accuracy of the model increase according to the number of reference species

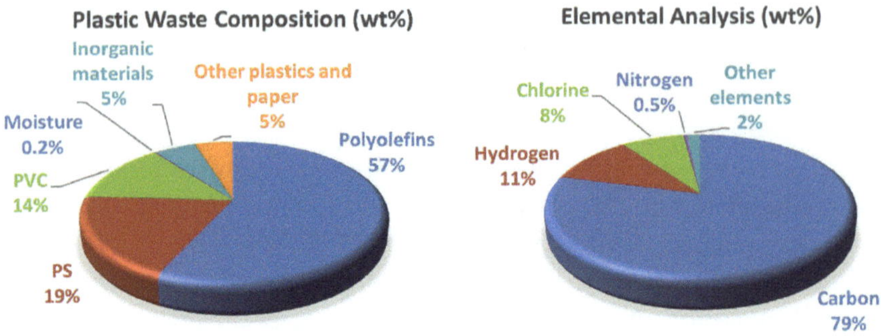

Fig. 1 Plastic waste composition and the corresponding elemental analysis from household waste. *Taken from Dogu O, Pelucchi M, Van de Vijver R, et al.: The chemistry of chemical recycling of solid plastic waste via pyrolysis and gasification: State-of-the-art, challenges, and future directions,* Prog Energy Combust Sci *84:100901, 2021. Adapted from Kaminsky W: Chemical recycling of mixed plastics of pyrolysis,* Adv Polym Technol *14:337–344, 1995.*

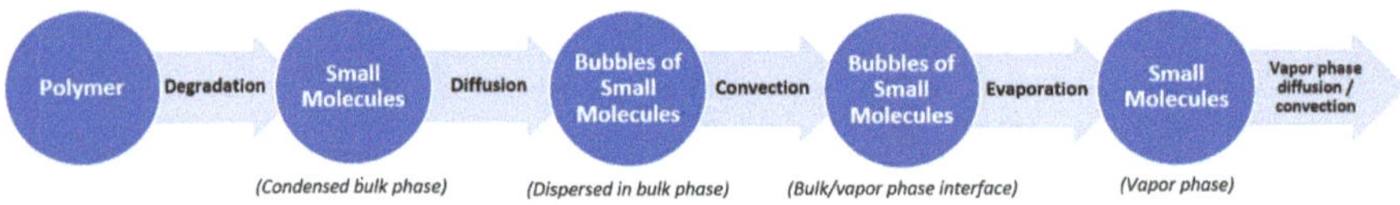

Fig. 2 Different phenomena taking place in the volatilization process of polymers. *Adapted from Dogu O, Pelucchi M, Van de Vijver R, et al.: The chemistry of chemical recycling of solid plastic waste via pyrolysis and gasification: State-of-the-art, challenges, and future directions,* Prog Energy Combust Sci *84:100901, 2021.*

accounted for, including polymers, impurities and additives, and the level of accuracy through which the mixture is characterized also affects the level of accuracy in modeling its evolution.

The degradation of the reacting mixture involves several distinct physico-chemical transformations as schematically represented in Fig. 2. The SPW mixture upon heating melts (e.g., $T=140$–$280\,°\mathrm{C}$ for the most common commercial polymers and polyolefins) and decomposes (e.g., $T=200$–$500\,°\mathrm{C}$) forming smaller species. The released intermediates cover a wide distribution of molecular weights, resulting in simultaneous formation of solids, liquids, vapors and supercritical gases. The coexistence of multiple phases requires estimation of their contact areas and their characteristic times. For instance, formation of bubbles in the liquid melt can enhance mass and heat exchange in the reactor, while proper quantification of contact times with solid ashes present in the feedstock is necessary for evaluating catalytic effects. Secondary gas-phase cracking or cross-linking reactions also contribute to alter the expected product distribution. The multi-phase nature can result in additional transport limitations, which can lead to formation of by-products that further increase mass and heat transfer resistance in a complex interplay. As with the multi-component nature of the system, the multi-phase system can be simplified considering only global behaviors that reduce the complexity at the expenses of heavy losses in accuracy of model predictions (Madanikashani et al., 2022).

The description of the entire process therefore requires proper characterization of different time- and length-scales due to the different physico-chemical phenomena involved. Indeed, the smallest scale involved are chemical reactions, taking place at the Angstrom scale, and physical transformation which occur at the nanoscale as depicted in Fig. 3. These molecular transformations concur with bubble and particle evolution at the micro- or millimeter scale and with the reactor itself that is at the meter scale. Simultaneous description of all the scales involved requires modeling the largest scale (10^0 m) with the characteristic length-scale of the smallest one (10^{-10} m) (Horstemeyer, 2010). To model the process, the different scales must be bridged in a multi-scale framework, where the reactor and particle scale employ subgrid models able to represent the complex physical behavior (Hughes, 1995). Kinetic models can be seen as a bridging strategy between molecular and reactor scale. The aim of these models is

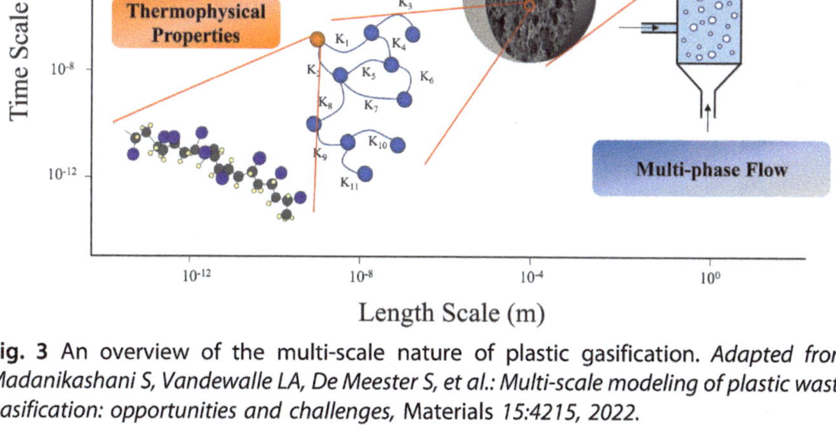

Fig. 3 An overview of the multi-scale nature of plastic gasification. *Adapted from Madanikashani S, Vandewalle LA, De Meester S, et al.: Multi-scale modeling of plastic waste gasification: opportunities and challenges,* Materials *15:4215, 2022.*

representing the stochastic nature of the physico-chemical transformation with more or less simplified terms that can be employed in particle and reactor simulations (Maestri and Cuoci, 2013; Wehinger et al., 2022).

A kinetic mechanism is a network of chemical reactions describing the transformation of reactants into products and the various intermediates involved. The kinetic model, to describe the dynamics of the reactive system, requires also defining species thermodynamic and transport properties and the kinetic constants of reaction rates (Dente et al., 2007). According to the level of detail, the reaction network can span from 10^0 to $>10^9$ number of reactions (De Oliveira et al., 2016). The peculiarity of polymeric systems stems from the single polymer high molecular weight and polydispersity which, coupled with the complex free-radical mechanism, results in enormous and hardly manageable networks of reactions (Ranzi et al., 2014). The broad product spectrum from plastic pyrolysis (Dogu et al., 2021) is consequence of the various isomers and classes of compounds that form during the process which, in theory, must all be considered. In general, the kinetic models can be classified in two macro-kinds (De Oliveira et al., 2016): global and mechanistic models. The first class aims at representing the overall behavior through few simplified global reactions, while the latter tries to describe the underlying physico-chemical phenomena to achieve a more comprehensive description although at higher computational cost. The different approaches will be discussed in the following sections, starting with SPW pyrolysis as hierarchically relevant also in gasification conditions. Nevertheless, further considerations are required on pyrolysis kinetics in gasifying conditions, as discussed in the last sections of the present chapter.

2.1 Global kinetic models for plastics pyrolysis

The focus of global models is to evaluate the overall decomposition rate of the reacting mixture. These models are generally obtained fitting rate equations to the experimental weight loss profiles. While these models are extremely simple and flexible, they yield case-specific reaction parameters as they are generally obtained from thermogravimetric analyses (TGA) with no possibility to decouple reaction kinetics from transport phenomena (Aguado et al., 2003; Neeft et al., 1996; Zhao et al., 2020). Indeed, since weight-loss occurs only due to evaporation of sufficiently small fragments, one-step global models describe in truth the global evaporation rate (Westerhout et al., 1997) without characterizing the sequential occurrence of cracking and evaporation, nor the chemical dynamics of products release.

Several authors report significantly different kinetic parameters even for the same single polymers (Bockhorn et al., 1999a; Bouster et al., 1980; Brems et al., 2011; Flynn and Wall, 1966; Harmon et al., 2021; Poutsma, 2003). Some differences arise from variation in the properties and characteristic of the polymers investigated (Marongiu et al., 2007) while others are due to the presence of different additives, impurities or weak links resulting from the specific production process (Kruse et al., 2002). Nevertheless, these discrepancies result also from the diversity in experimental conditions investigated as these models are valid only in a small range of operating conditions and conversion levels (Dogu et al., 2021). In general, one-step models assume a power-law dependence on the mass conversion, α, of the kind:

$$\frac{da}{dt} = k(T)f(a) \quad . \tag{1}$$

where $k(T)$ is the kinetic constant, usually evaluated with an Arrhenius form, and $f(\alpha)$ is the conversion dependence. For polymers such as PE, PP, PS, a simple first-order law can represent the overall behavior, although in a narrow range of operating conditions. On the other hand, for more complex systems as PVC, PU or plastic mixtures, more complex functions are required to capture the multi-step degradation. A different approach instead of defining a conversion function consists in the "distributed activation energy models" (DAEM) (Wang et al., 2018). These models assume the process as characterized by a continuous distribution of activation energy representative of the different constituents involved. In general, a Gaussian distribution is assumed and parameters are determined by experimental fitting (Tilghman and Mitchell, 2015). Despite the main interest of plastic polymers degradation is products formation, one-step global models focus only on the weight-loss profiles. To this aim, more complex global models have been developed that consider multiple reactions. The Independent Parallel Reactions (IRP) formulation has been employed to model different polymers and mixtures (Sorum et al., 2001; Yousef et al., 2020), while other authors (Koo and Kim, 1993) employ a more complex network accounting also for formation of gas, liquid and solid compounds. In general, more intricate models can be developed introducing global stoichiometric coefficients accounting for the condition dependent (i.e., mainly temperature and residence time) products distribution (Harmon et al., 2021). The main advantage of these models is their simplicity, which however still results in an extremely limited range of applicability.

2.2 Mechanistic kinetic models for SPW pyrolysis

The aim of mechanistic kinetic models is to represent the physico-chemical transformations to describe the evolution of the initial mixture. Thermochemical reactivity is described by a complex radical and molecular reactivity which results in a considerable number of elementary reactions. Nevertheless, these schemes are characterized by a modular and hierarchical structure (Dente et al., 1979) which implies that the formulation for simpler systems (e.g., from natural gas up to diesel fuels) can be extended through the definition of appropriate reaction classes and by assigning reference kinetic parameters. In the following paragraphs the degradation mechanism of the main polymer constituents of SPW mixtures is discussed. Then, studies on the chemistry of mixture interactions are considered and the families of mechanistic models proposed in literature are presented.

2.2.1 Solid plastic wastes reaction mechanisms

Different reaction classes must be considered for each polymer. The number and type of reaction classes depends strictly on molecular features of each polymer such as strength of chemical bonds along the backbone, the substituent groups and the presence of hetero-atoms or weak links along the chain (Dogu et al., 2021).

2.2.2 Backbone degradation mechanism: Polyethylene, polypropylene and polystyrene

Polyethylene (PE), polypropylene (PP) and polystyrene (PS) are characterized by side-groups with chemical bonds stronger than the chain backbone C—C bonds. For this reason, upon heating these polymers tend to decompose in smaller fragments which retain the same chemical structure of the starting polymer (Marongiu et al., 2007). Their decomposition mechanism can therefore be related almost entirely to a radical backbone degradation. Studies on thermal degradation kinetics of these polyolefins started in the late 1940s (Dickens, 1982), although only in the late 1990s the first mechanistic models were proposed (Ranzi et al., 1997; Woo and Broadbelt, 1998). Table 1 reports the families of reactions (i.e. reaction classes) typically considered in different literature studies on these polyolefins (Harmon et al., 2021; Marongiu et al., 2007).

In general, pyrolysis of long hydrocarbon chains can be broadly represented through initiation, propagation and termination reactions. Backbone initiation reactions are high-temperature random scission reactions that form the first radicals triggering the radical chain decomposition mechanism. These reactions are characterized by high activation energies

Table 1 Reaction families in the depolymerization mechanism for a generic vinyl polymer. The blue circle represents H, CH_3, and C_6H_5 for PE, PP, and PS respectively.

Reaction families (backbone degradation mechanism)	
Initiation reactions	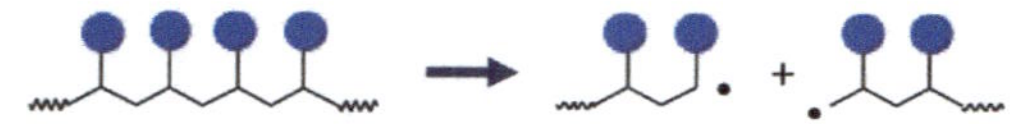
Initiation from "weak links"	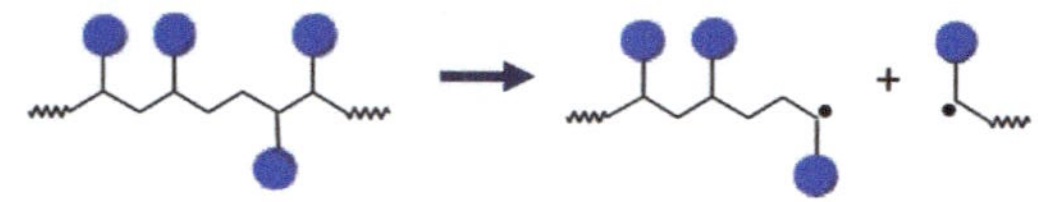
Recombination	
Specific scission (e.g., allylic position)	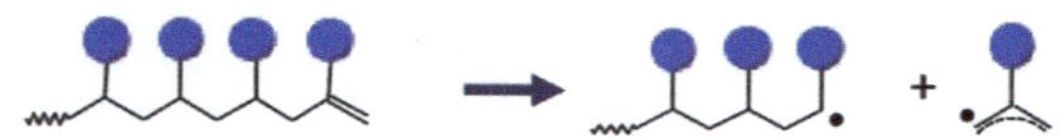
Disproportionation	
H-abstractions (end-chain radicals)	
H-abstractions (mid-chain radicals)	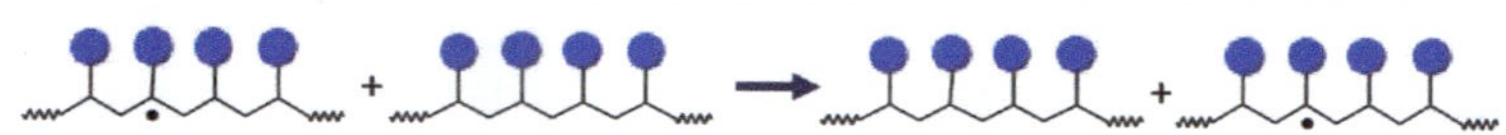

Continued

Table 1 Reaction families in the depolymerization mechanism for a generic vinyl polymer. The blue circle represents H, CH_3, and C_6H_5 for PE, PP, and PS respectively.—cont'd

Reaction families (backbone degradation mechanism)	
H-abstractions (specific position, e.g., allylic)	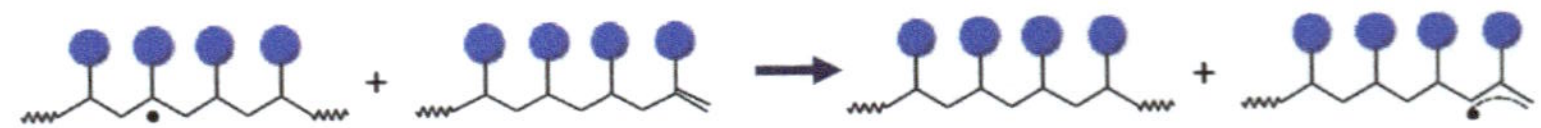
β-scission	
β-scission (unzipping)	
Isomerization (backbiting, mid-chain radical, (1,4), (1,5), (1,6))	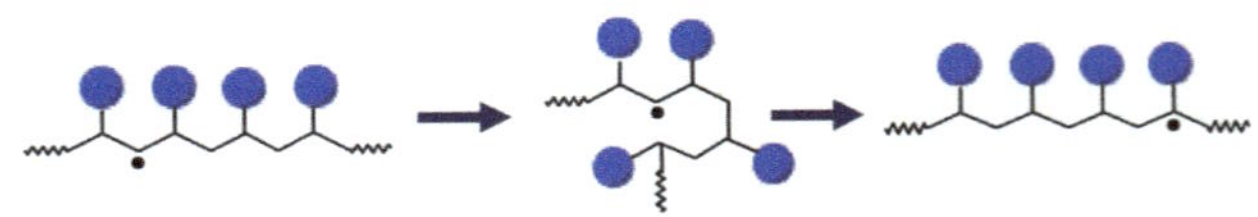
Isomerization (backbiting, end-chain radical, (1,4), (1,5), (1,6), (1,7))	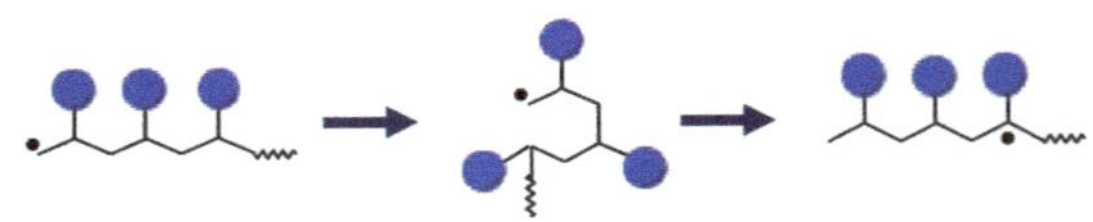
Addition reactions	
Cyclization of allyl radicals	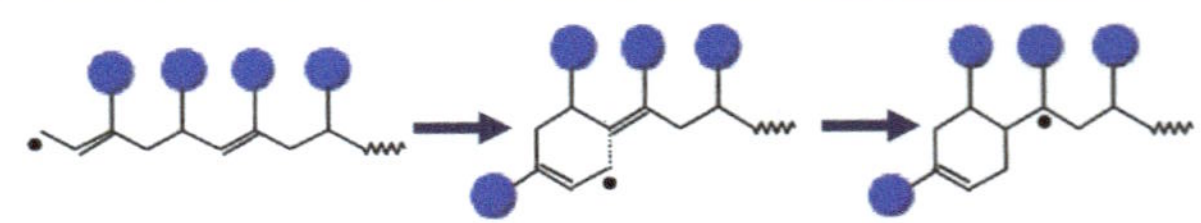

($E_a=250-360$ kJ/mol) (Kruse et al., 2002; Ranzi et al., 1997), but they significantly affect the degradation process as they control the onset of the radical pool. Presence of weak links such as peroxides impurities (Wong et al., 2000) or defects from the specific polymerization process (Moens et al., 2020) facilitate the onset of polymer degradation. The presence of double bonds (e.g., allyl terminal positions) or catalyst impurities also contributes to increasing the initiation rates (Faravelli et al., 1999). On the contrary, the radical pool is inhibited by termination reactions, i.e., bimolecular reactions that convert two highly reactive radicals in stable molecules. The main termination pathways are radical-radical recombination and disproportionation reactions, where the first results in a single longer molecule while the latter in one unsaturated and one saturated molecule. While these reactions are characterized by significantly lower activation energies ($E_a=8-60$ kJ/mol) (Faravelli et al., 2001; Levine and Broadbelt, 2009), the low radical concentrations result in reaction rates that are comparable to those of random scissions. While initiation and termination reactions affect the overall radical concentration, propagation reactions are the rate determining steps dominating both degradation process and product selectivity. Radical species can undertake different propagation channels such as isomerization and β-scission. The latter forms one unsaturated molecule and a terminal radical of shorter length and are characterized by activation energies in the range of $E_a=90-140$ kJ/mol (Kruse et al., 2003a; Mastral et al., 2007). The radical can also undergo unimolecular or bimolecular H-transfer and addition reactions. Compared to gas-phase pyrolysis, liquid-phase reactions are characterized by higher concentrations, resulting in higher importance of bimolecular reactions and lower alkene yields with respect to the gas-phase counterparts (Ranzi et al., 1997). Bimolecular H-transfer are characterized by low activation energies ($E_a=30-60$ kJ/mol) (Faravelli et al., 2001; Levine and Broadbelt, 2009), while bimolecular radical additions are the inverse of β-scission and have extremely low activation energies $E_a=4-24$ kJ/mol (Beuermann et al., 2022; Levine and Broadbelt, 2009). For terminal radicals, intramolecular radical additions can result in the formation of cyclic compounds, while terminal β-scission, i.e., unzipping, results in monomer formation. Backbiting isomerization corresponds to intramolecular H-transfer reactions proceeding through the formation of 3 to 7 member rings transition states with activation energies in the range $E_a=40-120$ kJ/mol (Harmon et al., 2021; Marongiu et al., 2007) according to the strain involved in ring formation. While formation of small rings results in high strain energies, formation of high-membered rings requires a specific arrangement of multiple atoms in place and therefore is entropically limited. The most favored isomerization is 1–5 backbiting, i.e., formation of a 6-member ring able to preserve bond angles at low entropic cost, which is responsible for preferential formation of specific compounds such as n-hexene from PE and trimers from PP and PS (Faravelli et al., 2001, Faravelli et al., 1999; Kruse et al., 2003b). The presence of double bonds plays an important role in the reaction mechanism as formation of allyl radicals is highly favored at the temperatures of interest due to resonance stabilization increasing product stability.

The degradation mechanisms of the most diffused polyolefins PE, PP and PS differ mainly because of the kinetic parameters involved (Marongiu et al., 2007). The higher stability of benzyl radicals results in lower decomposition temperature for PS and a higher importance of unzipping reactions. Indeed, as terminal radicals undergo unzipping before being stabilized, the yield to monomer formation for PS is 70–90% in the 500–800 °C temperature range (Faravelli et al., 2001; Kruse et al., 2002). PP has a higher thermal recalcitrance than PS, although lower than PE due to the presence of tertiary radicals. A key difference also for PP and PS with respect to PE is the distinction between head and tail configurations (Faravelli et al., 2001), which results in significantly more complex kinetic mechanisms and spiked product distributions. The similarity between reaction mechanisms is a consequence of assuming the side-groups as inert, although at higher temperatures this assumption fails and additional compounds are experimentally observed (Audisio and Bertini, 1992). To overcome this simplification, the side-group mechanism discussed in the following paragraph could be considered also for these polymers, resulting however in significantly more complex reaction networks. The presence of impurities can also result in additional reactive pathways that alter products selectivity. Several shortcomings are observed when these mechanisms are employed. The main difference with experimental data is these reaction pathways do not explain char formation and predict equimolar amounts of paraffins and α-ω dialkenes. While the first can be attributed to catalytic effects of ashes or side-group reactions not considered in the pure model polymers the kinetic models are developed for, further modeling and experimental studies can aid in improving the understanding of the underlying physico-chemical phenomena (Poutsma, 2003).

2.2.2.1 Side-group degradation mechanism: Polyvinylchloride, polyvinylbromide and polychloroprene

Halogenated vinyl-polymers as polyvinylchloride (PVC), polyvinylbromide (PVBr) and polychloroprene (PCP) are characterized by side-groups chemical bonds that are weaker compared to backbone C—C bonds. PVC is among the main commodity polymers and has been extensively studied in the past decades (Yu et al., 2016), while other halogenated polymers have received less consideration due to their niche application (e.g., flame retardants such as PVBr). These polymers are of relevance as halogenated compounds requires specific reactors, purification units and, in general, downstream equipment. However, few modeling activities have been performed due to the complexity resulting from the side-group degradation mechanism. This reaction network is also composed of radical pathways which can be distinguished in initiation, propagation and termination steps, although additional molecular pathways must be accounted for, as briefly summarized in Table 2. Radical initiation mainly occurs through backbone—side-group bond cleavage, i.e., C-Cl for PVC (E_a~210 kJ/mol), forming a carbon chain radical and a side-group radical. Propagation reactions are side-group radicals' stabilization through H-abstractions on the polymer chain and consequent evaporation. In PVC and PVBr, this dehydrohalogenation step is responsible for the initial >50% mass loss (Mehl et al., 2004) and leads to formation of stable polyene structures. Indeed, mid-chain radicals tend to undergo β-scission forming mid-chain double bonds which enhance the formation of other neighboring mid-chain radicals through allyl-resonance stabilization. The dehydrohalogenation process can also take place through a 4-centered molecular reaction. The resulting polyenes then undergo intra- and inter-molecular Dies-Alder reactions that form aromatic and alkyl-aromatic groups. Further condensation and dealkylation of these structures lead to the second mass-loss step forming a char residue and releasing small olefins and aromatic compounds (Marongiu et al., 2003). Chlorine and bromine can also interact with the polyene and aromatic structures leading to formation of toxic species as chlorobenzenes. For polychloroprene (PCP), the dehydrochlorination and condensation step partially overlap due to the stronger vinyl–Cl bond (Mehl et al., 2004).

The side-group mechanism involves reaction classes that significantly alter the polymer structure leading to a broad range of classes of compounds (Dogu et al., 2021). The same reaction classes can be coupled with the backbone mechanism to improve model predictions although at higher computational cost. For instance, for PS formation benzene and polyaromatics can be accounted for with these mechanisms, while for PMMA formation of CO, CO_2, and small olefins. Similarly, the backbone degradation mechanism must be considered in PCP degradation due to the lower difference in backbone and side-groups bond strength (Mehl et al., 2004).

Table 2 Reaction families of the side-group mechanism for a generic vinyl polymer.
Reaction families ("side group" mechanism)

Initiation reactions	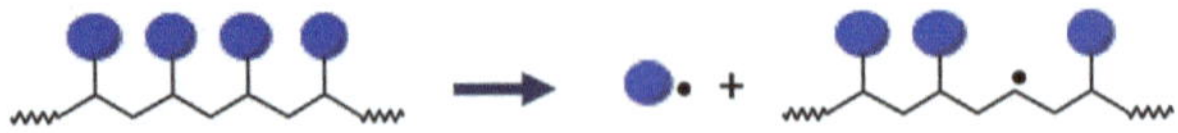
Specific scission (*e.g.* allylic position)	
Recombination	
H-abstractions	
H-abstractions (specific position, e.g., allylic)	
β-scission	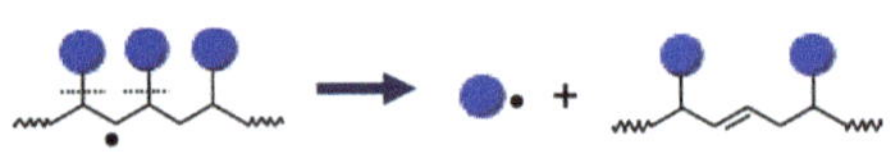
Cyclization through intramolecular Diels-Alder	
Cyclization through intermolecular Diels-Alder	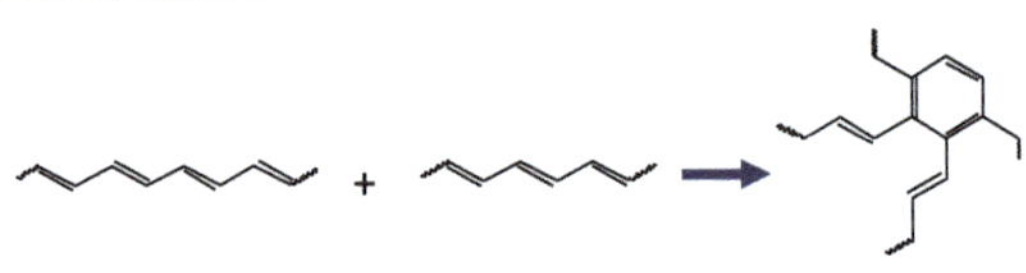
Addition to polyenes	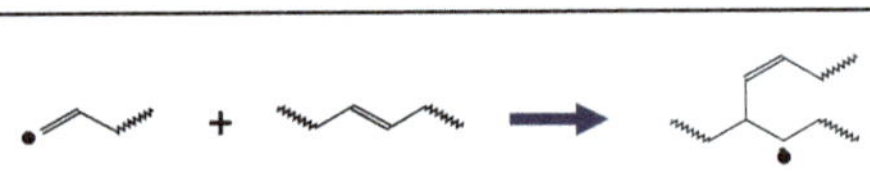
β-Scission of unsaturated resonance stabilized radicals	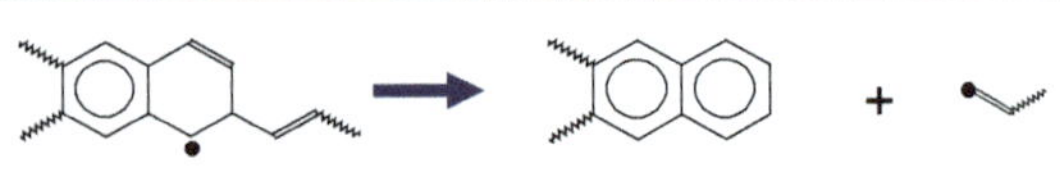
Four-center molecular decomposition	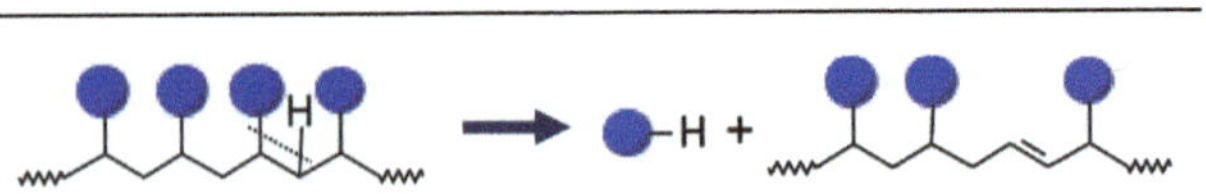

2.2.2.2 Molecular degradation mechanism: Polyethyleneterephthalate, polyamides and polyurethanes

The reactivity of heteroatoms-containing polymers is not fully described by radical decomposition mechanisms. In general, the presence of non-carbon atoms has a weakening effect on the chemical bonds resulting in competitive four- or six-centered molecular reactions (Dogu et al., 2021). As discussed for halogenated polymers, dehydrohalogenation can occur by both a radical mechanism and a low-temperature 4-centered concerted molecular pathway. Similarly, molecular reactions are responsible for cyclization, dealkylation and further dehydration of polyene structures. If heteroatoms are present along the backbone as in the case of polyethyleneterephthalate (PET), polyamides (PA) and polyurethanes (PU), their weakening effect is more prominent. For these polymers, no mechanistic kinetic model has been proposed to the authors' knowledge, but several possible reaction pathways have been discussed (Bockhorn et al., 1999a; Dhahak et al., 2020; Ohtani et al., 1987). In general, these functionalities allow intra- and inter-molecular concerted H-abstractions that lead to formation of vinyl ends and functional specific groups. For instance, ester functionalities in PET are hypothesized to undergo intramolecular concerted decompositions through a six-membered ring cyclization forming a vinyl and a carboxyl end as shown in Fig. 4A. These reactions are assumed the main pathways leading to monomer, dimer and trimer formation, while additional decarboxylation reactions are considered responsible for formation of other aromatic compounds as benzoic acid, small olefins and acetylene (Dhahak et al., 2019). PET pyrolysis leads to formation of non-negligible amounts of char (>15% wt) which is thought to originate from molecular condensation reactions similar to PVC. Concerted molecular reactions are also hypothesized as the main depolymerization reactions for polyamides. Indeed, four-member interactions among nitrogen, hydrogen and ketone functionalities lead to formation of cyclic monomers and dimers as shown in Fig. 4B. The selectivity towards this pathway is extremely high (>95%), and no formation of char is generally observed (Bockhorn et al., 1999b). The term "polyurethanes" refers to a broad class of polymers with significantly different chemical structures, usually involving ester or ether linkages and aromatic or aliphatic segments (Kumagai et al., 2017). While the degradation mechanism depends on the polyurethane monomers, the first degradation step can be related to molecular decomposition of the urethane linkages, either forming isocyanate, amine or imine ends (Chattopadhyay and Webster, 2009; Ravey and Pearce, 1998). According to the molecular

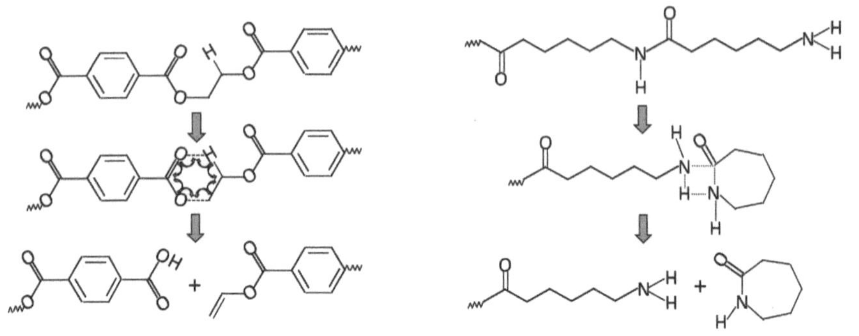

Fig. 4 Hypothetical mechanisms for degradation to monomer formation for (A) PET and (B) Nylon-6. *Adapted from Dogu O, Pelucchi M, Van de Vijver R, et al.: The chemistry of chemical recycling of solid plastic waste via pyrolysis and gasification: State-of-the-art, challenges, and future directions,* Prog Energy Combust Sci *84:100901, 2021.*

structure, significant amounts of char are experimentally observed (>40%) which may result from further dehydrogenation and condensation reactions similar to PVC and PET (Grassie et al., 1980). While molecular reactions are hypothesized as the dominant pathways, further studies are required to investigate the possibility of concurring radical degradation mechanisms.

2.2.2.3 Other polymers: Polyisobutylene, polyalphamethylstyrene, polymethylmethacrylate and polystyreneperoxide

The discussed degradation mechanisms have been employed to also describe additional polymer pyrolysis processes. Among the different hydrocarbon polymers, polyisobutylene (PIB) (Poutsma, 2005) and polyacrylamide copolymers (PAMs) (Woo and Broadbelt, 1998) have been described through a backbone degradation mechanism involving similar reaction classes as in Table 1. Polystyrene peroxide (PSP) and polymethylmethacrylate (PMMA) are heteroatom-containing polymer which however behave similarly to polystyrene. With respect to PMMA, the presence of oxygen in the side-group further enhances formation of tertiary backbone radicals through resonance stabilization. This results in high selectivity towards unzipping reactions that lead to MMA yields >95% (Kaminsky and Eger, 2001). Compared to PE, PP and PS, studies on PMMA pyrolysis have started to emerge only in the past decades (Moens et al., 2020) and further improvement is still required. Also for PSP the presence of oxygen does not modify the backbone degradation mechanism but enhances radical formation due to the weaker chemical bonds (Vinu et al., 2012).

2.2.2.4 Solid Plastic Waste mixtures

To model pyrolysis of real mixtures, the interactions among the main constituents must be considered, i.e. the interactions between polymers, additives and inorganics. Most of the studies targeting this topic in the scientific literature are limited to PE, PP and PS. In general, independent decompositions are observed if the extent of mixing is poor, but significant interactions are observed if molecular-scale mixing is present (Faravelli et al., 2003). Their interactions result from the similar radical pathways involving the same small radicals as reactants or products. For instance, radicals formed from PS pyrolysis can diffuse in the PE/PP melt enhancing their degradation if a PS/PE/PP mixture was to be considered. Nevertheless, few interactions are observed because of the low solubility and molecular diffusivities (Kruse et al., 2005) and accurate description of mixing phenomena are required to model the co-pyrolysis (Madanikashani et al., 2022). Considering other polymer binary or higher order mixtures, no modeling activity has been performed to the authors' knowledge. Experimental investigation on PVC interactions with PE and PS show that interactions are negligible, and the polymers can be assumed reasonably independent (Miranda et al., 2001). On the other hand, experimental results on PVC-PET-PA mixtures show that PET and PA decomposition are significantly enhanced (Bhaskar et al., 2004; Czégény and Blazsó, 2001; Knümann and Bockhorn, 2010; Kulesza and German, 2003). Similar considerations as to the polyolefin pairs discussed above cannot be employed in this case as experimental solubility of the two polymers is difficult to measure. Indeed their similar glass transition temperatures and the low degradation temperature of PVC (Marongiu et al., 2003) complicates this investigation. While aliphatic esters are reported as miscible with PVC (Prud'homme, 1982), further studies are required to evaluate their solubilities (Ziska et al., 1981). The synergistic behavior could also stem from the interactions between HCl released from PVC and PET ester groups, but liquid-phase concentrations of the first are expected to be negligible because of its high evaporation rates.

Similarly, interactions with additives and inorganics require further in-depth characterization, both from an experimental and modeling point of view. In general, experimental results show cellulosic impurities increase char yield, formation of an aqueous phase and CO and CO_2 formation, resulting in lower heating-values for gases and higher separation costs for liquids (Ragaert et al., 2017). $CaCO_3$ was found to affect the pyrolysis of PA and polycarbonates (PC), possibly due to the interactions with amide groups (Avella et al., 2006). Calcium carbonate was also found to be an efficient chlorine-capture agent (Bhaskar et al., 2004), although increasing the chlorine content in the liquid fraction (López et al., 2011b).

2.2.3 Kinetic models for solid plastic waste pyrolysis

A detailed description of the proposed reaction mechanisms requires the characterization of the reactivity of all compounds involved, i.e., reactants, intermediates and products. This detailed approach results in a network of elementary reactions which, in principle, is independent of the feedstock and allows for extrapolation. Nevertheless, according to the previous definition, the detailed model of a single polymer necessitates describing all chain-lengths and their potential isomers. Such reaction networks require a number of species $N_s > 10^7$ for each polymeric constituent, resulting in mechanisms that are unsuitable for systems characterized by complex fluid dynamics such as industrial reactors. To this aim, researchers have developed several approaches to simplify the involved chemistry while preserving the desired accuracy as discussed in the next paragraphs.

2.2.3.1 Discrete sections

The first detailed kinetic models employed for polymer pyrolysis date to the late 1990s (Ranzi et al., 1997). These models consider all stable chain-species introducing the quasi steady-state assumption (QSSA) for propagating radicals. The resulting kinetic models involve $N_s \sim 10^6$ but allow to employ non-stiff algorithms for their numerical solutions resulting in computational costs comparable to other approaches. These models have been successfully validated for PE, PP and PS (Faravelli et al., 2001, Faravelli et al., 1999; Ranzi et al., 1997) as shown in Fig. 5, both in terms of mass-loss profiles and products distribution. A free-radical backbone degradation mechanism is assumed, and the side-groups for PP and PS are assumed inert (Harmon et al., 2021; Marongiu et al., 2007). Faravelli et al. (1999) (Fig. 5D) also highlighted the importance of considering separate models for reaction kinetics and evaporation rates. Indeed, not considering appropriate evaporation models can result in dumping mass-transfer limitations on kinetic parameters resulting in less general or generalized predictions (Madanikashani et al., 2022).

The issue lies however in the scalability of these models to complex reactors and to more complex reaction mechanisms. To this aim, a discrete section lumping approach (Ranzi et al., 2001) has been employed to the models for PE (Marongiu et al., 2007) and PS (Faravelli et al., 2001). This method, employed also to describe particulate matter formation in combustion

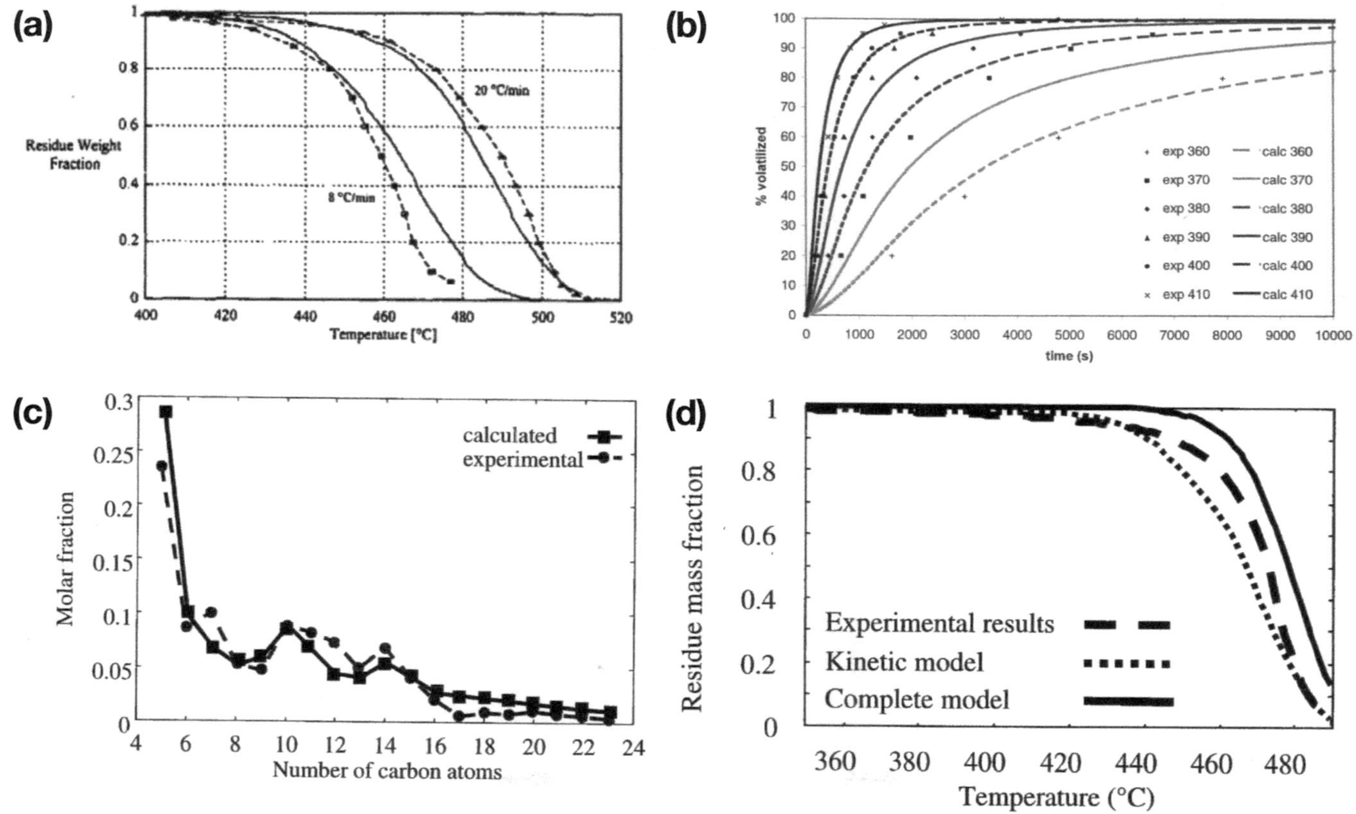

Fig. 5 Validation of discrete section models for pyrolysis of PE, PP and PS. Comparison of: (A) PE mass-loss profiles; (B) PS mass-loss profiles; (C) PE gas products distribution; (D) PE mass-loss profiles considering different evaporation models. *Adapted from Faravelli T, Pinciroli M, Pisano F, et al.: Thermal degradation of polystyrene,* J Anal Appl Pyrolysis *60:103–121, 2001; Faravelli T, Bozzano G, Scassa C, et al.: Gas product distribution from polyethylene pyrolysis,* J Anal Appl Pyrolysis *52:87–103, 1999; Ranzi E, Dente M, Faravelli T, et al.: Kinetic modeling of polyethylene and polypropylene thermal degradation,* J Anal Appl Pyrolysis *40–41:305–319, 1997.*

devices (Nobili et al., 2022), describes the chain distribution through few carbon numbers while the other chain-lengths are assumed as mixture of the chosen carbon numbers. This approach allows to significantly reduce the number of species involved without significant losses in terms of accuracy (Dente et al., 2007). The complete distribution is recovered in post-processing through appropriate delumping techniques (Ranzi et al., 2001).

2.2.3.2 Method of moments

The Method of Moments (MoM) approach was first introduced in the 1950s (Bamford and Tompa, 1954). Population balances are coupled to the theory of reactions in "continuous" mixtures (Wang et al., 1995) to represent the chain-length distribution through its statistical moments. This allows to simplify a theoretically infinite number of mass balance equations in a significantly smaller subset (Mastan and Zhu, 2015). Nevertheless, a distribution must be defined for chains with different structural characteristics, e.g., unsaturation and branching. In general, only the first three statistical moments are tracked per distribution, which however requires appropriate closure techniques (Kruse et al., 2003b; Marongiu et al., 2007). This approach is extremely powerful in describing the polymer chain-distribution evolution and has been successfully employed for PE, PP, PS and polyacrylamides (PAMS) degradation (Kruse et al., 2003b, Kruse et al., 2003a; Levine and Broadbelt, 2009; Woo and Broadbelt, 1998) as shown in Fig. 6. Marongiu et al. (2007) compared the MoM and discrete section models applied to HDPE and PS pyrolysis obtaining similar results as shown in Fig. 6D. This method has been employed also for soot predictions (Mueller et al., 2009) and shares the same limitations as the discrete sections. Indeed, while this approach yields extremely detailed results at low computational costs, the main issue lies in the need of a distribution for each different structural chain. For polymers as PVC, where significant formation of polyene and aromatic structures occurs, the Method of Moments requires a significant increase in number of equations to be considered (Marongiu et al., 2003).

2.2.3.3 Functional groups

Chemical reactions are local phenomena that involve only the nearest neighbors (Benson, 1976). Therefore, above a certain monomer-number, the elementary step does not depend on the chain-length but only on the neighboring chemical groups. The evolution of the polymer melt can be related to the reactivity of the chemical moieties of the chain and not to the

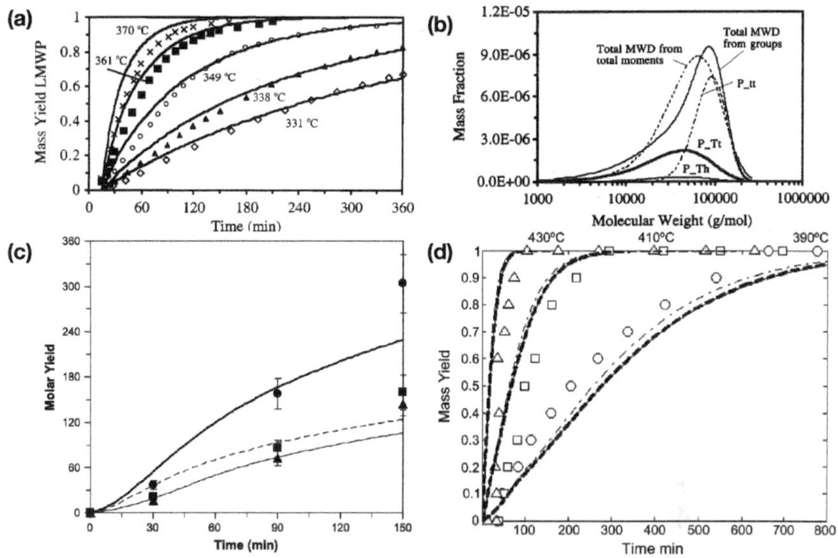

Fig. 6 Validation of Method of Moments models for pyrolysis of PE, PP and PS. Comparison of: (A) PS mass-loss profiles; (B) PS molecular weights distribution evolution; (C) PE total gas, alkane and alkene yields; (D) PE mass-loss profiles for MoM (- . -) and discrete section methods (—). *Adapted from Kruse TM, Wong HW, Broadbelt LJ: Modeling the evolution of the full polystyrene molecular weight distribution during polystyrene pyrolysis,* Ind Eng Chem Res *42:2722–2735, 2003a; Kruse TM, Woo OS, Wong HW, et al.: Mechanistic modeling of polymer degradation: a comprehensive study of polystyrene,* Macromolecules *35:7830–7844, 2002; Levine SE, Broadbelt LJ: Detailed mechanistic modeling of high-density polyethylene pyrolysis: low molecular weight product evolution,* Polym Degrad Stab *94:810–822, 2009; Marongiu A, Faravelli T, Ranzi E: Detailed kinetic modeling of the thermal degradation of vinyl polymers,* J Anal Appl Pyrolysis *78:343–362, 2007.*

distribution itself. This approach discards information about the chain distribution but allows to track different chemical structures at a low computational cost. The only chain-distribution information required is the distinction between mid- and end-chain chemical moieties because of their different selectivity to low molecular weight products. This distinction also allows to recover the number average molecular weight (Locaspi et al., 2022). This approach has been successfully employed to model PE, PP and halogenated polymers as PVC, PVBr and PCP (Locaspi et al., 2022; Loffredo, 2021; Marongiu et al., 2003; Mehl et al., 2004). To the authors' knowledge, this is the only mechanistic approach developed for PVC pyrolysis. Fig. 7 shows the models' predictions for different polymers in terms of mass-loss profiles, although satisfactory results are obtained also for gas-phase profiles.

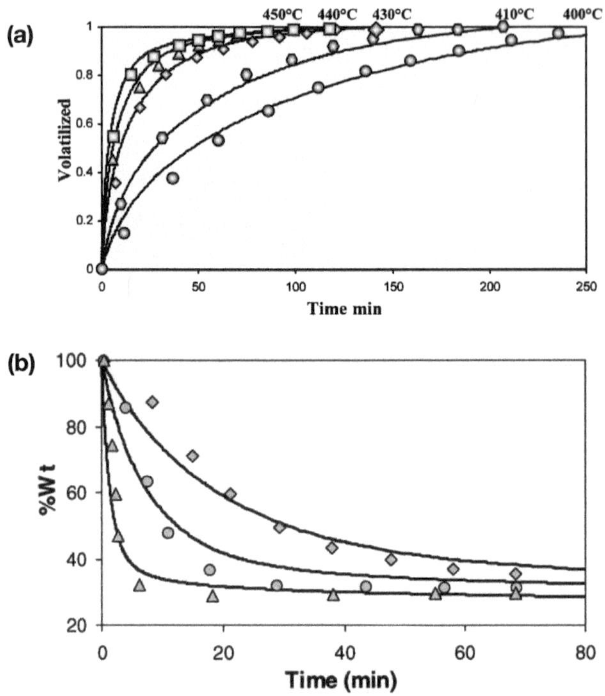

Fig. 7 Validation of functional group approach for pyrolysis of PVC and PCP. Comparison of: (A) PVC conversion and (B) PCP mass-loss profiles. *Adapted from Marongiu A, Faravelli T, Bozzano G, et al.: Thermal degradation of poly(vinyl chloride),* J Anal Appl Pyrolysis *70:519–553, 2003; Mehl M, Marongiu A, Faravelli T, et al.: A kinetic modeling study of the thermal degradation of halogenated polymers,* J Anal Appl Pyrolysis *72:253–272, 2004.*

2.2.3.4 Kinetic Monte Carlo

The kinetic Monte Carlo (kMC) algorithm employs a stochastic approach in contrast to the continuum deterministic modeling approaches previously described (Gillespie, 2007, Gillespie, 1977, Gillespie, 1976). This approach is a proven solution of the Chemical Master Equation (CME) and is able to describe explicitly every species in the system through a microscale homogeneous reaction volume (Vinu and Broadbelt, 2012). The drawback is the computational cost which is unfeasible for polymeric systems due to the huge number of atoms involved. Nevertheless, hybrid kMC methodologies able to reduce the computational cost exist and have been successfully implemented for PE, PS, PIB, PAMS and PSP (Poutsma, 2009, Poutsma, 2007, Poutsma, 2005, Poutsma, 2003; Vinu et al., 2012) as shown in Fig. 8.

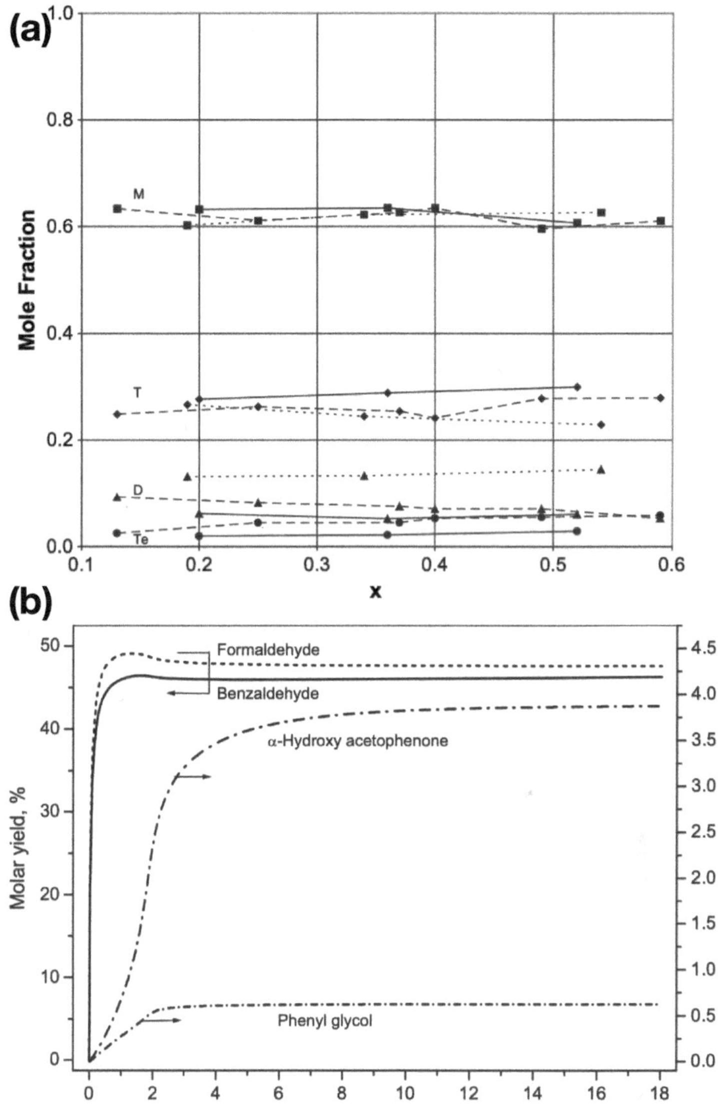

Fig. 8 Validation of kinetic Monte Carlo models. Comparison for: (A) PS gas molar fractions; (B) PSP molar yields. *Adapted from Poutsma ML: Further considerations of the sources of the volatiles from pyrolysis of polystyrene,* Polym Degrad Stab *94:2055–2064, 2009; Vinu R, Levine SE, Wang L, et al.: Detailed mechanistic modeling of poly(styrene peroxide) pyrolysis using kinetic Monte Carlo simulation,* Chem Eng Sci *69:456–471, 2012.*

2.2.4 Liquid-phase kinetic constants

To the authors' knowledge, no theoretical method to evaluate liquid-phase kinetic constants has been developed. On the other hand, several simplified approaches have been proposed in the literature based on gas-phase or structural analogies (Ranzi et al., 1997; Woo and Broadbelt, 1998). The structural analogy is based on Evans and Polanyi (1938) or Blowers and Masel (1999) relationships. These relations require estimation of reaction enthalpies, intrinsic barriers and transfer coefficients to obtain the reaction activation energies (Dogu et al., 2021). Corrections with respect to gas-phase must be included to account for the different entropic and enthalpic contributions resulting from interactions with surrounding compounds. In general, the kinetic parameters for liquid-phase reaction can be derived through the solvation free-energy corrections (Benson, 1976). The enthalpic corrections to the activation energy are related to the number of carbon atoms characterizing the polymer flow unit as per free volume theory, while the entropic corrections result from the loss of translation and rotational contributions. The flow unit can be experimentally estimated from the energy required for polymer segment mobility as a function of temperature starting from the evaporation enthalpy. For gas-phase barrierless reactions, such as radical recombination, the analogy requires further consideration as the rate determining step is not the reaction itself but the diffusive step (Ranzi et al., 1997). This approach affects mainly high activation energy reactions like chain initiations, while "caging effects" are less important for propagation reactions due to their lower energy barriers (Marongiu et al., 2007). Table 3 reports the kinetic parameters for PE pyrolysis evaluated according to the structural and gas-phase analogies. Significant differences are observed in terms of kinetic constants of initiation and termination reactions, while comparable values are obtained for the propagation steps. Although the reaction-class-based approach is consistent and robust, integration with rate constant measurements and other estimates should be included to improve the models predictions (Dogu et al., 2021).

2.3 Gasification and secondary reactions

The pyrolytic mechanisms discussed in the previous sections characterize the polymer degradation. Nevertheless, the low molecular weight compounds that form can nucleate homogeneously or heterogeneously to form bubbles that grow and break-up on the liquid surface (Faravelli et al., 1999). Secondary gas-phase reactions can play an important role both in

Table 3 Arrhenius rate parameters and rate coefficient (k) values at $T = 500\,°C$ for polyethylene degradation. Units are L, mol, cal, s. $k = A \cdot \exp(-E_a/RT)$.

	Levine and Broadbelt (2009)			Marongiu *et al.* (2007)		
Reaction type (units)	***A***	***E_a***	**k (T = 500 °C)**	***A***	***E_a***	**k (T = 500 °C)**
Initiation reactions (s^{-1})	$1.0 \cdot 10^{16}$	89.8	$4.4 \cdot 10^{-10}$	$8.0 \cdot 10^{14}$	78.0	$7.6 \cdot 10^{-8}$
Specific scission (allylic position) (s^{-1})	$1.0 \cdot 10^{16}$	75.3	$5.5 \cdot 10^{-6}$	$8.0 \cdot 10^{14}$	73.2	$1.7 \cdot 10^{-6}$
H-abstraction end-chain ($L\,mol^{-1}\,s^{-1}$)	$2.8 \cdot 10^{8}$	11.2	$1.9 \cdot 10^{5}$	$3.0 \cdot 10^{8}$	12.0	$1.3 \cdot 10^{5}$
H-abstraction mid-chain ($L\,mol^{-1}\,s^{-1}$)	–	–	–	$3.0 \cdot 10^{8}$	13.2	$5.8 \cdot 10^{4}$
Mid-chain β-scission (s^{-1})	$5.4 \cdot 10^{14}$	28.9	$3.8 \cdot 10^{5}$	$1.5 \cdot 10^{14}$	30.1	$4.6 \cdot 10^{5}$
End-chain β-scission (s^{-1})	$1.3 \cdot 10^{13}$	28.4	$1.7 \cdot 10^{5}$	$3.5 \cdot 10^{14}$	30.1	$1.1 \cdot 10^{6}$
Radical addition ($L\,mol^{-1}\,s^{-1}$)	$2.9 \cdot 10^{7}$	5.9	$6.2 \cdot 10^{5}$	–	–	–
1,4 end-Hydrogen shift (s^{-1})	$1.6 \cdot 10^{11}$	20.8	$2.1 \cdot 10^{5}$	$1.0 \cdot 10^{11}$	20.6	$1.6 \cdot 10^{5}$
1,5 end-Hydrogen shift (s^{-1})	$1.8 \cdot 10^{10}$	13.7	$2.4 \cdot 10^{6}$	$1.6 \cdot 10^{10}$	14.4	$1.4 \cdot 10^{6}$
1,6 end-Hydrogen shift (s^{-1})	$1.1 \cdot 10^{10}$	18.3	$7.1 \cdot 10^{4}$	$5.0 \cdot 10^{9}$	20.6	$7.7 \cdot 10^{3}$
1,7 end-Hydrogen shift (s^{-1})	$3.0 \cdot 10^{9}$	18.3	$2.0 \cdot 10^{4}$	–	–	–
1,3 mid-Hydrogen shift (s^{-1})	$1.0 \cdot 10^{11}$	21.2	$1.0 \cdot 10^{5}$	$1.0 \cdot 10^{11}$	20.6	$1.6 \cdot 10^{5}$
1,4 mid-Hydrogen shift (s^{-1})	$1.3 \cdot 10^{10}$	14.7	$8.8 \cdot 10^{5}$	$1.6 \cdot 10^{10}$	14.4	$1.4 \cdot 10^{6}$
1,5 mid-Hydrogen shift (s^{-1})	$7.2 \cdot 10^{9}$	18.1	$5.5 \cdot 10^{4}$	$5.0 \cdot 10^{9}$	20.6	$7.7 \cdot 10^{3}$
Recombinations ($L\,mol^{-1}\,s^{-1}$)	$1.0 \cdot 10^{11}$	2.3	$2.5 \cdot 10^{10}$	$5.0 \cdot 10^{7} \cdot T^{1}$	6.0	$7.9 \cdot 10^{8}$
Disproportionations ($L\,mol^{-1}\,s^{-1}$)	$1.0 \cdot 10^{10}$	2.3	$2.5 \cdot 10^{9}$	–	–	–

bubble-phase and in the gas-phase exiting the reactor altering the product selectivity, and therefore proper quantification of bubble and gas-phase residence times are also required (Madanikashani et al., 2022). SPW gasification can be regarded as a combination of polymer pyrolysis and secondary reactions, where the products further react with gas-phase oxidizing agents. The discussed pyrolysis models could be therefore coupled with gasification subsets to extend the range of operating conditions where the model could be applied. However, the high temperatures involved in gasification can decrease model reliability due to the increased importance of alternative chemistry pathways, e.g., the side-groups cannot be assumed inert (Audisio and Bertini, 1992). Furthermore, experimental studies show that the concentration of oxygen in the gasifying agent can affect also melt decomposition and evaporation (Kashiwagi and Ohlemiller, 1982). With respect to the gasification reactions, several detailed models have been developed for gas-phase kinetics (Ranzi et al., 2001), although care must be given to obtain mechanisms compact enough for reactor scale CFD simulations. On the other hand, char gasification is significantly more complex and still requires further work. Indeed, several model families have been developed to describe coal- and bio-char oxidation and gasification (Locaspi et al., 2021; Tilghman and Mitchell, 2015), but, to the authors' knowledge, no model has been assessed with plastic char oxidation/gasification. Nevertheless, char gasification is usually the rate determining step, impacting product selectivity significantly and thereof downstream processing costs (Tognotti et al., 1991). As with SPW, mechanistic kinetic models, coupled with structural information on pore-evolution and diffusion limitations, are required to model the overall process (Madanikashani et al., 2022).

2.4 Perspectives and challenges

Non-catalytic thermochemical conversion is an interesting pathway for SPW chemical recycling. The main advantages of these processes are their flexibility to treat complex contaminated mixtures and their scalability. Nevertheless, to optimize the wide product distribution, pollutant release and energy requirements, describing the waste conversion is of paramount importance. In this context, kinetic models coupled to CFD simulations allow for a quantitative description of the feedstock degradation. Kinetic models are either global or mechanistic. On the one hand, global models are simple and computationally inexpensive, but they represent the mass-loss profiles only in a narrow range of operating conditions on which they are fitted from experimental observations. On the other hand, mechanistic models are computationally expensive but can describe the underlying phenomena over a wide range of operating conditions in terms of both polymer degradation and product formation. While several models describing pyrolysis of several single polymers have been developed, no mechanistic model for solid plastic waste mixtures pyrolysis or gasification exists. With respect to pyrolysis, further work is required to expand the range of investigated polymers and their interactions. Additional studies are also required regarding the catalytic effect of inorganics in the mixture and secondary gas-phase cracking or gasification reactions. Solid Plastic Waste gasification can also benefit from these studies, as pyrolysis is the first hierarchical step occurring in both pyrolysis and gasification. Nevertheless, further validation of the pyrolysis models at gasification temperatures is required, as additional high-temperature exothermic competing pathways can significantly alter product selectivity. Furthermore, appropriate coupling of SPW pyrolysis models and gas- and solid-gas gasification models are required to completely characterize the process also describing char conversion.

Additionally, a general method for computing rate constants in liquid phase is missing in scientific literature. The state-of-the-art computational techniques for evaluating thermodynamics and kinetics involves implicit solvent models (Marenich et al., 2009). However, these models usually break down when unimolecular decomposition reactions are investigated because of the formation of a singularity inside the solvent accessible surface. For this reason, explicit solvent models are required for studying this kind of chemical reactions. The most recent developments in this field are originated by the Free Energy Perturbation Method (Zwanzig, 2004). Preliminary work is being carried out at Politecnico di Milano exploiting standard metadynamics coupled with Mean Force Integration algorithm (Marinova and Salvalaglio, 2019) for studying the unimolecular decomposition (or β-scission) of $CH_3CHClCH_2\dot{C}HCl$ into $CH_2CHCl + CH_3\dot{C}HCl$ (and the reverse, propagation) in water with the PM6-D3 quantum Hamiltonian, which aims at reproducing the rate limiting step of polyvinylchloride synthesis by suspension polymerization. This reaction is corroborated by experimental and computational data that can be found in the literature (Burnett and Wright, 1954; Kajiwara and Kamachi, 2000) allowing to test the validity and accuracy of the newly proposed methodology in the evaluation of kinetic parameters. Many other data are available from the IUPAC joint effort as reported in (Beuermann et al., 2022) paving the way for the definition of a general method for calculating rate parameters of unimolecular reactions in explicit solvent for application in polymerization and depolymerization (i.e. chemical recycling through pyrolysis) processes.

3. Catalytic thermochemical conversion

Catalytic pyrolysis differs from thermal pyrolysis, which relies uniquely on the thermal energy to degrade the polymers, for the addition of an active surface that influences both the reaction temperature and the product distribution. The introduction of a catalyst in the pyrolysis process allows to lower the reaction temperature from 450–500 °C to 300–350 °C thus reducing energy and production costs. In this sense, the effect of the catalyst is to lower the activation energy of cracking proportionally to its own activity. The higher the activity of the catalyst, the lower the cracking temperature (Aguado et al., 2007).

The second main advantage of catalytic cracking over thermal cracking is a higher selectivity to specific products with higher added value. The use of a cracking catalyst facilitates the production of a narrower product distribution (Lin et al., 2010; López et al., 2011a; Serrano et al., 2012) which can be tuned towards molecules of industrial interest, e.g., light olefins, gasoline or diesel by selecting a suitable catalyst and changing the operating conditions. In general, these products are lighter than those from thermal pyrolysis, which still contain impurities and residues. As a result, the products of catalytic pyrolysis show a larger fraction of gaseous species and a lower fraction of liquids, yet with a higher quality. As an example, Fig. 9 shows the composition of pyrolysis yield for thermal and catalytic cracking of HDPE, highlighting the difference between gaseous and liquid composition.

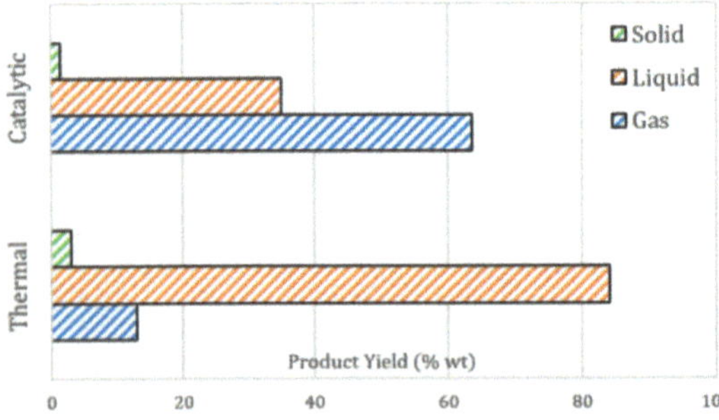

Yield (%wt)	Thermal	Catalytic
Gas	**13**	**63.5**
Liquid	**84**	**35**
C_6-C_{12}	*56.55*	*99.92*
C_{13}-C_{23}	*37.79*	*0.08*
$>C_{23}$	*5.66*	*0.0*
Solid	**3.0**	**1.5**

Fig. 9 Product yield comparison between thermal and catalytic pyrolysis of HDPE. *Figure taken from Daligaux V, Richard R, Manero MH: Deactivation and regeneration of zeolite catalysts used in pyrolysis of plastic wastes—a process and analytical review,* Catalysts *11, 2021. Adapted from the data of Seo Y, Lee K, Shin D: Investigation of catalytic degradation of high- density polyethylene by hydrocarbon group type analysis,* J Anal Appl Pyrolysis *70, 2003.*

Conversion values close to 100% can be achieved via the catalytic cracking of plastic waste (Sahu et al., 2013). However, despite this high conversion value, the main disadvantage lies in the need for pure polymers to avoid contaminants present in the mixed waste plastic steam, such as chloride or nitrogen components, which can lead to catalyst deactivation. (Achilias et al., 2007; Almeida and de Marque, 2016).

A pre-treatment of the waste is then often required to prevent contamination. Deactivation effects arise also from unwanted gas/vapor-solid reactions, poisoning of chemisorbed species that block the active sites, thermal degradation and coke deposition.

3.1 Classes of catalysts

Catalysis for the pyrolysis of plastics can be either homogeneous or heterogeneous, although the former is no longer used nowadays as it poses the problem of separating the catalyst from the products. In heterogeneous catalysis, the fluid product mixture is easily separated by the solid catalyst providing an economically viable solution for the reuse of an expensive catalyst.

According to the definition of Buekens and Huang (1998), catalytic cracking reactions can be classified into two groups: liquid-phase and vapor-phase. On the one hand, the catalyst is placed in the reactor, coming in direct contact with the molten polymer phase. On the other hand, the vapor formed during cracking are brought into contact with the catalyst in an independent catalytic bed.

Three families of catalysts are commonly employed in the pyrolysis of plastics: zeolites, fluid catalytic cracking catalysts and silica-alumina catalysts (Anuar Sharuddin et al., 2016; Daligaux et al., 2021).

Zeolites are aluminosilicate with a well-defined crystal structure, which differ from the silica-alumina catalyst which instead have an amorphous structure. The crystal structure of zeolites consists in a three-dimensional framework composed of T-O4 tetrahedral structural unit, where T is the central atom (i.e., Si or Al) surrounded by O atoms connecting the units together. The interconnection of these building blocks gives rise to different three-dimensional microporous networks and specific structural properties (Chester and Derouane, 2009). The Si/Al ratio determines the number of surface acid sites, which affects the overall reactivity of the zeolite. The most used zeolite catalyst is the HZSM-5 which has been employed for the pyrolysis of PE (Elordi et al., 2011; Sakata et al., 1999; Schirmer et al., 2001; Takuma et al., 2000), HDPE (del Remedio Hernandez et al., 2007; Lin et al., 2004; Liu et al., 2014; Mastral et al., 2006; Miskolczi et al., 2009,

Miskolczi et al., 2004; Navarro et al., 2009), LPDE (Gobin and Manos, 2004; Zhang et al., 2015; Zhang and Lei, 2016) and PP (Elordi et al., 2011), due to its high catalytic activity, better selectivity and limited deactivation. Other commonly adopted zeolites are HUSY (del Remedio Hernandez et al., 2007; Huang et al., 2010; Lin and Yang, 2008a, b; Navarro et al., 2009), Hβ (Elordi et al., 2009; Renzini et al., 2011), HY (Elordi et al., 2009; Kunwar et al., 2016; Seo et al., 2003), HZSM-11 (Renzini et al., 2011) and HMOR (Garforth et al., 1998).

Fluid catalytic cracking (FCC) catalysts consist in a zeolite crystal and non-zeolite acid matrix known as silica-alumina with the binder (Anuar Sharuddin et al., 2016). FCC units are commonly employed in the petrochemical refining industry to crack heavy oil to lighter products. Catalysts that have been used in refinery processes, known as spent FCC catalysts, can be reused in the plastic pyrolysis process. The plastic feed is solved in a suitable solvent and the solution is then pyrolyzed. Lee et al. (2002) report that the spent FCC catalyst is still active for the plastic pyrolysis with a liquid yield above 80 wt% for samples of HDPE, LDPE, PP and PS. The usage of FCC catalyst is encouraged to maximize the liquid product, with the condition of keeping the catalyst/polymer ratio lower than 20 wt% to minimize formation of coke and gaseous product (Abbas-abadi et al., 2014; Olazar et al., 2009). Silica-alumina catalysts are amorphous acid catalysts containing Bronsted and Lewis acid sites, with an ionizable hydrogen atom and an electron accepting site, respectively. The acidity of the catalyst is determined by the SiO_2/Al_2O_3 mole ratio oppositely to zeolite, i.e., high ratios indicate high acidity. The degree of acidity strongly influences the end product of plastic pyrolysis (Sakata et al., 1996). Another frequently used catalyst in the pyrolysis of plastics is MCM-41, a non-zeolite mesoporous material of the family of silicate and aluminosilicate characterized by a regular arrangement of cylindrical mesopores (Aguado et al., 2007; Artetxe et al., 2015; van Grieken et al., 2001; Huang et al., 2010; Lin et al., 2004). The review of Lopez (Lopez et al., 2017) summarizes the product distribution obtained in the catalytic pyrolysis of polyolefins over different catalysts and different reactors.

3.2 Kinetics of the catalytic cracking

While thermal pyrolysis proceeds via a free-radical mechanism, the catalytic degradation reaction has a very complex ionic mechanism (Miskolczi, 2006). When the plastic feed is introduced into the reactor, the polymer melts and spreads over the catalyst surface, where it reacts with the acid active sites initiating the degradation reaction. The initial step is agreed to be the adsorption of the reactant on the acid site and the production of a carbenium ion, which promotes the cracking of the molecule. The carbenium ion is produced either by abstraction of hydride ion at the Lewis acid site or by the proton addition to the polymer by Bronsted acid site. The acidity of the catalyst strongly affects the reaction rate and mechanism. When the acidity is strong, the catalytic cracking proceeds by end-chain scission, forming olefins. Conversely, in weak acidic catalysts, the cracking process occurs via random scission, leading to the formation of waxes. These initially formed products react further to produce lighter molecule, as shown in Fig. 10. The final product composition is strongly influenced by the operating parameters.

The kinetic models reported in the literature for catalytic cracking can be divided into two groups: (i) global kinetic models, based on thermal analysis of the weight loss curves, (ii) models with lumping schemes. In this contribution we aim to provide a review of these methods focusing on their application in the field of the catalytic cracking of plastics.

3.2.1 Models based on the thermal analysis of the weight loss curves

Thermogravimetric analysis (TGA) is used to determine the weight loss of the sample with increasing temperature at a constant heating rate. A thermogravimetric apparatus continuously measures the mass while the temperature is increased over time. When the product molecules evaporate in the gas fraction, the mass loss of the sample is detected.

To describe the kinetics of this non-isothermal solid-state reactions the kinetic equations of isothermal homogeneous gas or liquid phase

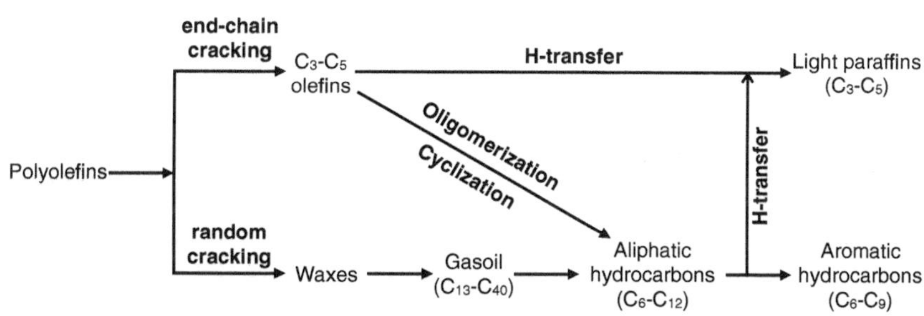

Fig. 10 Reaction pathways for catalytic pyrolysis of polyolefins. *Adapted from Aguado J, Serrano DP, Sotelo JL, et al.: Influence of the operating variables on the catalytic conversion of a polyolefin mixture over HMCM-41 and nanosized HZSM-5,* Appl Catal B Environ *25:5696–5704, 2001.*

(Baker, 1978) reactions for non-catalytic conversion is used as in Eq. (1). Nevertheless, thermogravimetric experiments are conducted at different heating rates $\beta = \frac{dT}{dt}$, hence Eq. (1) can be rearranged as:

$$\frac{da}{f(a)} = k_0 \, exp\left(-\frac{E_{act}}{RT}\right)\frac{1}{\beta}dT \tag{2}$$

Isoconversional methods allow to process TGA data, determining in the most accurate way the E_{act}-α dependence without the explicit knowledge of $f(\alpha)$, analyzing TGA curves at different heating rates for the same value of conversion (Vyazovkin et al., 2011). From the slope of the logarithmic heating rate curves it is possible to extract the activation energy. The underlying principle is that the reaction rate at a constant extent of conversion is only a function of the temperature (Sbirrazzuoli, 2020). Integral and differential methods to solve Eq. (2) are explained comprehensively in the review of Mortezaeikia (Mortezaeikia et al., 2021).

Fernandes and coworkers (Fernandes et al., 1997) computed the activation energies of thermal and catalytic degradation of HDPE on HZSM-5 zeolite catalyst integrating Eq. (2) using the formula of Doyle (1962). The activation energy for the HDPE degradation was found to be 170.2 kJ/mol for the thermal degradation and 67.0 kJ/mol in the presence of the catalyst, showing a remarkable lowering of the barrier.

Filho et al. (2005) applied the isoconversional model-free integral method of Vyazovkin (Vyazovkin and Wight, 1999) in the study of PP degradation on ZSM-12 and ZSM-5 catalysts. The ZSM-12 catalyst was found to be more efficient than ZSM-5 in the PP cracking, due to the larger diameter of the channel which allows larger fragment of the polymer to enter the channel and be cracked. The integral method of Vyazovkin was used by Fernandes et al. (2002) to study the kinetic parameters of the catalytic degradation of HDPE over a silicoaluminate SAPO-37 catalyst. The activation energy in absence of the catalyst was 290 kJ/mol, which is lowered to 220 kJ/mol when the catalyst is added.

Durmuş et al. (2005) integrated Eq. (2) with the Kissinger method (Kissinger, 1957), which assumes a generic n-th order conversion to approximate $f(\alpha) = (1 - \alpha)^n$. The activation energy can be determined regardless of the reaction order by TGA analysis of different heating rates. The authors investigated the thermo-catalytic pyrolysis of PP over BEA, ZSM-5 and MOR zeolite catalysts, finding that $E_{act}(ZSM) > E_{act}(BEA) > E_{act}(MOR)$. However, the catalyst with the lowest computed activation energy is not the one with the highest measured catalytic activity, probably due to the pore structure of MOR which favors the formation of coke, lowering the degradation rate.

Chandrasekaran et al. (2015) analyzed the TGA data of thermal and catalytic cracking of different plastics waste over zeolites (ZSM-5 and Y), $MgCO_3$, $Zr(SO_4)_2$, MCM-41 and spent FCC catalyst. The activation energies are reported in Table 4. Y-Zeolites were found to be more suitable for PP and PE cracking, while spent FCC and $Zr(SO_4)_2$ for PS and polyurethane foams (PUF), respectively.

Khedri (2017) computed the kinetic parameters of catalytic cracking of HDPE over ZSM-5 using the kinetic study previously developed by their group for thermal cracking (Khedri and Elyasi, 2016). In the presence of the catalyst the rate coefficient increases, especially at lower temperature, whereas coke formation and catalyst deactivation are reduced.

In the recent work of Kremer et al. (2021) the isoconversional model-free method of Friedman (1964) has been used to analyze TGA data of non-recyclable waste plastic mixture, using Fe-ZSM and FCC catalysts. The activation energy E_{act} as a function of the conversion α is reported in Fig. 11. Two regions with different E_{act} can be distinguished for all the samples, indicating a two-step decomposition mechanism.

The first step can be attributed to diffusion processes and the corresponding activation energy is lowered from 144 kJ/mol in the thermal case to 95 and 85 kJ/mol in the presence of FCC and Fe-ZSM catalyst, respectively. The second step is attributed to a Avrami-Erofeev reaction type, whose activation energy is reduced by 6.5% and 8.3% when the catalysts are added.

Al-Yaari and Dubdub (2020) tested three isoconversional methods, namely Friedman (1964), Flynn-Wall-Qzawa (FWQ) and Kissinger-Akahira-Suose (KAS) and two non-isoconversional methods, i.e., Arrenhius and Coats-Redfern to compute the activation energy of the catalytic pyrolysis of

Table 4 Activation energy of thermal and catalytic pyrolysis of different plastic feedstock (Chandrasekaran et al., 2015).

Feedstock	E_{act}^{therm} (kJ/mol)	E_{act}^{cat} (kJ/mol)
PP	182	113
PE	294	169
PUF	30	11
PS	182	45
HDPE	294	87

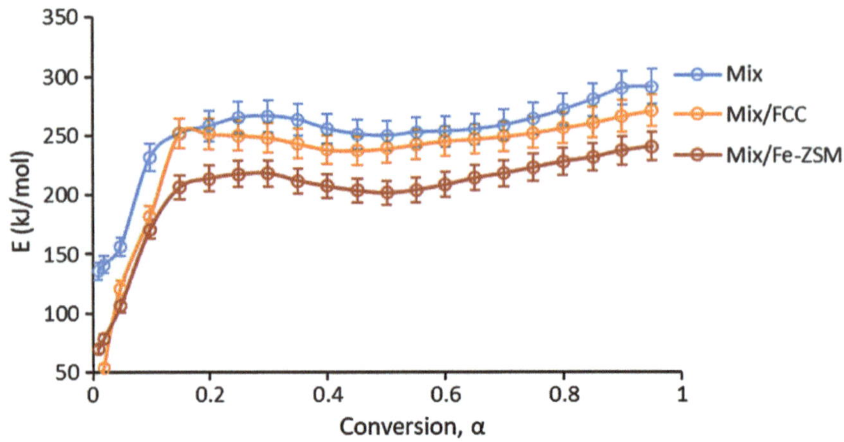

Fig. 11 Activation energy computed with the Friedmann method as a function of the conversion (Kremer et al., 2021).

Table 5 Activation energy (kJ/mol) of the catalytic cracking of HDPE over HZSM-5 predicted by isoconversional and non-isoconversional methods (Al-Yaari and Dubdub, 2020).

	HZSM-5/HDPE = 0.5	HZSM-5/HDPE = 0.77	HZSM-5/HDPE = 1
Friedmann	186	135.5	178.7
FWO	196	178	219.75
KSA	194.5	177.25	184.5
Coats-Redfern	144.7	131.7	131.7
Arrhenius	145	139	140

HDPE over HZSM-5, at different catalyst/polymer ratio. As shown in Table 5, the activation energies predicted by isoconversional methods share the same trend as a function of the catalyst/polymer ratio while the absolute value of the activation energy depends on the specific model.

The group of Marcilla has published several works in which TGA data were used to fit the parameters of a kinetic model of increasing complexity (Marcilla et al., 2007, Marcilla et al., 2003a; Marcilla et al., 2003b, Marcilla et al., 2001b, Marcilla et al., 2001b). The first version of the model was introduced to interpret the TGA data of LDPE pyrolysis over different catalysts, such as MCM-41, USY and ZSM-5 (Marcilla et al., 2001a). The model assumes that in the presence of the catalyst there are two parallel pathways for the decomposition:

$$P \xrightarrow{k_1} V_1 \tag{W1}$$

$$P + C \xrightarrow{k_v} [P - C] \xrightarrow{k_3} V_2 \tag{W2}$$

where P represents the polymer, C the catalyst, $[P-C]$ a reaction intermediate and V_1 and V_2 the volatile species formed during the thermal ($W1$) and the catalytic pathway ($W2$), respectively. The kinetic model of the thermal decomposition is a nth order Arrhenius law, while the k_v constant has a saturating dependence on the concentration of the catalyst. The model has 8 parameters, with a well-defined physical meaning, to be optimized, namely three pre-exponential factors, three activation energies, one reaction order and one parameter K_F representing the amount of the catalyst needed to obtain an acceleration of the reaction which is half of the maximum that can be obtained at a definite temperature. The parameters are obtained minimizing an objective function, O.F., defined as:

$$O.F. = \sum_i \sum_j \left[\frac{\left(\frac{dW}{dt}\right)_{exp_{i,j}} - \left(\frac{dW}{dt}\right)_{calc_{i,j}}}{\left(\frac{dW}{dt}\right)^{\max}_{exp\ i,j}} \right]^2 \tag{3}$$

where i represents the experimental data at time t of the experiment j. The weight loss of the model is computed from the kinetic equations describing the evolution of the decomposition and the concentration of the species. The agreement with the experimental data is good, as shown in Fig. 12. The reaction order is close to zero (0.0624), in accordance with the reported data of thermal pyrolysis. The ratio between the rate constant of the thermal and the catalytic pathway lies between 0.00027 and 0.012 indicating that the thermal pathway is not kinetically relevant.

In a subsequent work (Marcilla et al., 2001b), the model was improved since the agreement with the experimental data of pyrolysis of several PE over MCM-41 was not satisfactory, with significant deviations at low temperature and high heating rate. The kinetic parameters were then optimized considering that the weight measured at any time is the addition of all the solid species present in the reactor and that a gap between nominal and actual temperature exists. The reactions are then modified as:

$$P \xrightarrow{k_1} V_1 \tag{W1}$$

$$P + aC \xrightarrow{k_v} (1 + a)[P - C] \xrightarrow{k_3} V_2 + aC \tag{W2}$$

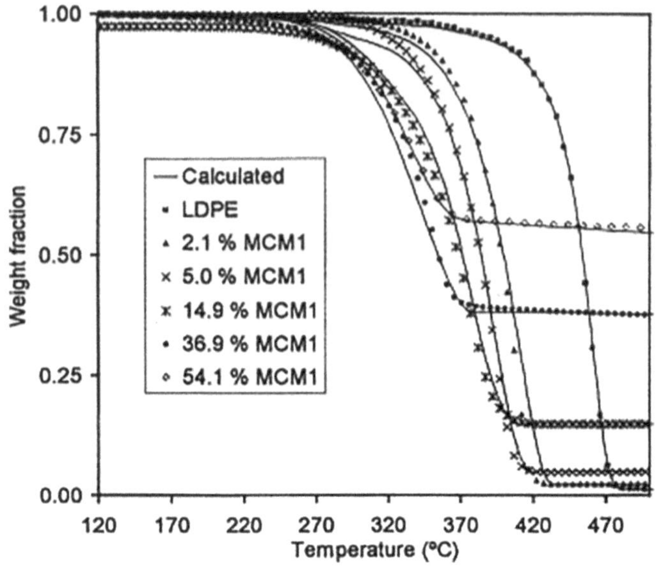

Fig. 12 Experimental and calculated TGA curves of the catalytic pyrolysis of LDPE over MCM-1 (Marcilla et al., 2001a).

while the actual temperature at the time t_j differs from the nominal one proportionally to the mass present in the sample

$$T_j = T_{j,exp} + a_i W_{calc} \tag{4}$$

In the above equations a represents the mass ratio (in grams) of catalyst per gram of polymer necessary to obtain $[P-C]$ while α_i is a parameter that depend on the initial mass. The introduction of these parameters improves the agreement with the experimental data.

The saturating effect within k_v was removed to give rise to a more general model that successfully reproduced the catalytic pyrolysis of PP over MCM-41 catalyst (Marcilla et al., 2003a) and over ZSM-5 and FCC catalyst (Marcilla et al., 2003b):

$$P \xrightarrow{k_1} V_1 \tag{W3}$$

$$P + aC \xrightarrow{k_2} (1+a)[P-C] \xrightarrow{k_3} V_2 + aC \tag{W4}$$

k_v was substituted with a generic k_2 and the concentration of each species is represented with a n-th order kinetics. When the eleven parameters are optimized, the model correctly reproduces the TGA curves of PP cracking. The activation energies of the three steps of the PP decomposition are reported in Table 6.

Table 6 Comparison of the activation energies of the three decomposition steps of PP over different catalysts (Marcilla et al., 2003b, Marcilla et al., 2003a).

Activation energy	MCM-41	FCC	ZSM-6
E_1 (kJ/mol)	204.4	204.4	246.3
E_2 (kJ/mol)	157.5	160.5	130.1
E_3 (kJ/mol)	40.8	53.5	121.8

The activation energies of the thermal reaction (E_1) are much higher than those in the catalyzed reaction. However, it is worth noticing that ZSM-5 shows a higher value of E_3 compared to the other catalysts which means that the [P-C] intermediate has a slower decomposition rate, hence a longer lifetime.

This scheme has proven to be effective in describing the catalytic cracking of polymers in single-step processes, yet it fails in reproducing the shoulders in the derivative TGA curves, which are related to the existence of two-step decomposition mechanisms as it is the case for the LDPE cracking over Hβ catalyst (Marcilla et al., 2007). To account for this experimental observation, the model is modified as follows:

$$P \xrightarrow{k_0} V_1 \tag{W5}$$

$$P + aC \xrightarrow{k_1} (1 + a)[P - C] \xrightarrow{k_2} (1 - b)V_2 + aC + bP^* \tag{W6}$$

$$P^* + dC \xrightarrow{k_3} (1 + d)\left[P^* - C\right] \xrightarrow{k_4} dC + V_3 \tag{W7}$$

The catalytic reaction proceeds through two different steps. First, the acid site causes partial cracking of the polymer (W6) and the oligomers formed in this stage are later cracked in the second step (W7). $[P-C]$ and $[P^* - C]$ are the activated complexes formed in the two catalytic pathways, while the parameters a and d represent the ratio between the gram of catalysts and the gram of polymer necessary to obtain the activated complexes and $(1 - b)$ is the fraction of volatile compounds. The kinetic parameters obtained in this way allow to reproduce correctly the experimental TGA curves, as shown in Fig. 13.

3.2.2 Models with lumping schemes

In lumped kinetic schemes, the large number of individual constituents in a complex feedstock are grouped into classes of compounds, known as lumps, with a simplified reaction network connecting them. Each lump is treated as

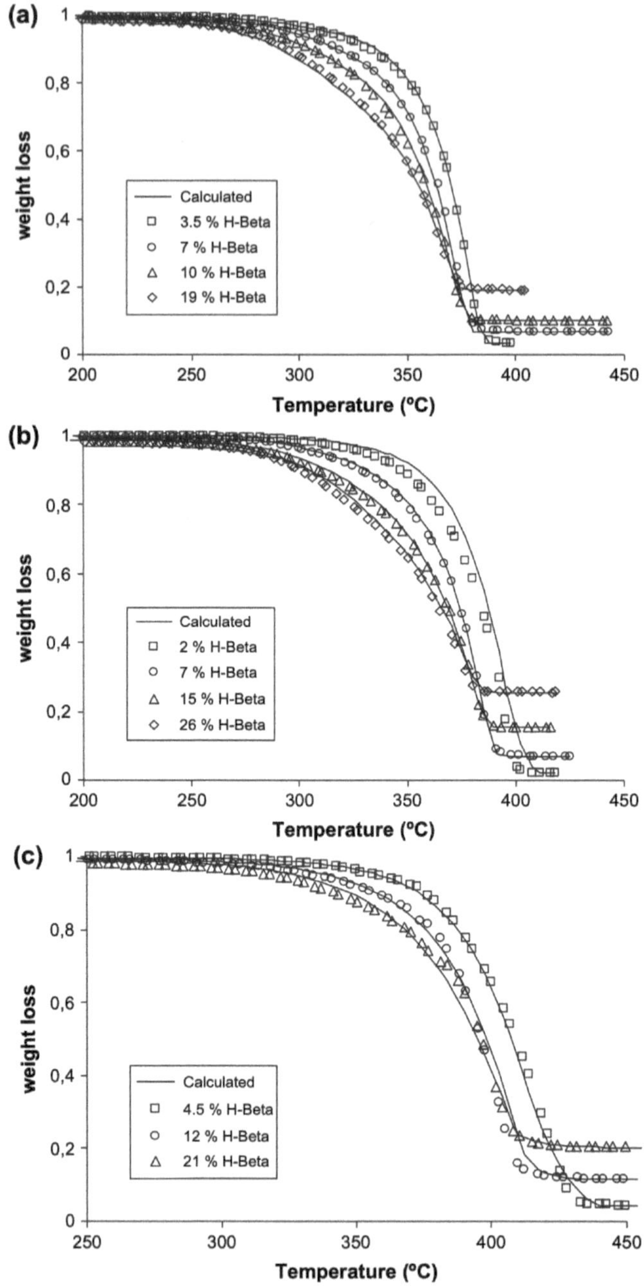

Fig. 13 Experimental and calculated TGA curves for the catalytic degradation of LDPE using Hβ zeolite catalyst at different heating rates: (A) 5 °C/min, (B) 10 °C/min, (C) 35 °C/min (Marcilla et al., 2007).

a single component. Lumped models generally rely on the following assumptions (Lin et al., 2001):

- Liquid phase polymer. The solid polymer introduced into the reactor immediately melts and disperses around the catalyst, forming a polymer/catalyst complex and starting the reaction at the surface. Melting and dispersion time of the polymer are negligible.
- Evolution of the intermediates. Scission reactions generate intermediates including long-chain olefins and intermediate precursors of carbenium ion. The number of precursors is limited by the number of active sites, and the carbenium ions concentration rapidly reaches the steady-state. Alkanes may be generated via hydrogen transfer.
- Evolution of products. Once the intermediates are produced, further reactions could be expected to produce smaller molecules, such as smaller-chain olefins, alkanes, benzene, toluene and xylene (BTX) and coke, in equilibrium with the surface carbenium ions.

Once a lumped model is proposed, the rate of production and consumption of each lump is written accordingly. The kinetic parameters are obtained from theoretical calculations, structural analysis or group contributions. However, because of the complexity of underlying phenomena, a simple least square minimization is employed to obtain the parameters that reduce deviations between the experiments and the calculated results.

In this context, Lin and coworkers (Lin et al., 2001) propose a generic lumped model, which provided a good representation of the catalytic degradation of HDPE and PP over fluidized acidic catalysts such as HZSM-5, HUSY and the silica-alumina SAHA. The model comprises six lumps: the first representing the liquid phase polymer species (W_{c*}), the second the intermediates of olefins and carbenium ions (W_I), and the other representing four different families of products (Fig. 14).

The rate of formation $r_{i,j}$ of the product i from the reactant r through reaction j is written as:

$$r_{i,j} = k_{i,j} W_r^{n_j} \eta_j \tag{5}$$

where $k_{i,j}$ is the rate constant of the product i from the j-th reaction, W_r the weight fraction of the reactant r on the acid sites and n_j the reaction order of the j-th reaction. η_j represents the catalyst activity decay due to presence of coke, which is assumed to be the same for all the reaction steps:

$$\eta_j = \eta = \exp\left(-\alpha W_r\right) \tag{6}$$

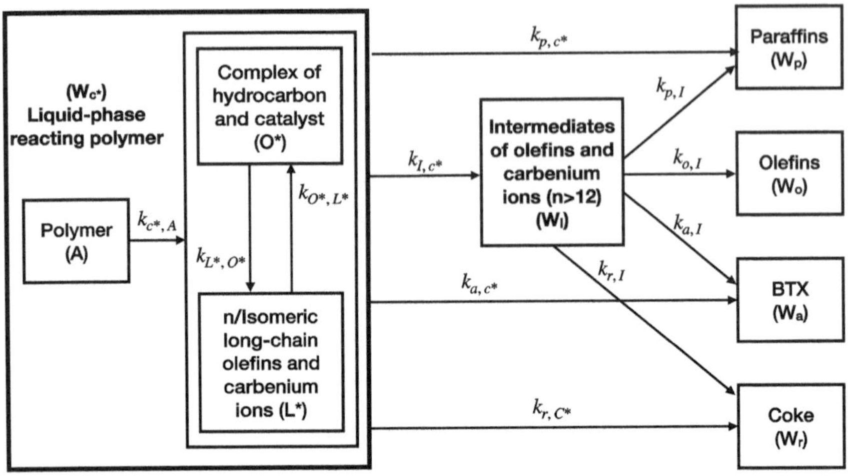

Fig. 14 Mechanistic and kinetic scheme representing polymer degradation over cracking catalysts. *Adapted from Lin YH, Hwu WH, Ger MD, et al.: A combined kinetic and mechanistic modelling of the catalytic degradation of polymers,* J Mol Catal A Chem *171:143–151, 2001.*

Table 7 Apparent rate constants for HDPE over various cracking catalyst at 360 °C.

Catalyst	$k_{i,\ c*}$	$k_{r,\ c*}$	$k_{p,\ c*}$	$k_{a,\ c*}$	$k_{O,\ i}$	$k_{r,\ i}$	$k_{p,\ i}$	$k_{a,\ i}$
HZSM-5	245.5	1.3	1.3	0,04	30.1	0.8	10.9	0.98
HUSY	329.3	2.8	0.86	0.03	18.6	2.3	·25.7	0.66
SAHA	182.2	5.1	0.06	<0.01	31.5	2.1	2.7	<0.01

The constants are expresses as (10^{-2} min^{-1}) (Lin et al., 2001).

where W_r is the mass of the coke deposited on the catalyst and α a constant found to be 0.163 $(\%w/w)^{-1}$ for HDPE. The rate of consumption and production of each lumped species is written assuming simple first-order reactions, which are numerically integrated to obtain the rate constants (Table 7).

The generation of intermediates has by far the highest rate constant, $k_{i,\ c*}$, which is higher for zeolite catalysts than SAHA, indicating that zeolites are more effective in converting the polymer into intermediates of volatile precursors than the amorphous SAHA. Zeolites show a faster rate of generation of olefinic ($k_{O,\ i}$) and paraffinic lumps ($k_{p,\ i}$) rather than BTX ($k_{a,\ i}$) and coke ($k_{r,\ i}$). In SAHA, the generation of the olefinic lump is much faster than all the other lumps.

This six-lump method was later applied in different works, extending the finding to different feedstocks. Lin and Yang adopted this method to describe the catalytic cracking of PP and PE over HSZM-5, equilibrium spent catalyst ECat-1 and SAHA (Lin and Yang, 2008a, b). Huang et al. (2010) and Lin et al. (2010) employed the six-lump method to study the cracking of hospital plastic waste, mixture of PE/PP/PS over zeolitic and non-zeolitic catalysts.

Miskolczi and coworkers (Miskolczi et al., 2004) modeled the cracking reaction rates of PE into different fuels over HZSM, FCC and a natural clinoptilolite containing rhyolite tuff (CRT). A scheme of the model is reported in Fig. 15.

The resulting model, in which each reaction is represented by a first-order kinetics, shows good correlation with the experimental data. The rate constant of catalytic paraffin formation k_{pi} is lower than in the thermal degradation, while the opposite occurs for internal olefins formation k_{ioi}. The values of k_{ri} and k_{iw} reflect the reduced solid residue formation in the presence of the catalyst.

A four-lump model was introduced by Lin and Yang (2008a, b) to model the catalytic cracking of a mixture of HDPE and LDPE using both zeolitic (HZSM-5, HUSY) and non-zeolitic (SAHA) catalysts in a fluidized bed reactor. The model consists in only four lumps, representing the polymer (A), gasoline (B), light gases (C) and coke (D). The kinetic equations are written consistently with the model proposed in Fig. 16. Rate constants are reported in Table 8.

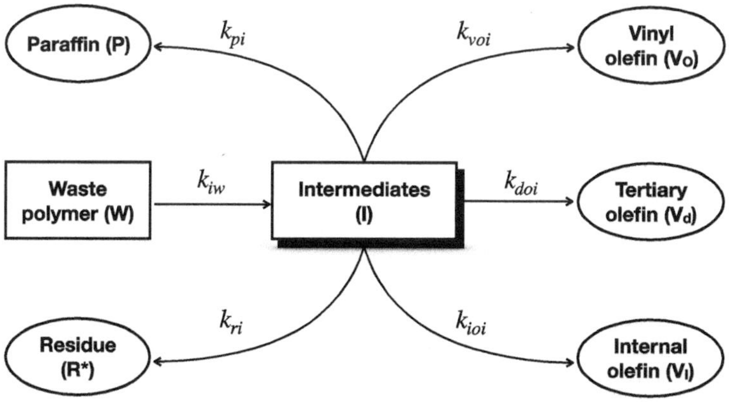

Fig. 15 Scheme of the polymer waste degradation. *Adapted from Miskolczi N, Bartha L, Deák G, et al.: Kinetic model of the chemical recycling of waste polyethylene into fuels,* Process Saf Environ Prot *82:223–229, 2004.*

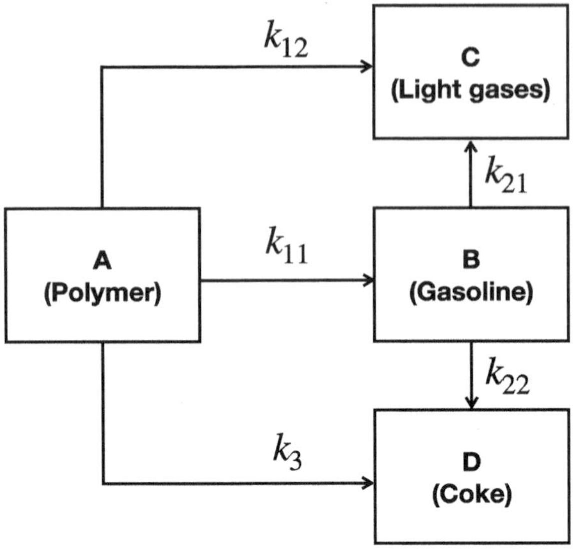

Fig. 16 Reaction scheme proposed in the four-lump model. *Adapted from Lin YH, Yang MH: A novel approach for the kinetic and mechanistic modeling of acid-catalyzed degradation of polymers,* Asia-Pacific J Chem Eng *4:147–153, 2008a; Lin YH, Yang MH: Tertiary recycling of polyethylene waste by fluidised-bed reactions in the presence of various cracking catalysts,* J Anal Appl Pyrolysis *83:101–109, 2008b.*

Table 8 Apparent rate constants and product selectivity for PE cracking over different catalysts at 360 °C.

Catalyst	k_{11}	k_{12}	k_3	k_{21}	S_B (%)	S_C (%)	S_D (%)
HUSY	24.3	13.9	1.34	0.15	57.7	37.1	3.5
HZSM-5	22.4	43.4	0.36	0.22	34.2	63.9	0.5
SAHA	16.9	9.3	0.80	0.03	58.5	34.2	2.7

The rate constants are expresses as (10^{-2} min^{-1}) (Lin and Yang, 2008a, b).

k_{11} and k_{12} are higher for zeolites than SAHA, suggesting that they depend on the acidity of the active sites. However, the selectivity to gasoline $S_B = k_{11}/(k_{11}+k_{12}+k_3)$ is similar for SAHA and HUSY and a lower value is found for HZSM-5, which, in contrast, has a much higher selectivity for light gases $S_C = k_{12}/(k_{11}+k_{12}+k_3)$. Moreover, the high value of the k_{21} constant indicates a gasoline overcracking towards light gases. The relative low value of coke selectivity $S_D = k_3/(k_{11}+k_{12}+k_3)$ suggests that the pore structure of HSZM-5 inhibits coke formation.

Artetxe et al. (2014) reported a kinetic method based on lumps which includes the catalyst deactivation to model cracking of HDPE pyrolysis volatiles on a HSZM-5 catalyst. The experiment was carried out in two steps: (i) a thermal pyrolysis in a CSBR at 500 °C and (ii) catalytic cracking of the volatiles (waxes, C_{12+}) formed in the first step in a fixed bed reactor. Consequently, the kinetic model consists in two steps as well: (i) a kinetic scheme established according to the results at zero time on stream (after thermal pyrolysis) and (ii) inclusion of the catalyst deactivation on the results obtained at zero time on stream. The first part of the model, reported in Fig. 17, has an increased

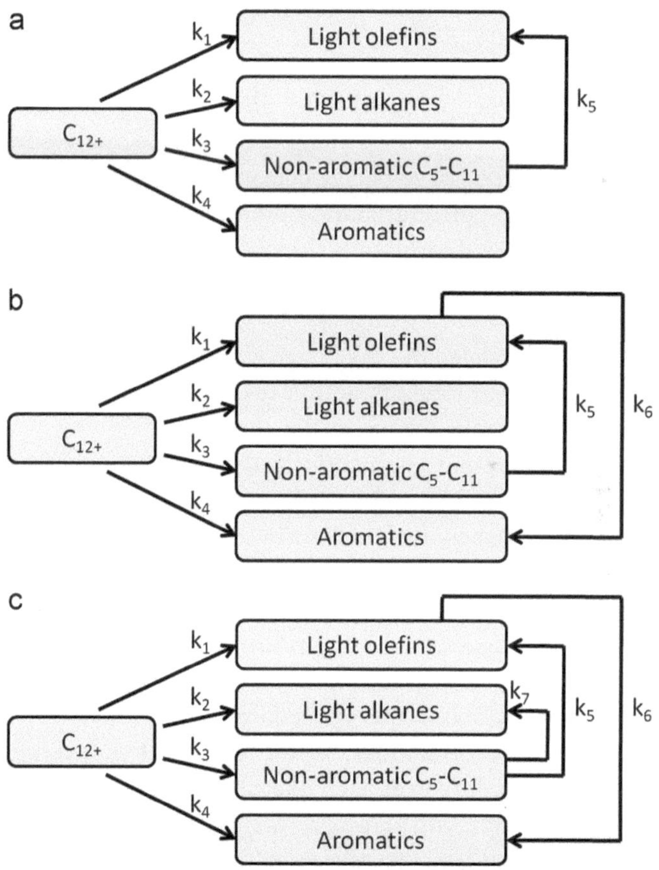

Fig. 17 Possible kinetic schemes for the catalytic cracking of the volatiles formed in the pyrolysis of HDPE. Reactions among the four primary products are gradually added through k_5 (A), k_6 (B) and k_7 (C). *Taken from Artetxe M, Lopez G, Amutio M, et al.: Kinetic modelling of the cracking of HDPE pyrolysis volatiles on a HZSM-5 zeolite based catalyst,* Chem Eng Sci *116:635–644, 2014.*

complexity in order to evaluate the effect of including the condensation reaction of light olefins to aromatic compound (B) and cracking of non-aromatic to yield light alkane (C). Among these, only the inclusion of the former was found to have a significant kinetic relevance.

The kinetic model at zero time of stream in Fig. 17 provides information about the potential reaction on the adopted catalyst, yet its deactivation must be included to quantify the evolution of component concentration over time. The proposed kinetic equation for catalyst deactivation considers the deactivation rate proportional to wax concentration:

$$\frac{da}{dt} = -k_d[C_{12+}]\alpha \tag{7}$$

where k_d is the kinetic constant of deactivation and α is the activity defined as the ratio between the reaction rate at time t and time zero. The kinetic equations of model B of Fig. 17 are then integrated with the catalyst deactivation and the kinetic parameters have been calculated by fitting the experimental results of product lump concentration at different space-times and time on streams. The product evolution with time of stream is in good agreement with the experiments and the activation energy associated to k_d was found to be 13.4 kJ/mol.

Till et al. (2018) modeled the two-stage pyrolysis of a HDPE/PP/LDPE mixture on zeolite-based catalysts with a reaction network with ten reactions connecting six lumps, shown in Fig. 18.

The six lumps represent the plastic feedstock (P), a cracked polymer intermediate (P-), heavy (L+) and light (L−) liquids, gas (G) and coke (C). First order reaction between the lumps were considered. The kinetic parameters were optimized introducing a novel algorithm which requires no a-priori knowledge about slow or fast reactions, nor temperature dependence of the rate coefficients. The algorithm consists in two identification and one approximation steps as reported in Fig. 18. We refer to the original work for more details about the procedure. The model can reproduce several experimental features, such as the gas yield increasing effect of Ni/Mo-Al_2O_3 catalyst, however the coke depletion on this catalyst is not captured by the model. Furthermore, the results suggest that thermo-catalytic pyrolysis takes place in parallel reactions and the effect of the catalyst of the further cracking of gaseous products is not imminent.

3.3 Perspectives and challenges

In the previous sections, we presented the two classes of kinetic models mostly employed to theoretically describe the catalytic pyrolysis of

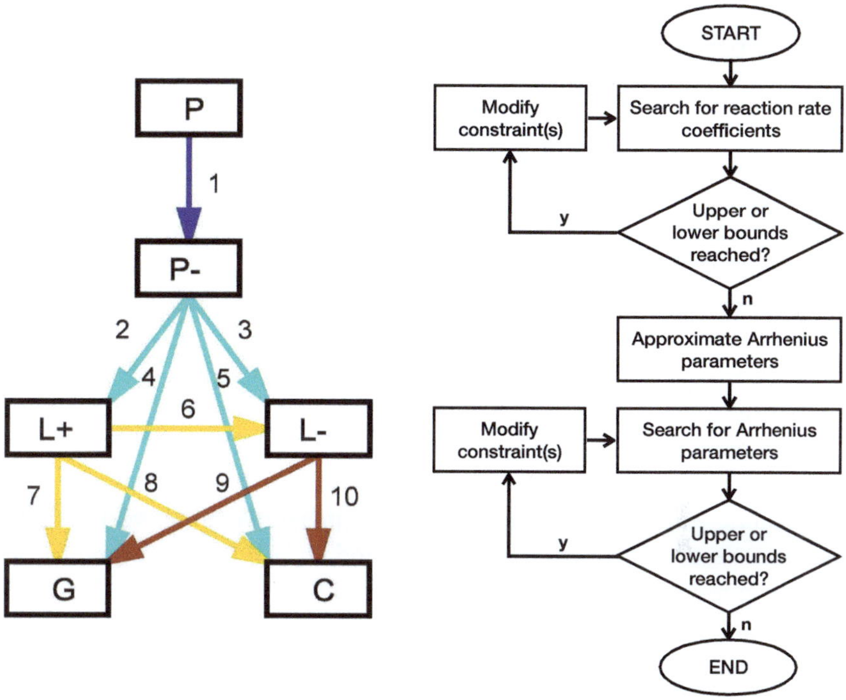

Fig. 18 Reaction network with ten reactions between six lumps to model the catalytic cracking of a mixture of HDPE/PP/LDPE (left). Identification strategy of the kinetic parameters (right). *Adapted from Till Z, Varga T, Só ja J, et al.: Kinetic identification of plastic waste pyrolysis on zeolite-based catalysts,* Energy Convers Manag *173:320–330, 2018.*

plastic waste. Although they provide a detailed description of the product distribution, they are still of a rather empirical nature. In this section, we briefly present other techniques that have been recently adopted in the study of catalytic pyrolysis of plastic waste, but which are not yet of common use.

Zavala-Gutierrez and coworkers (Zavala-Gutiérrez et al., 2019) reported the first mathematical model for the catalytic pyrolysis of HDPE based on the chemistry of the carbenium ion. The proposed mechanism goes beyond and form of empiricism, since it is founded on a solid knowledge of the kinetics of FCC of relatively smaller hydrocarbons. The mechanism consists of three steps: (i) initiation, which consists in a protonation reaction at the Bronsted site of olefins, resulting in the formation of carbenium ions, and paraffins. Alternatively, a hydride ion can be extracted from paraffins in a Lewis acid site. (ii) Propagation, in which the products obtained in the first step reduce their length through successive reactions with the acid center of the catalyst. (iii) Termination, where products are desorbed from the surface, returning a proton to restore the active site. Other secondary reactions are not included. A large set of ordinary differential equations (from 2000 to 50,000) is derived from the mass balance of all the species that participate the reaction in a plug-flow reactor. The model is compared to catalytic cracking data from the literature, showing a good agreement, making it a promising tool for future studies.

De la Flor-Barriga and Rodríguez-Zúñiga (2022) simulated the catalytic pyrolysis of HDPE over ZSM-5 catalyst in a fluidized bed reactor using computational fluid dynamics (CFD) simulations. The kinetic mechanism of Artexte (Artetxe et al., 2014) was employed. The main outcome of the work is that a feed rate of 1 kg/m^2 s at 500 °C yields the best gasoline production while consuming the lowest amount of energy per kg of product. This work shows that it is possible to obtain operative information from CFD simulation, which may help in the setting up the experiments.

Machine learning based tools, such as artificial neural network (ANN), have been recently employed by Al-Yaari and Dubdub (2020) to predict the catalytic pyrolysis of HDPE using non-isothermal TGA data. An ANN scheme with two hidden layers has been developed to predict the mass left percentage, based on a set of 900 experimental TGA data points taken from the literature. 70% of the data were used for training, 15% for validation and 15% for testing. The predicted values of the ANN showed a very good agreement with the experimental data ($R^2 > 0.999$) and an efficient performance for new input data.

Saha and Ghoshal (2007) adopted a hybrid genetic algorithm (HGA) to minimize the objective function that determines the kinetic parameters of the catalytic decomposition of LDPE over ZSM-5 from TGA data. HGA adopts the probabilistic approach of standard genetic algorithms, improving their convergence to global optima via a local optimization algorithm. The ZSM-5 catalyzed reaction shows multiple peaks in the $\frac{d\alpha}{dt}$ versus T plot, indicating the existence of a multistep reaction, different from the thermal cracking, where a single peak is observed. The HGA approach predicts kinetic parameters in agreement with the experimental data and with a model-free approach, coupled with local optimization algorithm. In a later work, HGA was employed to estimate the kinetic parameters of the catalytic cracking of PP over MCM-41, based on 15 decomposition model and multi-heating experimental data (Saha et al., 2008). According to the HGA analysis, the degradation mechanism of PP changes from initiation-propagation to random scission method after the consumption of weak links. Similarly, Reddy et al. (2008) explored parallel approaches based on genetic algorithms to study the catalytic pyrolysis of PP over HZSM-5, find that GA can predict the variations of kinetic parameters depending on the catalyst used. These works represent a first application of machine learning tools to model the data obtained from catalytic pyrolysis (Armenise et al., 2021).

To the best of the author's knowledge, quantum mechanical method such as density functional theory have not been employed in the study of the kinetics of catalytic cracking of solid plastic waste both due to the complexity of the mechanism and the prohibitive computational cost. Efforts in this direction would be beneficial for speeding up model development and to achieve a higher degree of predictive capabilities.

4. Conclusions and outlooks

Catalytic and non-catalytic thermochemical conversions allow to recycle complex mixtures of SPW. These processes are advantageous because of their flexibility in treating complex contaminated mixtures and their scalability. On the other hand, the wide product distributions, pollutant release and energy requirements are among the main drawbacks. Chemical kinetic models and CFD tools aid in describing waste conversion to quantitatively optimize the process.

Kinetic models are either global or mechanistic. On the one hand, global models are simple and computationally inexpensive, but they represent the mass-loss profiles only in a narrow range of operating conditions on which they are fitted from experimental observations. On the other hand, mechanistic models are computationally expensive but can describe the underlying phenomena over a wide range of operating conditions in terms of both polymer degradation ad product formation. Nevertheless, to the best of the authors' knowledge, no mechanistic model describing SPW mixture conversion exists, neither for catalytic nor non-catalytic valorization.

Several kinetic models describing pyrolysis of several single polymers have been developed, but future works is still required, starting with expanding the investigated range of polymers and their interactions. Model development is performed in a hierarchical fashion, where pyrolysis is the first hierarchical step while gasification and catalytic pyrolysis are secondary reactions. With respect to gasification, for gas-phase reactions several models have been proposed but further studies are required to describe the heterogeneous plastic-char conversion. In this context, studies on coal-char and bio-char can be employed as reference, although with appropriate consideration. On the other hand, studies on catalytic pyrolysis are at a significantly less mature stage. The presence of a catalyst allows to tailor the product distribution and lower process temperatures but introduces several complexities in modeling activities. With respect to vapor-phase cracking, as for gasification a modular approach can be employed considering gas-solid cracking kinetic models already developed. On the other hand, for liquid-cracking, to the best of the authors' knowledge, no mechanistic model has been developed and further work is required starting with the single polymer-catalyst pair.

In this context, estimation of kinetic constants and reaction pathways is of paramount importance both in catalytic and non-catalytic conversion. At present, kinetic constants are obtained from experiments or simplified approaches based on gas-phase or structural analogies. While accurate quantum methods have been developed for gas- and, at a lower extent, gas-solid reactions, further work is required to consider liquid-phase phenomena.

References

Abbas-abadi MS, Nekoomanesh M, Yeganeh H, et al.: Evaluation of pyrolysis process parameters on polypropylene degradation products, *J Anal Appl Pyrolysis* 109:272–277, 2014.

Achilias DS, Roupakias C, Megalokonomos P, et al.: Chemical recycling of plastic wastes made from polyethylene (LDPE and HDPE) and polypropylene (PP), *J Hazard Mater* 149:536–542, 2007.

Aguado R, Olazar M, Gaisán B, et al.: Kinetics of polystyrene pyrolysis in a conical spouted bed reactor, *Chem Eng J* 92:91–99, 2003.

Aguado J, Serrano DP, Miguel GS, et al.: Catalytic activity of zeolitic and mesostructured catalysts in the cracking of pure and waste polyolefins, *J Anal Appl Pyrolysis* 78:153–161, 2007.

Almeida D, de Marque MF: Thermal and catalytic pyrolysis of polyethylene plastic waste in semi, *Polimeros* 26:1–8, 2016.

Al-Salem SM, Lettieri P, Baeyens J: Recycling and recovery routes of plastic solid waste (PSW): a review, *Waste Manag* 29:2625–2643, 2009.

Al-Yaari M, Dubdub I: Application of artificial neural networks to predict the catalytic pyrolysis of hdpe using non-isothermal TGA data, *Polymers (Basel)* 12:1813, 2020.

Anuar Sharuddin SD, Abnisa F, Wan Daud WMA, et al.: A review on pyrolysis of plastic wastes, *Energy Convers Manag* 115:308–326, 2016.

Arena U: Process and technological aspects of municipal solid waste gasification. A review, *Waste Manag* 32:625–639, 2012.

Armenise S, Wong S, Ramírez-Velásquez JM, et al.: Application of computational approach in plastic pyrolysis kinetic modelling: a review, *React Kinet Mech Catal* 134:591–614, 2021.

Artetxe M, Lopez G, Amutio M, et al.: Kinetic modelling of the cracking of HDPE pyrolysis volatiles on a HZSM-5 zeolite based catalyst, *Chem Eng Sci* 116:635–644, 2014.

Artetxe M, Lopez G, Amutio M, et al.: Styrene recovery from polystyrene by flash pyrolysis in a conical spouted bed reactor, *Waste Manag* 45:126–133, 2015.

Audisio G, Bertini F: Molecular weight and pyrolysis products distribution of polymers. I. Polystyrene, *J Anal Appl Pyrolysis* 24:61–74, 1992.

Avella M, Cosco S, Di Lorenzo ML, et al.: Nucleation activity of nanosized CaCO3 on crystallization of isotactic polypropylene, in dependence on crystal modification, particle shape, and coating, *Eur Polym J* 42:1548–1557, 2006.

Baker R: Kinetic parameters from the non-isothermal decomposition of a multi-component solid, *Thermochim Acta* 23:201–212, 1978.

Bamford CH, Tompa H: The calculation of molecular weight distributions from kinetic schemes, *Trans Faraday Soc* 50:1097–1115, 1954.

Benson SW: *Thermochemical kinetics*, 1976, John Wiley & Sons.

Beuermann S, Harrisson S, Hutchinson RA, et al.: Update and critical reanalysis of IUPAC benchmark propagation rate coefficient data, *Polym Chem* 13:1891–1900, 2022.

Bhaskar T, Kaneko J, Muto A, et al.: Pyrolysis studies of PP/PE/PS/PVC/HIPS-Br plastics mixed with PET and dehalogenation (Br, Cl) of the liquid products, *J Anal Appl Pyrolysis* 72:27–33, 2004.

Blowers P, Masel RI: An extension of the marcus equation for atom transfer reactions, *J Phys Chem A* 103:7047–7054, 1999.

Bockhorn H, Hornung A, Hornung U: Mechanisms and kinetics of thermal decomposition of plastics from isothermal and dynamic measurements, *J Anal Appl Pyrolysis* 50:77–101, 1999a.

Bockhorn H, Hornung A, Hornung U, et al.: Kinetic study on the non-catalysed and catalysed degradation of polyamide 6 with isothermal and dynamic methods, *Thermochim Acta* 337:97–110, 1999b.

Bouster C, Vermande P, Veron J: Study of the pyrolysis of polystyrenes: I. Kinetics of thermal decomposition, *J Anal Appl Pyrolysis* 1:297–313, 1980.

Brems A, Baeyens J, Beerlandt J, et al.: Thermogravimetric pyrolysis of waste polyethylene-terephthalate and polystyrene: a critical assessment of kinetics modelling, *Resour Conserv Recycl* 55:772–781, 2011.

Buekens AG, Huang H: Catalytic plastics cracking for recovery of gasoline-range hydrocarbons from municipal plastic wastes, *Resour Conserv Recycl* 23:163–181, 1998.

Burnett GM, Wright WW: The photosensitized polymerization of vinyl chloride in tetrahydrofuran solution III. Determination of the kinetic coefficients, *Proc R Soc London Ser A Math Phys Sci* 221:41–53, 1954.

Chandrasekaran SR, Kunwar B, Moser BR, et al.: Catalytic thermal cracking of postconsumer waste plastics to fuels. 1. Kinetics and optimization, *Energy Fuel* 29:6068–6077, 2015.

Chattopadhyay DK, Webster DC: Thermal stability and flame retardancy of polyurethanes, *Prog Polym Sci* 34:1068–1133, 2009.

Chen D, Yin L, Wang H, et al.: Pyrolysis technologies for municipal solid waste: A review, *Waste Manag* 34:2466–2486, 2014.

Chester AW, Derouane EG: *Zeolite characterization and catalysis*, 2009, Springer. https://link.springer.com/book/10.1007/978-1-4020-9678-5.

Czégény Z, Blazsó M: Thermal decomposition of polyamides in the presence of poly(vinyl chloride), *J Anal Appl Pyrolysis* 58–59:95–104, 2001.

Daligaux V, Richard R, Manero MH: Deactivation and regeneration of zeolite catalysts used in pyrolysis of plastic wastes—a process and analytical review, *Catalysts* 11, 2021.

De la Flor-Barriga LA, Rodríguez-Zúñiga UF: Numerical analysis on a catalytic pyrolysis reactor design for plastic waste upcycling using CFD modelling, *RSC Adv* 12:12436–12445, 2022.

De Oliveira LP, Hudebine D, Guillaume D, et al.: A review of kinetic modeling methodologies for complex processes, *Oil Gas Sci Technol – Rev d'IFP Energies Nouv* 71:45, 2016.

Dente M, Ranzi E, Goossens AG: Detailed prediction of olefin yields from hydrocarbon pyrolysis through a fundamental simulation model (SPYRO), *Comput Chem Eng* 3:61–75, 1979.

Dente M, Bozzano G, Faravelli T, et al.: Kinetic modelling of pyrolysis processes in gas and condensed phase, *Adv Chem Eng* 32:51–166, 2007.
Dhahak A, Hild G, Rouaud M, et al.: Slow pyrolysis of polyethylene terephthalate: online monitoring of gas production and quantitative analysis of waxy products, *J Anal Appl Pyrolysis* 142, 2019, 104664.
Dhahak A, Grimmer C, Neumann A, et al.: Real time monitoring of slow pyrolysis of polyethylene terephthalate (PET) by different mass spectrometric techniques, *Waste Manag* 106:226–239, 2020.
Dickens B: Thermally degrading polyethylene studied by means of factor-jump thermogravimetry, *J Polym Sci Polym Chem Ed* 20:1065–1087, 1982.
Dogu O, Pelucchi M, Van de Vijver R, et al.: The chemistry of chemical recycling of solid plastic waste via pyrolysis and gasification: State-of-the-art, challenges, and future directions, *Prog Energy Combust Sci* 84:100901, 2021.
Doyle CD: Estimating isothermal life from thermogravimetric data, *J Appl Polym Sci* 6:639–642, 1962.
Durmuş A, Koç SN, Pozan GS, et al.: Thermal-catalytic degradation kinetics of polypropylene over BEA, ZSM-5 and MOR zeolites, *Appl Catal B Environ* 61:316–322, 2005.
Elordi G, Olazar M, Lopez G, et al.: Catalytic pyrolysis of HDPE in continuous mode over zeolite catalysts in a conical spouted bed reactor, *J Anal Appl Pyrolysis* 85:345–351, 2009.
Elordi G, Olazar M, Lopez G, et al.: Continuous polyolefin cracking on an HZSM-5 zeolite catalyst in a conical spouted bed reactor, *Ind Eng Chem Res* 50:6061–6070, 2011.
Evans MG, Polanyi M: Inertia and driving force of chemical reactions, *Trans Faraday Soc* 34:11–24, 1938.
Faravelli T, Bozzano G, Scassa C, et al.: Gas product distribution from polyethylene pyrolysis, *J Anal Appl Pyrolysis* 52:87–103, 1999.
Faravelli T, Pinciroli M, Pisano F, et al.: Thermal degradation of polystyrene, *J Anal Appl Pyrolysis* 60:103–121, 2001.
Faravelli T, Bozzano G, Colombo M, et al.: Kinetic modeling of the thermal degradation of polyethylene and polystyrene mixtures, *J Anal Appl Pyrolysis* 70:761–777, 2003.
Fernandes VJ, Araujo AS, Fernandes GJT: Catalytic degradation of polyethylene evaluated by TG, *J Therm Anal* 49:255–260, 1997.
Fernandes GJT, Fernandes VJ, Araujo AS: Catalytic degradation of polyethylene over SAPO-37 molecular sieve, *Catal Today* 75:233–238, 2002.
Filho JGAP, Graciliano EC, Silva AOS, et al.: Thermo gravimetric kinetics of polypropylene degradation on ZSM-12 and ZSM-5 catalysts, *Catal Today* 107–108:507–512, 2005.
Flynn JH, Wall LA: General treatment of the thermogravimetry of polymers, *J Res Natl Bur Stand Sect A Phys Chem* 70A:487, 1966.
Friedman HL: Kinetics of thermal degradation of char-forming plastics from thermogravimetry. Application to a phenolic plastic, *J Polym Sci C* 6:183–195, 1964.
Garforth AA, Lin YH, Sharratt PN, et al.: Production of hydrocarbons by catalytic degradation of high density polyethylene in a laboratory fluidised-bed reactor, *Appl Catal A Gen* 169:331–342, 1998.
Gillespie DT: A general method for numerically simulating the stochastic time evolution of coupled chemical reactions, *J Comput Phys* 22:403–434, 1976.
Gillespie DT: Exact stochastic simulation of coupled chemical reactions, *J Phys Chem* 81:2340–2361, 1977.
Gillespie DT: Stochastic simulation of chemical kinetics, *Annu Rev Phys Chem* 58:35–55, 2007.
Gobin K, Manos G: Polymer degradation to fuels over microporous catalysts as a novel tertiary plastic recycling method, *Polym Degrad Stab* 83:267–279, 2004.
Grassie N, Zulfiqar M, Guy MI: Thermal degradation of a series of polyester polyurethanes, *J Polym Sci Polym Chem Ed* 18:265–274, 1980.

Harmon RE, Sribala G, Broadbelt LJ, et al.: Insight into polyethylene and polypropylene pyrolysis: global and mechanistic models, *Energy Fuel* 35:6765–6775, 2021.
Horstemeyer MF: Multiscale modeling: a review, *Pract Asp Comput Chem Methods Concepts Appl*:87–135, 2010.
Huang WC, Huang MS, Huang CF, et al.: Thermochemical conversion of polymer wastes into hydrocarbon fuels over various fluidizing cracking catalysts, *Fuel* 89:2305–2316, 2010.
Hughes TJR: Multiscale phenomena: Green's functions, the Dirichlet-to-Neumann formulation, subgrid scale models, bubbles and the origins of stabilized methods, *Comput Methods Appl Mech Eng* 127:387–401, 1995.
Kajiwara A, Kamachi M: ESR observation of propagating radical and estimation of propagation rate constant in the radical polymerization of vinyl chloride, *Macromol Chem Phys* 201:2165–2169, 2000.
Kaminsky W, Eger C: Pyrolysis of filled PMMA for monomer recovery, *J Anal Appl Pyrolysis* 58–59:781–787, 2001.
Kashiwagi T, Ohlemiller TJ: A study of oxygen effects on nonflaming transient gasification of PMMA and PE during thermal irradiation, *Symp Combust* 19:815–823, 1982.
Khedri S: *Kinetic and thermodynamic studies on pyrolysis of waste polymers (HDPE) (PhD thesis)*, 2017, Lakehead University.
Khedri S, Elyasi S: Corrigendum to kinetic analysis for thermal cracking of HDPE: a new, *Polym Degrad Stab* 133:330–338, 2016.
Kissinger HE: Reaction kinetics in differential thermal analysis, *Anal Chem* 303:1702–1706, 1957.
Knümann R, Bockhorn H: Investigation of the kinetics of pyrolysis of PVC by TG-MS-analysis, *Combust Sci Technol* 101:285–299, 2010.
Koo JK, Kim SW: Reaction kinetic model for optimal pyrolysis of plastic waste mixtures, *Waste Manag Res* 11:515–529, 1993.
Kremer I, Tomić T, Katančić Z, et al.: Catalytic pyrolysis of mechanically non-recyclable waste plastics mixture: kinetics and pyrolysis in laboratory-scale reactor, *J Environ Manag* 296, 2021.
Kruse TM, Woo OS, Wong HW, et al.: Mechanistic modeling of polymer degradation: a comprehensive study of polystyrene, *Macromolecules* 35:7830–7844, 2002.
Kruse TM, Wong HW, Broadbelt LJ: Modeling the evolution of the full polystyrene molecular weight distribution during polystyrene pyrolysis, *Ind Eng Chem Res* 42:2722–2735, 2003a.
Kruse TM, Wong HW, Broadbelt LJ: Mechanistic modeling of polymer pyrolysis: polypropylene, *Macromolecules* 36:9594–9607, 2003b.
Kruse TM, Levine SE, Wong HW, et al.: Binary mixture pyrolysis of polypropylene and polystyrene: a modeling and experimental study, *J Anal Appl Pyrolysis* 73:342–354, 2005.
Kulesza K, German K: Chlorinated pyrolysis products of co-pyrolysis of poly(vinyl chloride) and poly(ethylene terephthalate), *J Anal Appl Pyrolysis* 67:123–134, 2003.
Kumagai S, Motokucho S, Yabuki R, et al.: Effects of hard- and soft-segment composition on pyrolysis characteristics of MDI, BD, and PTMG-based polyurethane elastomers, *J Anal Appl Pyrolysis* 126:337–345, 2017.
Kunwar B, Moser BR, Chandrasekaran SR, et al.: Catalytic and thermal depolymerization of low value post-consumer high density polyethylene plastic*, *Energy* 111:884–892, 2016.
Lee K, Noh N, Shin D, et al.: Comparison of plastic types for catalytic degradation of waste plastics into liquid product with spent FCC catalyst, *Polym Degrad Stab* 78:539–544, 2002.
Levine SE, Broadbelt LJ: Detailed mechanistic modeling of high-density polyethylene pyrolysis: low molecular weight product evolution, *Polym Degrad Stab* 94:810–822, 2009.

Lin YH, Yang MH: A novel approach for the kinetic and mechanistic modeling of acid-catalyzed degradation of polymers, *Asia-Pacific J Chem Eng* 4:147–153, 2008a.
Lin YH, Yang MH: Tertiary recycling of polyethylene waste by fluidised-bed reactions in the presence of various cracking catalysts, *J Anal Appl Pyrolysis* 83:101–109, 2008b.
Lin YH, Hwu WH, Ger MD, et al.: A combined kinetic and mechanistic modelling of the catalytic degradation of polymers, *J Mol Catal A Chem* 171:143–151, 2001.
Lin HT, Huang MS, Luo JW, et al.: Hydrocarbon fuels produced by catalytic pyrolysis of hospital plastic wastes in a fluidizing cracking process, *Fuel Process Technol* 91:1355–1363, 2010.
Lin YH, Yang MH, Yeh TF, et al.: Catalytic degradation of high density polyethylene over mesoporous and microporous catalysts in a fluidised-bed reactor, *Polym Degrad Stab* 86:121–128, 2004.
Liu M, Zhuo JK, Xiong SJ: Catalytic degradation of high-density polyethylene over a clay catalyst compared with other catalysts, *Energy Fuels* 28(9):6038–6045, 2014. https://doi.org/10.1021/ef501326k.
Locaspi A, Debiagi P, Pelucchi M, et al.: A predictive physico-chemical model of biochar oxidation, *Energy Fuel* 35:14894–14912, 2021.
Locaspi A, Pelucchi M, Mehl M, et al.: Towards a lumped approach for solid plastic waste gasification: polyethylene and polypropylene pyrolysis, *Waste Manag*, 2022.
Loffredo M: A lumped kinetic model for the pyrolysis of polyethylene (Master Thesis), *Politecnico di Milano*, 2021.
Lombardi L, Carnevale E, Corti A: A review of technologies and performances of thermal treatment systems for energy recovery from waste, *Waste Manag* 37:26–44, 2015.
López A, de Marco I, Caballero BM, et al.: Deactivation and regeneration of ZSM-5 zeolite in catalytic pyrolysis of plastic wastes, *Waste Manag* 31:1852–1858, 2011a.
López A, De Marco I, Caballero BM, et al.: Dechlorination of fuels in pyrolysis of PVC containing plastic wastes, *Fuel Process Technol* 92:253–260, 2011b.
Lopez G, Artetxe M, Amutio M, et al.: Thermochemical routes for the valorization of waste polyolefinic plastics to produce fuels and chemicals. A review, *Renew Sust Energ Rev* 73:346–368, 2017.
Madanikashani S, Vandewalle LA, De Meester S, et al.: Multi-scale modeling of plastic waste gasification: opportunities and challenges, *Materials* 15:4215, 2022.
Maestri M, Cuoci A: Coupling CFD with detailed microkinetic modeling in heterogeneous catalysis, *Chem Eng Sci* 96:106–117, 2013.
Marcilla A, Beltran M, Conesa JA: Catalyst addition in polyethylene pyrolysis: thermogravimetric study, *J Anal Appl Pyrolysis* 58–59:117–126, 2001a.
Marcilla A, Gómez A, Reyes-Labarta JA: MCM-41 catalytic pyrolysis of ethylene-vinyl acetate copolymers: Kinetic model, *Polymer (Guildf)* 42:8103–8111, 2001b.
Marcilla A, Gómez A, Reyes-Labarta JA, et al.: Catalytic pyrolysis of polypropylene using MCM-41: kinetic model, *Polym Degrad Stab* 80:233–240, 2003a.
Marcilla A, Gómez A, Reyes-Labarta JA, et al.: Kinetic study of polypropylene pyrolysis using ZSM-5 and an equilibrium fluid catalytic cracking catalyst, *J Anal Appl Pyrolysis* 68–69:467–480, 2003b.
Marcilla A, Gómez-Siurana A, Valdés F: Catalytic cracking of low-density polyethylene over H-Beta and HZSM-5 zeolites: Influence of the external surface. Kinetic model, *Polym Degrad Stab* 92:197–204, 2007.
Marenich AV, Cramer CJ, Truhlar DG: Universal solvation model based on solute electron density and on a continuum model of the solvent defined by the bulk dielectric constant and atomic surface tensions, *J Phys Chem B* 113:6378–6396, 2009.
Marinova V, Salvalaglio M: Time-independent free energies from metadynamics via mean force integration, *J Chem Phys* 151, 2019, 164115.
Marongiu A, Faravelli T, Bozzano G, et al.: Thermal degradation of poly(vinyl chloride), *J Anal Appl Pyrolysis* 70:519–553, 2003.

Marongiu A, Faravelli T, Ranzi E: Detailed kinetic modeling of the thermal degradation of vinyl polymers, *J Anal Appl Pyrolysis* 78:343–362, 2007.

Mastan E, Zhu S: Method of moments: A versatile tool for deterministic modeling of polymerization kinetics, *Eur Polym J* 68:139–160, 2015.

Mastral JF, Berrueco C, Gea M, et al.: Catalytic degradation of high density polyethylene over nanocrystalline HZSM-5 zeolite, *Polym Degrad Stab* 91:3330–3338, 2006.

Mastral JF, Berrueco C, Ceamanos J: Theoretical prediction of product distribution of the pyrolysis of high density polyethylene, *J Anal Appl Pyrolysis* 80:427–438, 2007.

Mehl M, Marongiu A, Faravelli T, et al.: A kinetic modeling study of the thermal degradation of halogenated polymers, *J Anal Appl Pyrolysis* 72:253–272, 2004.

Miranda R, Yang J, Roy C, et al.: Vacuum pyrolysis of commingled plastics containing PVC I. Kinetic study, *Polym Degrad Stab* 72:469–491, 2001.

Miskolczi N: Kinetic Model of the Chemical and Catalytic Recycling of Waste Polyethylene into Fuels, *Feed Recycl Pyrolysis Waste Plast Convert Waste Plast into Diesel Other Fuels* 225–47, 2006.

Miskolczi N, Bartha L, Deák G, et al.: Kinetic model of the chemical recycling of waste polyethylene into fuels, *Process Saf Environ Prot* 82:223–229, 2004.

Miskolczi N, Angyal A, Bartha L, et al.: Fuels by pyrolysis of waste plastics from agricultural and packaging sectors in a pilot scale reactor, *Fuel Process Technol* 90:1032–1040, 2009.

Moens EKC, De Smit K, Marien YW, et al.: Progress in reaction mechanisms and reactor technologies for thermochemical recycling of poly(methyl methacrylate), *Polymer* 12:1667, 2020.

Mortezaeikia V, Tavakoli O, Khodaparasti MS: A review on kinetic study approach for pyrolysis of plastic wastes using thermogravimetric analysis, *J Anal Appl Pyrolysis* 160:105340, 2021.

Mueller ME, Blanquart G, Pitsch H: Hybrid method of moments for modeling soot formation and growth, *Combust Flame* 156:1143–1155, 2009.

Navarro R, Marcilla A, Beltra MI: Thermal and catalytic pyrolysis of polyethylene over HZSM5 and HUSY zeolites in a batch reactor under dynamic conditions, *Appl Catal B Environ* 86:78–86, 2009.

Neeft JPA, Hoornaert F, Makkee M, et al.: The effects of heat and mass transfer in thermogravimetrical analysis. A case study towards the catalytic oxidation of soot, *Thermochim Acta* 287:261–278, 1996.

Nobili A, Pejpichestakul W, Pelucchi M, et al.: Modeling soot particles as stable radicals: a chemical kinetic study on formation and oxidation. Part II. Soot oxidation in flow reactors and laminar flames, *Combust Flame*:112072, 2022.

Ohtani H, Kimura T, Okamoto K, et al.: Characterization of polyurethanes by high-resolution pyrolysis-capillary gas chromatography, *J Anal Appl Pyrolysis* 12:115–133, 1987.

Olazar M, Lopez G, Amutio M, et al.: Influence of FCC catalyst steaming on HDPE pyrolysis product distribution, *J Anal Appl Pyrolysis* 85:359–365, 2009.

Poutsma ML: Reexamination of the pyrolysis of polyethylene: data needs, free-radical mechanistic considerations, and thermochemical kinetic simulation of initial product-forming pathways, *Macromolecules* 36:8931–8957, 2003.

Poutsma ML: Comparison of literature models for volatile product formation from the pyrolysis of polyisobutylene at mild conditions: Data analysis, free-radical mechanistic considerations, and simulation of initial product-forming pathways, *J Anal Appl Pyrolysis* 73:159–203, 2005.

Poutsma ML: Mechanistic analysis and thermochemical kinetic simulation of the products from pyrolysis of poly(α-methylstyrene), especially the unrecognized role of phenyl shift, *J Anal Appl Pyrolysis* 80:439–452, 2007.

Poutsma ML: Further considerations of the sources of the volatiles from pyrolysis of polystyrene, *Polym Degrad Stab* 94:2055–2064, 2009.

Prud'homme RE: Miscibility phenomena in polyester/chlorinated polymer blends, *Polym Eng Sci* 22:90–95, 1982.

Ragaert K, Delva L, Van Geem K: Mechanical and chemical recycling of solid plastic waste, *Waste Manag* 69:24–58, 2017.

Ranzi E, Dente M, Faravelli T, et al.: Kinetic modeling of polyethylene and polypropylene thermal degradation, *J Anal Appl Pyrolysis* 40–41:305–319, 1997.

Ranzi E, Dente M, Goldaniga A, et al.: Lumping procedures in detailed kinetic modeling of gasification, pyrolysis, partial oxidation and combustion of hydrocarbon mixtures, *Prog Energy Combust Sci* 27:99–139, 2001.

Ranzi E, Frassoldati A, Stagni A, et al.: Reduced kinetic schemes of complex reaction systems: Fossil and biomass-derived transportation fuels, *Int J Chem Kinet* 46:512–542, 2014.

Ranzi E, Debiagi PEA, Frassoldati A: Mathematical modeling of fast biomass pyrolysis and bio-oil formation. Note II: secondary gas-phase reactions and bio-oil formation, *ACS Sustain Chem Eng* 5:2882–2896, 2017.

Ravey M, Pearce EM: Flexible polyurethane foam. I. Thermal decomposition of a polyether-based, water-blown commercial type of flexible polyurethane foam, *J Appl Polym Sci* 63:47–74, 1998.

Reddy PK, Chowlu ACK, Ghoshal AK: Hybrid genetic algorithm and model-free coupled direct search methods for kinetics of nanocrystalline ZSM-5-catalyzed decomposition of PP, *Appl Catal A Gen* 351:195–203, 2008.

del Remedio Hernandez M, Gomez A, Garcia A, et al.: Effect of the temperature in the nature and extension of the primary and secondary reactions in the thermal and HZSM-5 catalytic pyrolysis of HDPE, *Appl Catal A Gen* 317:183–194, 2007.

Renzini MS, Lerici LC, Sedran U, et al.: Stability of ZSM-11 and BETA zeolites during the catalytic cracking of low-density polyethylene, *J Anal Appl Pyrolysis* 92:450–455, 2011.

Saha B, Ghoshal AK: Hybrid genetic algorithm and model-free coupled direct search methods for pyrolysis kinetics of ZSM-5 catalyzed decomposition of waste low-density polyethylene, *Ind Eng Chem Res* 46:5485–5492, 2007.

Saha B, Chowdhury P, Ghoshal AK: Al-MCM-41 catalyzed decomposition of polypropylene and hybrid genetic algorithm for kinetics analysis, *Appl Catal B Environ* 83:265–276, 2008.

Sahu JN, Mahalik KK, Nam K, et al.: Feasibility study for catalytic cracking of waste plastic to produce fuel oil with reference to malaysia and simulation using ASPEN plus, *AIChE* 00:1–10, 2013.

Sakata Y, Uddin A, Koizumi K, et al.: Thermal degradation of polyethylene mixed with poly(viny1 chloride) and poly(ethyleneterephthalate), *Polym Degrad Stab* 53:111–117, 1996.

Sakata Y, Uddin A, Muto A: Degradation of polyethylene and polypropylene into fuel oil by using solid acid and non-acid catalysts, *J Anal Appl Pyrolysis* 51:135–155, 1999.

Sbirrazzuoli N: Interpretation and physical meaning of kinetic parameters obtained from isoconversional kinetic analysis of polymers, *Polymers* 12(6):1280, 2020. https://doi.org/10.3390/polym12061280.

Schirmer J, Kim JS, Klemm E: Catalytic degradation of polyethylene using thermal gravimetric analysis and a cycled-spheres-reactor, *J Anal Appl Pyrolysis* 60:205–217, 2001.

Seo Y, Lee K, Shin D: Investigation of catalytic degradation of high- density polyethylene by hydrocarbon group type analysis, *J Anal Appl Pyrolysis* 70, 2003.

Serrano DP, Aguado J, Escola JM: Developing advanced catalysts for the conversion of polyolefinic waste plastics into fuels and chemicals, *ACS Catal* 2:1924–1941, 2012.

Solis M, Silveira S: Technologies for chemical recycling of household plastics—a technical review and TRL assessment, *Waste Manag* 105:128–138, 2020.

Sommariva S, Grana R, Maffei T, et al.: A kinetic approach to the mathematical model of fixed bed gasifiers, *Comput Chem Eng* 35:928–935, 2011.

Sorum L, Gronli MG, Hustad JE: Pyrolysis characteristics and kinetics of municipal solid wastes, *Fuel* 80:1217–1227, 2001.

Takuma K, Uemichi Y, Ayame A: Product distribution from catalytic degradation of polyethylene over H-gallosilicate, *Appl Catal A Gen* 192:273–280, 2000.

Tilghman MB, Mitchell RE: Coal and biomass char reactivities in gasification and combustion environments, *Combust Flame* 162:3220–3235, 2015.

Till Z, Varga T, Sója J, et al.: Kinetic identification of plastic waste pyrolysis on zeolite-based catalysts, *Energy Convers Manag* 173:320–330, 2018.

Tognotti L, Longwell JP, Sarofim AF: The products of the high temperature oxidation of a single char particle in an electrodynamic balance, *Symp Combust* 23:1207–1213, 1991.

van Grieken R, Serrano DP, Aguado J, et al.: Thermal and catalytic cracking of polyethylene under mild conditions, *J Anal Appl Pyrolysis* 59:127–142, 2001.

Vinu R, Broadbelt LJ: Unraveling reaction pathways and specifying reaction kinetics for complex systems, *Annu Rev Chem Biomol Eng* 3:29–54, 2012.

Vinu R, Levine SE, Wang L, et al.: Detailed mechanistic modeling of poly(styrene peroxide) pyrolysis using kinetic Monte Carlo simulation, *Chem Eng Sci* 69:456–471, 2012.

Vyazovkin S, Wight CA: Model-free and model-fitting approaches to kinetic analysis of isothermal and nonisothermal data, *Thermochim Acta* 341:53–68, 1999.

Vyazovkin S, Burnham AK, Criado JM, et al.: ICTAC Kinetics Committee recommendations for performing kinetic computations on thermal analysis data, *Thermochim Acta* 520:1–19, 2011.

Wang M, Smith JM, McCoy BJ: Continuous kinetics for thermal degradation of polymer in solution, *AICHE J* 41:1521–1533, 1995.

Wang L, Chai M, Liu R, et al.: Synergetic effects during co-pyrolysis of biomass and waste tire: a study on product distribution and reaction kinetics, *Bioresour Technol* 268:363–370, 2018.

Wehinger GD, Ambrosetti M, Cheula R, et al.: Quo vadis multiscale modeling in reaction engineering?—A perspective, *Chem Eng Res Des* 184:39–58, 2022.

Westerhout RWJ, Waanders J, Kuipers JAM, et al.: Kinetics of the low-temperature pyrolysis of polyethene, polypropene, and polystyrene modeling, experimental determination, and comparison with literature models and data, *Ind Eng Chem Res* 36:1955–1964, 1997.

Wong HW, Kruse TM, Woo OS, et al.: Tertiary resource recovery from waste polymers via pyrolysis: polypropylene, *ACS Div Fuel Chem Prepr* 45:480–481, 2000.

Woo OS, Broadbelt LJ: Recovery of high-valued products from styrene-based polymers through coprocessing: experiments and mechanistic modeling, *Catal Today* 40:121–140, 1998.

Yousef S, Eimontas J, Striugas N, et al.: Modeling of metalized food packaging plastics pyrolysis kinetics using an independent parallel reactions kinetic model, *Polymer* 12:1763, 2020.

Yu J, Sun L, Ma C, et al.: Thermal degradation of PVC: a review, *Waste Manag* 48:300–314, 2016.

Zavala-Gutiérrez J, Pérez-Camacho O, Villarreal-Cárdenas L, et al.: Mathematical modeling of the catalytic pyrolysis of high-density polyethylene in a plug-flow tubular reactor, *Ind Eng Chem Res* 58:19050–19060, 2019.

Zhang X, Lei H: Synthesis of high-density jet fuel from plastics via catalytically integral processes, *RSC Adv* 6:6154–6163, 2016.

Zhang X, Lei H, Yadavalli G, et al.: Gasoline-range hydrocarbons produced from microwave-induced pyrolysis of low-density polyethylene over ZSM-5, *Fuel* 144:33–42, 2015.

Zhao D, Wang X, Miller JB, et al.: The chemistry and kinetics of polyethylene pyrolysis: a process to produce fuels and chemicals, *ChemSusChem* 13:1764–1774, 2020.

Ziska JJ, Barlow JW, Paul DR: Miscibility in PVC-polyester blends, *Polymer (Guildf)* 22:918–923, 1981.

Zwanzig RW: High-temperature equation of state by a perturbation method. i. nonpolar gases, *J Chem Phys* 22:1420, 2004.

Further reading

Aguado J, Serrano DP, Sotelo JL, et al.: Influence of the operating variables on the catalytic conversion of a polyolefin mixture over HMCM-41 and nanosized HZSM-5, *Appl Catal B Environ* 25:5696–5704, 2001.

Kaminsky W: Chemical recycling of mixed plastics of pyrolysis, *Adv Polym Technol* 14:337–344, 1995.

CHAPTER THREE

Kinetic experiments for pyrolytic recycling of solid plastic waste

Ankush Jain and Ravikrishnan Vinu*
Department of Chemical Engineering, Indian Institute of Technology Madras, Chennai, India
*Corresponding author: e-mail address: vinu@iitm.ac.in

Contents

Abstract

Pyrolysis is a promising strategy to convert plastic wastes to valuable liquid products that can be used as fuels, chemicals and platform molecules. An understanding of the pyrolysis chemistry and kinetics of neat polymers and contaminated waste plastics is valuable to design and optimize the pyrolysis reactors for maximum recovery and selectivity to valuable organics in the liquid pyrolysis oil. Thermogravimetric analysis (TGA) is a popular analytical technique to obtain mass loss data, which can then be used to obtain the decomposition regimes and kinetics of polymer pyrolysis. Owing to the limitations of TGA such as slow heating rate and the influence of heat and mass transfer on kinetics and pyrolysate evolution, a number of analytical pyrolyzers have captured the attention of researchers for accurate estimation of product yields and fast pyrolysis kinetics. This article provides an overview of the different pyrolyzers including TGA, Pyroprobe, Curie point pyrolyzer, Frontier micropyrolyzer, Pyrola and PHASR (Pulse Heated Analysis of Solid Reactions), analyzes their capabilities, and pros and cons of each of them for the kinetics of plastic decomposition. Salient results from the literature on the use of the above pyrolysis systems are highlighted. While Pyroprobe and Frontier

Advances in Chemical Engineering, Volume 60
ISSN 0065-2377
https://doi.org/10.1016/bs.ache.2022.09.008

micropyrolyzer have extended capabilities to analyze the time evolution of pyrolysate composition in seconds timeframe and obtain non-isothermal kinetics using the evolved gas analysis, respectively, the PHASR reactor is gaining popularity owing to its accuracy in obtaining product evolution data in milliseconds timeframe. Finally, the challenges and opportunities in obtaining kinetic data in heat and mass transport-free conditions are outlined, which would eventually pave way for a mechanistic understanding of polymer pyrolysis.

1. Introduction

Hydrocarbon-derived plastics constitute an integral part of our everyday lives. Owing to their low cost, durability, versatility and high strength-to-mass ratio, they serve as key enablers of a range of sectors like packaging, construction, healthcare, electronics and transportation to thrive. According to statistics, the plastic production is expected to stand at 1124 million metric tons (MMT) by 2050, which is more than three times of its production of 367 MMT in 2020. More importantly, the plastics' share of global oil consumption is expected to rise to 20% in 2050 from the current 6% if there are no major interventions to either bring back the used plastics to the economy or minimize the utilization of hydrocarbon-derived plastics by switching to bio-degradable plastics (The New Plastics Economy: Rethinking the Future of Plastics, 2016). Plastic packaging represents 26% of the plastics produced by volume. It is estimated that only 14% of the plastic packaging wastes are collected for recycling globally, and about 95% of the material value of the plastic packaging is lost after the first use, amounting to 80–120 billion USD loss to the economy (The New Plastics Economy: Rethinking the Future of Plastics, 2016). Moreover, a significant 80% of marine plastic litter is supposed to have come from the land sources due to improper collection and recovery infrastructure in Asia. Therefore, creating an effective after-use economy for the plastics through recycling, reusing and remanufacturing is the only way to drastically minimize their leakage into natural bodies.

The intensity of the problem posed by the single-use packaging plastics, and the initiatives, regulations and policies of various nations in curbing the disposal of plastic packaging materials is outlined in the document by United Nations Environment Programme (2018). World over, the COVID-19 pandemic has exacerbated the disposal of single-use plastics, especially in the form of personal protection equipment (PPE) kits, disposable syringes and medical accessories, into the biomedical waste fraction, which are simply thermally treated or incinerated or landfilled thereby causing a huge stress to the ecosystem. The polymers used in the production of single-use packaging materials include the first six categories of plastics standardized by the Society of the Plastics Industry (SPI), viz., polyethyelene terephthalate (PET, #1), high density polyethylene (HDPE, #2), polyvinyl chloride (PVC, #3), low density polyethylene (LDPE, #4), polypropylene (PP, #5) and polystyrene (PS, #6). Chemical depolymerization of polymers is a promising strategy to improve the net recycling of waste plastics into the economy, either by converting them into feedstocks that can be used to make a similar or better materials (closed-loop recycling) or by converting them into products that have a lower-value application than the original waste plastic feedstock (cascaded recycling) (The New Plastics Economy: Rethinking the Future of Plastics, 2016).

Among the different depolymerization strategies, thermochemical conversion techniques have captured the attention of researchers and practitioners owing to their versatility and possibility of scale-up. These include processes like pyrolysis, gasification, combustion and hydrothermal liquefaction. Among these, pyrolysis is shown to be energetically and economically viable process to convert polymers to liquid products containing monomers, fine chemicals and fuel molecules. Unlike gasification that involves high temperatures (800–1000 °C) in air or reactive oxygen ambience or liquefaction that involves high pressures (>200 bar) in supercritical water phase, pyrolysis is conducted at moderate temperatures (400–600 °C) in inert ambience at an atmospheric pressure to yield pyrolysis oil, gas and solid residue as the major products. Gasification produces syn gas ($CO + H_2$) as the major product, which requires a second step of Fischer-Tropsch synthesis to produce liquid hydrocarbons and chemicals. Moreover, the high energy requirements for the process and higher tar content in the gaseous products are the major deterrents (Lopez et al., 2018). Combustion or incineration completely oxidizes the polymer to generate low value energy accompanied by the release of toxic gases, especially when PVC or halogenated plastics are present in the waste stream. Hydrothermal liquefaction is at its nascent stage for plastic recycling, even though it is shown to be a potential process to generate monomers (Seshasayee and Savage, 2020). Nevertheless, significant debottlenecking is required to scale-up the process owing to the pressure-related and safety-related challenges.

The olefinic polymers have been a subject of pyrolysis research through experimentation, both analytical and applied, and kinetic modeling studies in the past decades. More than 80 wt% oil yields are reported from pyrolysis of polymers like PE, PP and PS in the temperature range of 450–500 °C, with a gross calorific value of 40–43 MJ kg^{-1}, resembling hydrocarbon fuels (Sharma et al., 2022). The pyrolysis oils generated from PE and PP are shown to be rich in alkenes followed by alkanes and alkadienes in a wide carbon number range, and are found to be a potential steam-cracker feedstock (Vinu et al., 2016). On the other hand, pyrolysis of PS is shown to produce styrene monomer as the major product along with oligomers and other aromatics like α-methyl styrene, benzene, toluene and ethyl benzene (Ojha and Vinu, 2015b). The pyrolysis chemistry of olefinic polymers like PE, PP, PS and PVC is well known, and established to occur through a series of free radical elementary reactions involving C—C bond fission, recombination, hydrogen abstraction, chain β-scission, radical addition and intramolecular backbiting reactions (Marongiu et al., 2003, 2007; Vinu and Broadbelt, 2012; Zhou et al., 2016). Mechanistic kinetics of pyrolysis of these individual polymers to predict the pyrolysate composition and degradation rates are well known in the literature (Marongiu et al., 2007; Vinu and Broadbelt, 2012; Zhou et al., 2016). Nevertheless, the complexity of the emerging packaging wastes containing multiple layers of different polymers designed for better barrier properties poses challenges in the understanding of pyrolysis chemistry and kinetics of polymer blends. Moreover, the sheer volumes at which the plastic wastes are generated and disposed pose major challenge in segregating them based on SPI codes in order to recycle them through different pathways. For example, the contamination of PVC in olefinic plastic waste streams presents risks of HCl and polychlorinated dioxin liberation during pyrolysis. Therefore, it is imperative to understand the pyrolysis chemistry and kinetics of polymer mixtures so that smart strategies can be devised through catalysis or cascade processes to depolymerize the mixtures and generate valuable products without spending energy and economics in segregation processes. It is also necessary to understand the pyrolysis behavior of PET and SPI #7 category plastics such as polycarbonates (PC), poly(acrylonitrile-butadiene-styrene) (ABS), poly(methyl methacrylate) (PMMA), polyacrylates, polylactides, etc., for designing efficient recycling processes.

The past decades have relied greatly on thermogravimetric analysis (TGA) to obtain the pyrolysis kinetics of different polymers and mixtures of polymers mimicking waste plastics. TGA aids in obtaining the mass loss of the polymer when it is heated from ambient temperature to the pyrolysis temperature and beyond, and provides the typical decomposition regimes. The mass loss data can be further used to obtain the kinetic parameters of pyrolysis, viz., apparent activation energy and pre-exponential factor, based on an assumption of a reaction kinetic model. Several TG studies on virgin plastics (Aboulkas et al., 2010; Al-Salem and Lettieri, 2010; Anderson and Freeman, 1961; Encinar and González, 2008; Kayacan and Doğan, 2008; Osman et al., 2020), plastic blends (Costa et al., 2010; Wu et al., 1993; Yuan et al., 2020) and solid-waste plastics (Anene et al., 2018; Liu et al., 2020; Özsin and Pütün, 2019; Pan et al., 2021; Tang et al., 2017; Zhang et al., 2021) have been reported in literature. TGA is still the technique of choice to understand the decomposition regimes of plastics, new materials and complicated waste mixtures. However, the need to acquire comprehensive information on pyrolysis including pyrolysate composition, time evolution of pyrolysates, and the kinetics of pyrolysis under fast heating rates under kinetically limited conditions have prompted the researchers to develop novel hyphenated methods and sample heating strategies. These include TGA coupled with Fourier Transform Infrared Spectrometer (TG-FTIR) and Mass Spectrometer (TG-MS), analytical pyrolysis coupled with GC/MS (Py-GC/MS) and FTIR spectrometer (Py-FTIR). Besides, novel analytical microreactors based on different sample configurations and heating mechanisms have evolved that aim to deliver a holistic understanding of polymer pyrolysis at different reaction conditions. It is hence the aim of this state-of-the-art review to analyze the different analytical pyrolysis techniques in terms of their capabilities, advantages and disadvantages, and the studies that have been carried out by various researchers. The importance of understanding the transport limitations on pyrolysis of plastics is discussed from the viewpoint of Pyrolysis-Biot numbers. This is followed by a systematic analysis of the various systems by highlighting the results from the key research works. The chapter concludes by outlining the existing challenges in obtaining the kinetic information via experiments for a mechanistic understanding of pyrolysis of different types of plastics.

2. Transport and kinetic regimes for common plastics

Transport effects play a vital role when performing kinetic studies. Studies have proved that transport effects change product distribution obtained from pyrolysis of different plastics (Ceamanos et al., 2002). Sample mass, particle size, heating mechanism, residence time and carrier gas flow rate are among the few factors known to impart heat and mass transfer limitations within the plastic sample. Sample weight and particle size can influence the mass and heat transfer length scales (Levine and Broadbelt, 2009; Teixeira et al., 2016). Different residence times can facilitate secondary gas-phase reactions to different degrees depending on the design of the pyrolyzer (Sezgi et al., 1998; Williams and Williams, 1997). Thus, designing a kinetic experiment to limit transport effects can be a challenging task.

Guidelines to design kinetic experiments can be helpful to derive kinetic parameters free from transport effects. In an effort to understand the effect of sample dimension on pyrolysis, Paulsen et al. (2013) generated the Pyrolysis-Biot number or *Py*-*Bi* profiles to contrast cellulose pyrolysis kinetics with conductive and convective heat transfer. The Pyrolysis numbers, Py^I and Py^{II}, relate the reaction kinetics to conductive and convective heat transfer, respectively. The Biot number relates the conductive heat transfer within the particle to the external heat transfer by convection. The expressions for Py^I, Py^{II} and Bi are given by Eqs. (1)–(3).

$$Py^I = \frac{\lambda}{\rho C_p L^2 k} \tag{1}$$

$$Py^{II} = \frac{h}{\rho C_p L k} \tag{2}$$

$$Bi = \frac{hL}{\lambda} \tag{3}$$

The physical properties in the above equations are as follows:

- λ is the thermal conductivity (SI unit: $\mathrm{W\,m^{-2}\,K^{-1}}$)
- h is the convective heat transfer co-efficient (SI unit: $\mathrm{W\,m^{-1}\,K^{-1}}$)
- L is the characteristic length (SI unit: m)
- ρ is the density (SI unit: $\mathrm{kg\,m^{-3}}$)
- C_p is the specific heat capacity (SI unit: $\mathrm{kJ\,kg^{-1}\,K^{-1}}$)
- k is the first order rate constant (SI unit: $\mathrm{s^{-1}}$)

A plot of the Pyrolysis numbers against the Biot number for different characteristic lengths of the particle provides an understanding of the controlling mechanism. The pyrolysis process is controlled by kinetics when $Py \gg 1$ and heat transfer controlled when $Py \ll 1$. When $Bi \gg 1$, intra-particle conduction is limiting as compared to the external heat transfer by convection; while the external convection is limiting when $Bi \ll 1$. The *Py*-*Bi* profiles have not been reported for plastics before. In this contribution, the *Py*-Bi profiles at 800 K are generated for two commonly found plastics in solid plastic wastes, namely PP and PS. It is worthwhile to note that both the plastics do not remain in their polymeric forms and produce respective pyrolysis products including oligomers and low molecular weight products at 800 K. Thus, pyrolysate composition for PP and PS was obtained from the literature to determine the critical properties of the melt phase.

Table 1 shows the physical properties of the pyrolysis products from polypropylene and polystyrene that were used for developing the *Py–Bi* profiles. The products considered for PP pyrolysis accounted for 90% of the total products identified by Supriyanto et al. (2021). Studies have shown that PS pyrolysis at 800 K yielded about 94% of styrene. The density of the compounds was evaluated at their boiling points using boiling-point volume correlation given by the Tyn and Calus method (Tyn and Calus, 1975). The thermal conductivities and the specific heat capacities were calculated at 800 K using the Roy-Thodos method (Roy and Thodos, 1968, 1970) and Benson's method of group contribution (Benson et al., 1969; Eigenmann et al., 1973; O'Neal and Benson, 1970; Stein et al., 1977), respectively. The Roy-Thodos and Tyn-Calus methods require information on the critical properties of the compounds. Such properties were calculated using the method proposed by Marrero-Morejón and Pardillo-Fontdevila (1999).

Table 1 Physical properties of compounds at 800 K used to evaluate Pyrolysis and Biot numbers for PP and PS pyrolysis.

Compound	Boiling-point density (kg m^{-3})	$C_{p,800}$ (kJ kg^{-1} K^{-1})	λ_{800} (W m^{-1} K^{-1})
2,4-Dimethyl-1-heptene	643.90	3.24	0.05
Pentane	620.48	3.47	0.10
2-Methyl pentene	634.62	3.19	0.06
1-Docosene	616.16	3.27	0.02
2,4-Dimethyl-1-pentene	640.98	3.22	0.05
(Z)-3-Tetradecene	625.70	3.25	0.04
(E)-9-Eicosene	618.73	3.26	0.02
2,3,3-Trimethyl-1-hexene	643.90	3.28	0.05
(E)-3-Eicosene	618.73	3.26	0.02
4-Methyl-,(E)-2-heptene	640.63	3.21	0.05
2-Methyl pentane	618.77	3.49	0.10
Styrene	774.27	2.48	0.14[a]

[a]The methods available to estimate thermal conductivity of compounds yielded unrealistic values for styrene. The value tabulated is the liquid thermal conductivity of styrene at 322 K.

The physical properties of some compounds were obtained from NIST library (Linstrom, 2017) and Physical and Chemical Property Prediction (2010). The convective heat transfer co-efficient was assumed as $3160\,Wm^{-1}K^{-1}$ based on a study by Jin et al. (2018). The rate constants at 800 K for PP and PS pyrolysis were calculated using the rate parameters of the C—C bond fission step, which is usually rate limiting (Kruse et al., 2001, 2003; Vinu and Broadbelt, 2012). The *Py-Bi* plot for PS and PP pyrolysis for different particle sizes is depicted in Fig. 1.

The *Py-Bi* plot for PS and PP reveal an important aspect of their pyrolysis behavior. It can be seen that for the characteristic lengths considered in the analysis that both PP and PS pyrolysis at 800 K are in the kinetically controlled regime. The $Py \gg 1$ feature of the *Py-Bi* plot indicates that pyrolysis of these plastics should not be influenced by heat transfer effects. This is in contrast with the *Py-Bi* plot for cellulose pyrolysis presented by Paulsen et al. (2013). This can be explained, atleast partially, based on the fact that the polymers, PS and PP, are volatile-rich feedstocks without any fixed carbon in them, which leads to the formation of a homogeneous melt phase as they undergo cracking. The melt phase is typically characterized by the depolymerizing polymer containing oligomers of different chain lengths,

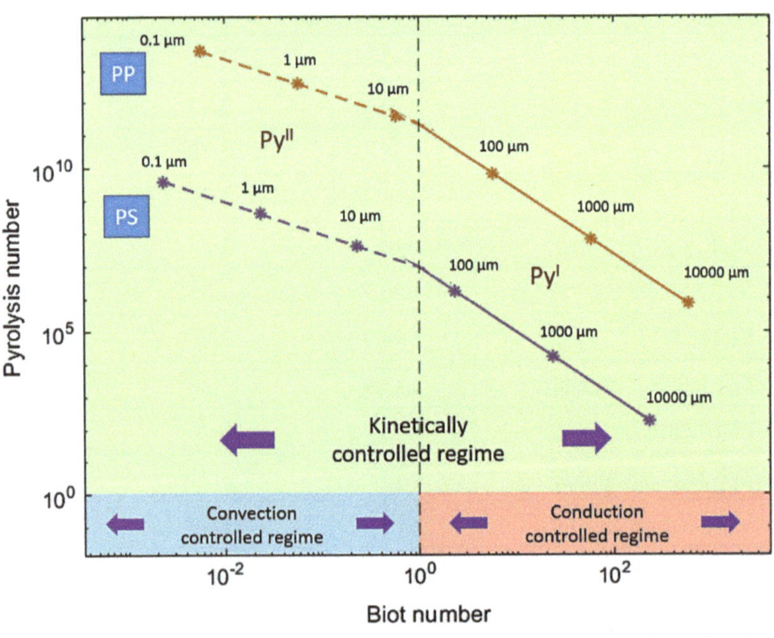

Fig. 1 *Py-Bi* plots for polypropylene (PP) and polystyrene (PS). Refer Table 1 for the data used to obtain the plots.

while the vapor phase product includes the low molecular weight pyrolysate. For a kinetically controlled experiment, it is vital to ensure that the temperature lags must be at the minimum, and to reliably calculate the rate constants from experiments, isothermal condition should be achieved in the least time possible. Maduskar et al. (2018) have proposed criteria to design an isothermal kinetic experiment. Although the criteria are specific to biomass, they can be adapted for measuring isothermal plastic pyrolysis kinetics. The following are the criteria:

(a) The *characteristic length* of the particle must be within the appropriate range (based on the type of plastic) to avoid internal thermal gradients.
(b) The *heating and cooling rates* must be high to avoid reactions in the transition time.
(c) Sample *temperature monitoring* system must be used, as non-uniform thermal distribution can influence the reaction rates.
(d) *Carrier gas flow* should be high to arrest secondary reactions with pyrolysis vapors.
(e) Time-resolved *analysis of products* should be performed to track the progress of pyrolysis.

Lédé and Authier (2015) showed that particle's temperature and heating rate are not the appropriate parameters to define fast pyrolysis of biomass particles, especially cellulose, and smaller particle size is neither sufficient not a necessary condition for fast pyrolysis. Importantly, the external heat transfer coefficient, which is a characteristic of the analytical reactor set-up and the faster rate of removal of primary pyrolysis products are shown to be necessary to achieve kinetically-limited conditions during pyrolysis. Considering that polymers pyrolyze through an intermediate melt phase, a general assumption based on which the *Py-Bi* plot can be understood is the melt phase is fully mixed and homogeneous across the length scale. Moreover, unlike biomass, the viscous nature of the polymers ensures that the melt phase is free of any particulate matter that usually is subject to heat and mass transport limitations.

3. Analytical pyrolysis systems to perform kinetic experiments

Different pyrolysis systems have different provisions to ensure that kinetic experiments are free from or performed with minimal transport effects. Among the many important points when considering a pyrolyzer are:

(a) Rapid sample heating rate for isothermal experiments
(b) Provisions to arrest secondary reactions
(c) Broad operating temperature range to avail multiple data

(d) Provision to perform pyrolysis in the "seconds" or even "milliseconds" time scale

(e) Capability to interface with an analytical equipment like GC/MS, FTIR spectrometer and/or a microbalance to monitor the evolution of pyrolysates and mass loss of the sample

(f) High pressure capability to perform pressure-dependent kinetic studies

Based on the above-mentioned points, Table 2 compares different pyrolyzers available in the market based on the chosen attributes. It is clear that the existing analytical pyrolyzers have the capability to interface with GC/MS, which has become a basic requirement for all analytical pyrolysis experiments. The product composition data is valuable to assess the quality of the plastic that is undergoing decomposition and the potential of pyrolysis products. The GC/MS total ion chromatogram (TIC) is used as a fingerprint to identify the type of plastic through a comparison with MS database of compounds. Interfacing analytical pyrolyzer with FTIR spectrometer is a unique capability as this requires the pyrolysis to be conducted in a special reaction cell that needs to be located in the FTIR spectrometer's sample compartment. The secondary reactions of the gas phase products may occur if sufficient residence time is allowed for the vapors to pass through a heated transfer line to the FTIR sample compartment. Only the Pyroprobe instrument has this capability. While obtaining mass loss data during pyrolysis is a unique aspect of conventional TGA at slow heating rates, such data has been extracted using other devices operating at faster heating rates. An important but not the only criterion to perform mass loss kinetic studies is the detachment of the pyrolyzer from the hyphenated analytical equipment, especially the GC/MS. The ability to unhook the pyrolyzer from the analytical equipment is considered sufficient enough to conduct mass loss experiments.

Aside from the physical distinctions between the pyrolyzers, they essentially differ in the heating mechanism. It is well known that heating mechanism can influence the distribution of heat within the sample thereby leading to heat transport effects. Resistive heating of the heating element can be seen as a common means of heating in most of the pyrolyzers. The Curie-point pyrolyzer offers the most unique way of heating through induction. As all the pyrolyzers can be coupled with a GC-based chromatographic equipment, the sample size for all pyrolyzers is limited to the range of 0.1–0.5 mg. This is to avoid the excessive amount of pyrolysis vapors affecting the filament of the mass spectrometer. An important requirement to perform isothermal kinetic analysis is the capability of achieving high heating rate. Isothermal experiments should satisfy the condition that $T(t) = T_{target}$,

Table 2 Comparison of different analytical pyrolyzers based on different attributes.

#	Reactor	Type of heating	Maximum claimed heating rate (°C s^{-1})	Typical sample mass (mg)	Temperature range (°C)	High pressure operation	Catalytic upgradation (in-situ/ex-situ/both)	Isothermal experiments	Non-isothermal experiments	Interface with GC/FID or GC/MS	Interface with FTIR	Possibility of mass-loss kinetics
1	Thermogravimetric analyzer (TGA)	Conductive + convective	5	1–50	150–1200	✓	In-situ	✗	✓	✓	✓	✓
2	Pyroprobe[a]	Resistive/ conductive	20,000	0.1–5	150–1400	✓	Both	✓	✓	✓	✓	✓
3	Curie-point pyrolyzer[b]	Inductive	3500	0.1–5	150–1040	✗	In-situ	✓	✗	✓	✗	✓
4	Micro-pyrolyzer[c]	Conductive + convective	>1000	0.2–0.5	150–1050	✓	Both	✓	✓	✓	✗	✓
5	Pyrola[d]	Resistive	>10,000	0.25–0.5	150–1400	✗	NR	✓	✗	✓	✗	✗
6	PHASR	Resistive	16,000	0.25–0.5	350–1000	✗	NR	✓	✗	✓	✗	✓

[a]C.D.S. Analytical (U.S.A.).
[b]Japan Analytical Industry *Co.* Ltd. (Japan).
[c]Frontier Lab (Japan).
[d]Pyrol AB (Sweden), NR—No Reports.

where *T(t)* is the temperature at any given time *t*, and T_{target} is the targeted pyrolysis temperature. The sample temperature in isothermal kinetic experiments is expected to reach the reaction temperature instantaneously (or) within the shortest amount of time to avoid non-isothermal behavior at the inception of the process. It is important to note in Table 2 that the values of heating rates tabulated are the values claimed by the manufacturers. Studies have shown that the actual heating rate experienced by the sample may be much lower that the rated values in different analytical pyrolyzers (Ojha et al., 2017; Proano-Aviles et al., 2017).

3.1 TGA

TGA provides the ability to measure the sample mass and temperature as a function of time. The experiments generate mass loss and differential thermogravimetric (DTG) curves, which essentially describe the rate of degradation of the material as the temperature increases at a constant rate. Isothermal experiments have been performed using the TGA in the past (Muravyev et al., 2017; Sichina, 2000; Vyazovkin and Wight, 1997). However, attaining the pyrolysis temperature in a short period of time is usually difficult in a conventional TGA as the maximum rate at which the sample can be heated is usually limited to 200 °C min^{-1}. This means that even during the initial heat up phase, significant mass loss is recorded and the sample characteristics are altered. Hence, TGA is preferred for dynamic experiments when the sample heating rate is usually within 20 °C min^{-1}, which allows minimum thermal lag between the furnace and the sample. While some studies use even up to 50 °C min^{-1} to obtain mass loss data and derive the rate parameters, the rate parameters are usually apparent or global in nature and do not correspond to the elementary reaction step (Bockhorn et al., 1999; Ceamanos et al., 2002; Gao et al., 2003). Moreover, significant heat transport effects and thermal lag may yield erroneous values of these kinetic parameters with a wide standard deviation.

Fig. 2 depicts the TG mass loss and DTG profiles of common polymeric components of a PPE kit, multilayers packaging (MLP) and expanded PS (EPS) waste. The mass loss profiles in Fig. 2A clearly point to the decomposition of the plastics occurring in the temperature range of 350–500 °C. Importantly, some polymers like PC, PET and MLP do not devolatilize completely and leave a solid residue, char, at the final pyrolysis temperature of 900 °C. This shows that these polymers cannot be completely converted

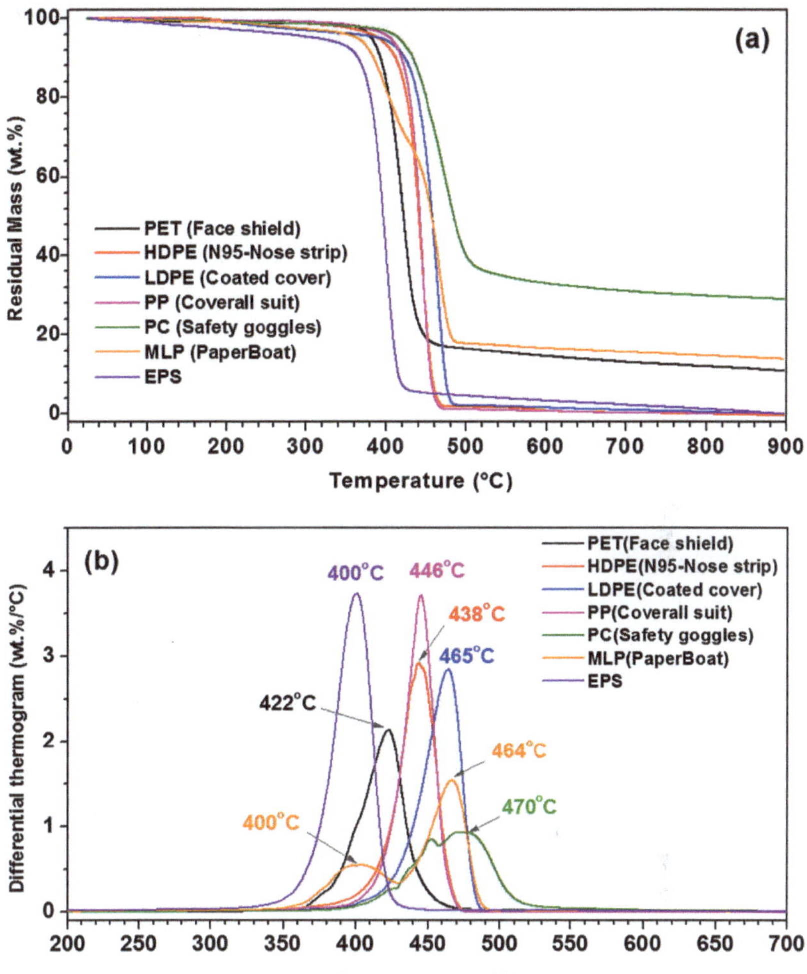

Fig. 2 (A) Thermogravimetric (TG) mass loss, and (B) differential TG profiles of common waste plastics including those from the COVID-19 PPE kits and packaging wastes. The experiments were performed in a SDT Q600 TG analyzer (T.A. Instruments) in IIT Madras at 5 °C min^{-1} heating rate with 100 mL min^{-1} flow of N_2 gas.

to liquid and gaseous products, and the char is essentially a culmination of the aromatic structures in the polymer converting to polyaromatic hydrocarbons during pyrolysis. Fig. 2B depicts the decomposition regimes of the polymers. It is evident that composite materials like MLP exhibit two distinct decomposition regimes corresponding to polymeric layers that degrade at different temperature regimes. Some complex SPI #7 plastics like PC also exhibits multiple decomposition peaks albeit in a very narrow and overlapping temperature zones.

The TG curves of pure and virgin plastics like HDPE, PP, PS, and LDPE have been used as benchmark to investigate the synergistic effects in polymer blends (Chowlu et al., 2009). The ease of degradation is given by the order: PS > PP > LDPE > HDPE. Dubdub and Al-Yaari (2020) blended common plastics in specific ratios to show that synergy exists between the constituent plastics. They performed the TGA of pure plastics along with their binary and ternary mixtures. The profiles showed that the temperature corresponding to maximum decomposition rate of the binary mixture, PS and HDPE, was greater than that of PS. In case of PP and PS, all blends had a higher degradation temperature than either of the plastics. In the ternary mixtures, the degradation temperature increased with increasing percentage of PE (LDPE or HDPE). Thus, a synergy has been claimed to exist between constituents of the mixture. Synergistic effects may arise due to immiscibility between constituent polymers (Miranda et al., 2001) or difference in their stability (Chowlu et al., 2009). Faravelli et al. (2003) probed this question of synergy in the pyrolysis of PS-PE blends by combining TGA experiments with kinetic models. They concluded that unless the mixing between the polymers is on a molecular scale, the polymers tend to retain their distinct pyrolysis behaviors in the blend. Kruse et al. (2005) modeled the interactions in binary blends of PP and PS by allowing the low molecular weight free radicals generated from PS to react with the PP melt phase, and observed the faster rate of decomposition of PP, in line with the experimental results. It is evident that TGA alone does not suffice in understanding the synergies in pyrolysis of polymer mixtures at a molecular level. It only provides the qualitative insights, which are outcomes of molecular interactions. Hence, combining the TGA data with product analysis and mechanistic modeling is a valuable exercise.

Korobeinichev et al. (2019) consolidated the pyrolysis of PMMA in the temperature range of 315–500 °C using different analytical pyrolysis units including TGA, microscale combustion calorimeter (MCC) and differential mass spectrometric thermal analysis (DMSTA). MCC involved heating rates up to 120 °C min^{-1}, while heating rates as high as 100 °C s^{-1} were achieved using the DMSTA. The intensity of the product peaks is tracked in DMSTA, which is proportional to the release rate of the respective products. In MCC, the products released from pyrolysis of the polymer are combusted using oxygen, and the oxygen consumption for complete combustion is calibrated against the change in crucible temperature in the pyrolysis chamber. Fig. 3A shows that the DTG and heat release rate measurements from MCC

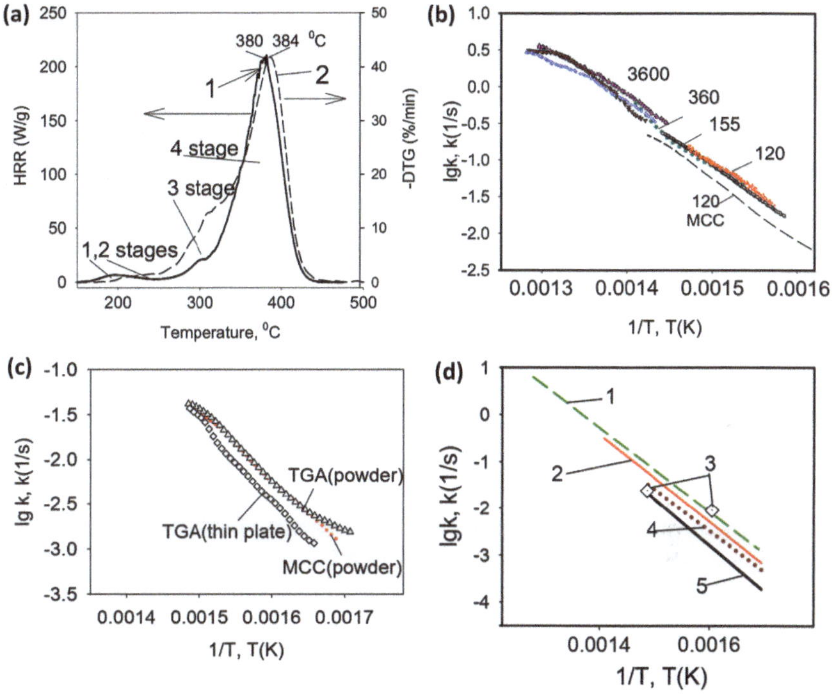

Fig. 3 (A) Variation of heat release rate (HRR) determined by microscale combustion calorimetry (MCC) and differential mass loss of PMMA powder at 30°C min^{-1} heating rate in N_2 ambience. (B) Arrhenius plot for the decomposition of PMMA powder measured at different heating rates (in °C min^{-1}) using differential mass spectrometric thermal analysis (DMSTA). (C) Arrhenius kinetic plot comparing the decomposition rates of PPMA in power and thin plate forms. (D) Arrhenius kinetic plots of PMMA powder using different methods (1)-TGA + DMSTA, (2)-MCC, (3)-fast pyrolysis, (4,5) TGA based on literature studies (Korobeinichev et al., 2019).

are in consonance with each other and capture the major step of decomposition with minimum difference in temperature corresponding to maximum degradation rate. From Fig. 3B, it is evident that the slopes of the Arrhenius plots are similar at different heating rates using MCC and DMSTA, which point to the similar activation energy of PMMA pyrolysis. The authors also studied the effect of sample dimension of PMMA (thin film vs powder) in TGA and MCC at similar heating rates and found that the activation energies are similar (Fig. 3C). Fig. 3D summarizes the Arrhenius plots obtained by using different techniques, and it was unequivocally shown that the first order activation energy and pre-exponential factor for PMMA decomposition are 171.4 kJ mol^{-1} and $10^{12.3}$ s^{-1}, respectively.

3.2 TG-FTIR and TG-MS

The hyphenation of TGA with FTIR spectrometer and MS sheds valuable insights on the nature of volatile species released from decomposition of the polymer at different temperatures and at different heating rates. This is usually accomplished by a heated transfer line that carries the pyrolysis vapors from the TGA furnace to either a gas cell of the FTIR spectrometer or directly to the MS without passing through a GC column. The transfer line is usually maintained at 150–200 °C to avoid condensation of the vapors. The time or temperature-based FTIR spectra of the vapors indicates the rate at which the key pyrolysate functional groups evolve, and the rate of change of absorbance of the peaks reflects the rate of formation and consumption of a functional group. Similarly, the evolution of intensity of a specific *m/z* peak in the MS shows the relative amount of a specific organic molecule with respect to others. It is important to note that both these techniques do not yield quantitative information of the release of pyrolysis vapors unless a calibration of absorbance of a specific FTIR peak or a specific m/z peak in MS is performed. The FTIR peaks corresponding to low molecular weight gases and vapors such as CO, CO_2, H_2O, HCl, HCN, C=O carbonyl/carboxyl are usually well indexed, and these are tracked by various studies. Yousef et al. (2020) studied the TGA, TG-FTIR and TG-MS of pyrolysis of metallized food packaging wastes of different types, and identified the key steps involved in the delamination and decomposition of the layers to form hydrocarbon oil products and char residue (Fig. 4A).

The TG-FTIR analysis was used to understand the effect of sample heating rate on the evolution of pyrolysates. It was found that at a slow heating rate of 5 °C min^{-1}, minor release of methane and carboxylic acid residues was observed in the form of peaks at 2960 cm^{-1}, 890 cm^{-1} (C-O-C stretch) and 1465 cm^{-1} (methylene bend) As the heat flux to the sample was increased, the intensity of the volatile functional groups increased signifying the delamination and decomposition of the inner layers of the sample (Fig. 4B and C). This was also corroborated by the evolution of hydrocarbons like propylene, pentane, methyl pentene, 2,4-dimethyl-1-heptene and furans from the decomposition of plastic layers like LDPE, PET and poly(ethylene-vinyl acetate). Zhou et al. (2014) studied the pyrolysis of individual and mixtures of components constituting the municipal solid waste like rice straw, poplar wood and PVC using TG-FTIR. Fig. 4D shows the typical FTIR spectrum of the gas phase from PVC, where the vibrations corresponding to HCl vapors and aromatic benzene-type

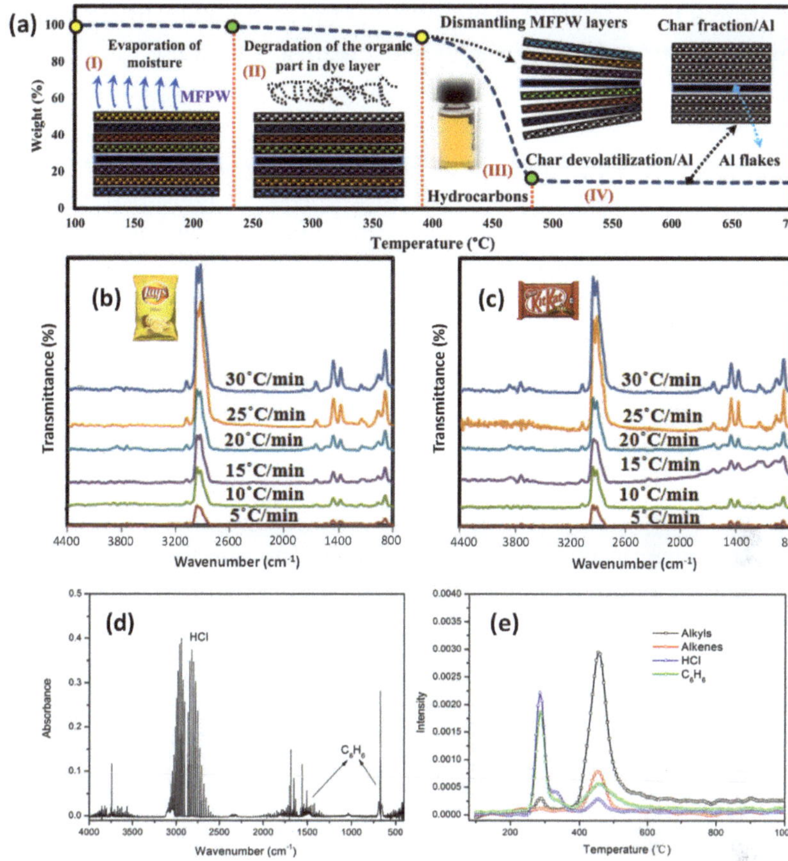

Fig. 4 (A) TG mass loss data of metallized food packaging wastes (MFPW) along with a schematic of the decomposition processes at different temperatures (Yousef et al., 2020). (B and C) FTIR spectra of vapor phase products from pyrolysis of two different MFPW wastes (Yousef et al., 2020). (D) FTIR profile of the pyrolysis vapors from PVC pyrolysis at the temperature corresponding to maximum mass loss rate (Zhou et al., 2014). (E) Time evolution of pyrolysis vapor species from PVC pyrolysis (Zhou et al., 2014).

molecules are evident. Fig. 4E clearly shows that HCl and aromatic hydrocarbon vapors evolve at low temperatures (250–350 °C) due to dehydrochlorination and bimolecular condensation reactions, while the generation of alkenes is prevalent at high temperatures. The authors also calculated the time resolved FTIR peak intensities from the pyrolysis of mixtures based on the area under the peaks of the pyrolysates from individual materials to understand the presence of interactions. In another study, Yang and Chen (2020) adopted the TGA-FTIR-GC/MS technique to probe the evolution of pyrolysates from different components of the automobile shredder residue containing plastic, rubber, leather, textile, foam and their mixtures in a wide temperature range of 200–700 °C. It was shown that the pyrolysates from plastics and textiles greatly influence the product distribution from the mixtures, and the major compounds were alkanes, alkenes, alcohols and aromatic hydrocarbons. In another study, Wu et al. (2014) employed TG-FTIR to understand the interactions among the olefinic plastics like PE, PVC and PS during pyrolysis, and concluded that the presence of PS accelerated the pyrolysis of PE, while that of PE and PVC were retarded when both were mixed together. Importantly, the methyl and methylene groups disappeared in the pyrolysate when PE and PVC were mixed.

3.3 Pyroprobe

Pyroprobe employs a resistively heated platinum coil to heat the sample, which is usually placed in a quartz tube inserted into the coil. While the platinum coil is reported to be heated at a maximum rate of 20,000 °C s^{-1}, the sample placed inside the quartz tube is usually heated at a much slower rate owing to the conductive heat transfer, which is the rate determining step. Krishna et al. (2019) simulated the unsteady temperature profiles within the sample when polymers such as PMMA, PE and ultra-high molecular weight PE (UHMWPE) are fast pyrolyzed in a Pyroprobe. Fig. 5A depicts the cross section of the quartz tube and the time evolution of temperature profiles. It is evident that the entire sample placed inside the quartz tube attains the set temperature of 500 °C within 4 s. This means the actual sample heating rate is 125–150 °C s^{-1}, which demonstrates that there is a lag time of few seconds until the sample reaches the set temperature. While this is much better than the TGA, the initial lag might still affect the kinetic measurements. Similar heating rates have been calculated for other samples like lignin and lignocellulosic biomass (Ojha et al., 2017). The sample mass taken in this pyrolyzer for Py-GC/MS studies is typically lesser than a mg, and the sample is usually held firm in the center portion of the quartz tube using quartz wool on both sides. Latest designs and models of the Pyroprobe have the capability to perform pyrolysis under high pressures (upto 30 bar) in reactive or inert ambience. There is also a provision of a temperature programmed catalytic reactor that can be used to carry out ex-situ vapor phase upgradation studies (Venkatesan et al., 2020). The provision of a Tenax cartridge aids in thermal desorption experiments where even low concentrations of pyrolysates can be analyzed accurately.

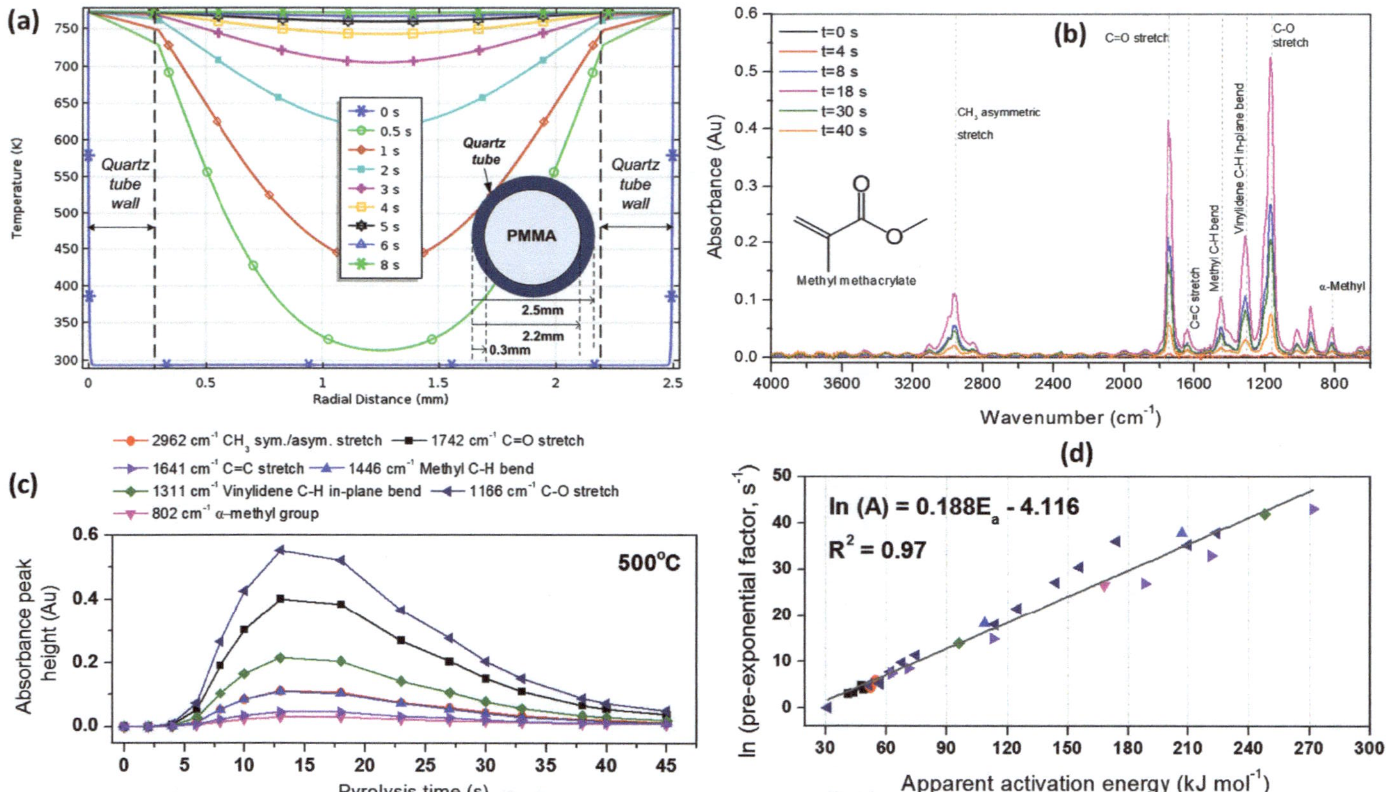

Fig. 5 (A) Simulated temperature profiles of the PMMA sample inside the quartz tube of the Pyroprobe at different time periods. (B) Time evolution of FTIR spectra when PMMA is pyrolyzed at 500 °C in the Pyroprobe-FTIR reactor. (C) Time evolution of absorbance of salient functional groups during pyrolysis of PMMA. (D) Kinetic compensation plot of PMMA pyrolysis based on a number of kinetic data from the literature (Krishna et al., 2019).

The Pyroprobe-GC/MS has been extensively utilized in understanding plastic pyrolysis in the past (Aguado et al., 2005; Blazsó, 1998; Conesa et al., 1997; Czégény et al., 2012; del Hernández et al., 2006; Fabbri et al., 1998, 2000; García et al., 1992, 1993; Krishna et al., 2019; Liebman et al., 1978; Liu et al., 2014; Okonsky et al., 2022; Zhou et al., 2020). Importantly, the Pyroprobe can be employed to obtain the cumulative amount of pyrolysates generated in specific temperature regimes. It can be very well used like a TGA by setting different heating rates in different temperature zones, and the pyrolysis vapors can be analyzed and quantified to probe the mechanism of decomposition. Viju et al. (2018) demonstrated the use of Pyroprobe to study the product evolution from individual biochemical components of algae in different temperature regimes. The formation of indole and phenol in the 30–350 °C range, aliphatic and aromatic hydrocarbons in 350–600 °C range and long chain oxygenates in the 600–1000 °C range confirmed the decomposition of proteins, lipids and condensation reactions mediated by char, respectively. In situ catalytic pyrolysis experiments can be performed in the Pyroprobe by either mixing the catalyst along with the polymer or by packing the catalyst as a layer slightly away from the polymer sample within the quartz tube (Ojha and Vinu, 2019). Okonsky et al. (2022) recently reported the use of Pyroprobe to study the in-situ catalytic co-pyrolysis of PET:LDPE of equal ratio with different zeolites. The interactions among the condensed phase intermediates from PET and LDPE decomposition in the presence of catalyst promoted the yield of alkylated benzenes and polyaromatics, while limiting the production of C_4-C_{10} paraffins and olefins.

The author's group is the first to demonstrate the application of the Pyroprobe to obtain mass loss data from isothermal fast pyrolysis experiments. A typical mass loss profile of polymer can be generated by performing multiple experiments by holding the sample at a specific pyrolysis temperature for different periods of time. Such an exercise has been performed for pyrolysis of polymers (PMMA, PS, UHMWPE) (Krishna et al., 2019), lignin (Ojha et al., 2017), microalgae (Gautam et al., 2017) and a variety of lignocellulosic biomasses (Ojha et al., 2021). It is worthwhile to note that the sample does not attain the set temperature instantaneously, but faces a lag time of a few seconds. Krishna et al. (2019) generated the isothermal mass loss profiles for the decomposition of PMMA, PS and UHMWPE using the Pyroprobe in the temperature range of 350–800 °C. At 600 °C, complete conversion of PMMA and PS volatile hydrocarbons was achieved in 10 s and 40 s, respectively, while only 70% conversion of UHMWPE was achieved at 750 °C even after 40 s of pyrolysis. Using the mass loss profiles at different temperatures, the apparent kinetic parameters can be determined by assuming different reaction models given by Vyazovkin et al. (2011). The diffusion model described the PMMA fast pyrolysis, while the Avrami model described the fast pyrolysis of PS and UHMWPE with good confidence. It is now well proven that the kinetic parameters generated using the isothermal pyrolysis experiments in the Pyroprobe reactor for a variety of feedstocks are lower in magnitude compared to the ones generated using TGA, which point to the diffusion-limited nature of the pyrolysis reactions in the Pyroprobe (Gautam et al., 2017; Krishna et al., 2019; Ojha et al., 2017, 2021). This is because, the Pyroprobe mimics a packed bed sample configuration where the pyrolysates are generally removed by the inert gas surrounding the sample tube. Nonetheless, the rate parameters determined from different studies follow a kinetic compensation effect, which is more statistical in nature. Such a plot for PMMA pyrolysis is shown in Fig. 5D. The black squares in the lower left end of the plot denote the kinetic parameters generated using the Pyroprobe.

3.3.1 Pyroprobe-FTIR

One of the unique features of the Pyroprobe is its ability to be interfaced with an FTIR spectrometer through a specially designed interface called the Brill cell, which is placed inside the sample compartment. The Brill cell is a cylindrical chamber with two ZnSe windows on both sides to allow the IR to scan the pyrolysis vapors as they escape the quartz tube of the Pyroprobe. The cell is usually heated to 200°C to make sure the vapors do not condense inside the cell, and an inert carrier gas is purged continuously to ensure the vapors do not stay inside the cell for a long time. The exact configuration of the sample in a Brill cell is given by Ojha et al. (2021). In this system, there is no need of a long, heated transfer line through which the gases and vapors have to pass through to get analyzed by the FTIR spectrometer. Our group has used this system significantly to study the pyrolysate evolution from different polymer and biomass samples at short time periods (Gautam et al., 2017; Krishna et al., 2019; Ojha et al., 2017, 2021; Ojha and Vinu, 2015a). Fig. 5B depicts the typical FTIR spectra collected at different time periods when PMMA is pyrolyzed at 500°C. A flat FTIR spectrum in the beginning ($t = 0$ s) marks the onset of pyrolysis reactions in the Pyroprobe. Fig. 5C shows the time evolution of absorbance of the major functional groups in the vapor phase. The maximum rate of evolution of vapors occurs at 12 s followed by a decline in the absorbance profile, which shows the complete conversion of the PMMA sample used in the experiments. Based on the functional groups evolved, and through independent Py-GC/MS analysis, the vapor phase is found to contain the monomer, methyl methacrylate. Compared to Py-GC/MS experiments where low sample mass is used, Py-FTIR experiments require a slightly larger amount of sample (few mg) to make sure the vapors in the gas phase are in sufficient concentration to be detected. This is because, the carrier gas tends to dilute the generated vapors.

The Pyroprobe-FTIR has also been demonstrated for fast pyrolysis of PP (Ojha and Vinu, 2015a), fast co-pyrolysis of PP with cellulose (Ojha and Vinu, 2015a), and to understand the role of flame retardants in decreasing the rate of PMMA pyrolysis (Krishna et al., 2020). Online gas analyzers can be integrated at the outlet of the Brill cell to evaluate the non-condensable gas composition. Recently, we have also employed the Pyroprobe-FTIR system to probe the fast hydropyrolysis kinetics of pine wood biomass in H_2 ambience at different temperatures and at high partial pressures of H_2 upto 15 bar (Venkatesan et al., 2020). In presence of H_2, the release of oxygenated organics containing C=O was found to be significant than in inert atmosphere. Importantly, the release of CO_2 was found to be enhanced, while the carbonyl (C=O) vibration decreased with increase in H_2 partial pressure to 15 bar. This was supposed to be due to the release of CO via decarbonylation pathway. The maximum rate of evolution of functional groups from the biomass decomposition was also found to be delayed at high pressures. The increase in methane evolution with increase in H_2 partial pressure at a high temperature of 700 °C was shown to be due to the decomposition of the methoxy groups in the lignin fraction of biomass. The existing results show that Pyroprobe-FTIR system can be good value addition to emerging studies on waste plastic upcycling by hydrogen-assisted conversion techniques.

3.4 Micropyrolyzer

Micropyrolyzer, originally developed by Frontier Laboratories, Japan, is a drop-cup-type of an analytical pyrolyzer used extensively for Py-GC/MS experiments. The sample to be pyrolyzed is taken in a deactivated stainless-steel cup, which is then dropped into a pre-heated quartz tube surrounded by a furnace. The pyrolysates are carried to the GC through a heated transfer line for the analysis. The system can be operated either in a single-shot mode (direct fast pyrolysis) or double-shot mode (thermal desorption followed by pyrolysis). Proano-Aviles et al. (2017) studied the heat and mass transport effects in this system through experiments and fluid dynamics simulations, and showed that conductive and convective heat transfer are equally important in pyrolyzing the sample. The experimentally determined heating rate of the sample in this pyrolyzer is shown to depend on the final pyrolysis temperature. At 995 °C, the sample heating rate was found to be maximum at 494 °C s^{-1}, while it was 180 °C s^{-1} at 500 °C. Moreover, the diffusion of pyrolysates from inside the sample cup to the gas phase was found to induce mass transfer limitations, especially when longer/deep cups were used as against shallow and perforated cups. The effect of temperature on pyrolysate vapor and gas composition can be very well studied using this system. Kannan et al. (2014) reported that low molecular weight gases like ethylene, propylene, 1-pentene and 1,3-butadiene are produced in significant amounts when LDPE is pyrolyzed beyond 800 °C.

Fig. 6A and B depicts the typical thermal desorption and pyrolysis chromatograms when an environmental plastic sample collected from the beach waste is subjected to Py-GC/MS. The thermal desorption step is usually a slow ramping stage where the physically adsorbed organics from the sample are removed leaving the polymer sample intact. The presence of fatty acids from the sample shows their presence in the form of oils sticking to the plastic packaging wastes and wrappers. From the pyrogram, the packaging material can be clearly identified as PE, owing to the presence of distinct triplets corresponding to alkane:alkene:alkadiene in approximately 1:3:1 ratio. The generation of hydrocarbon chains in the range of C_5-C_{22} and beyond is also evident from the pyrogram. While the area under the peaks can be used to assess the relative abundance of the hydrocarbons, quantification using pure compound standards yields accurate wt% yield of the pyrolysates. This technique can be used in the first place to identify the unknown waste plastic sample, and it also provides valuable information on the potential compounds that can be recovered through pyrolysis.

In-situ catalytic fast pyrolysis experiments can be performed in this system by mixing the catalyst with the sample, as shown by various studies (Nair and Vinu, 2016; Ojha and Vinu, 2015b). The effect of different zeolite catalysts of varying acidity, catalyst loading and temperature on catalytic reaction mechanism of PS was deciphered (Ojha and Vinu, 2015b). Recent models of this system are equipped with tandem microreactor, which involves a second stage catalyst bed though which the pyrolysates generated in the first reactor can be passed and upgraded. Similar to the Pyroprobe, hydropyrolysis experiments can also be performed at high partial

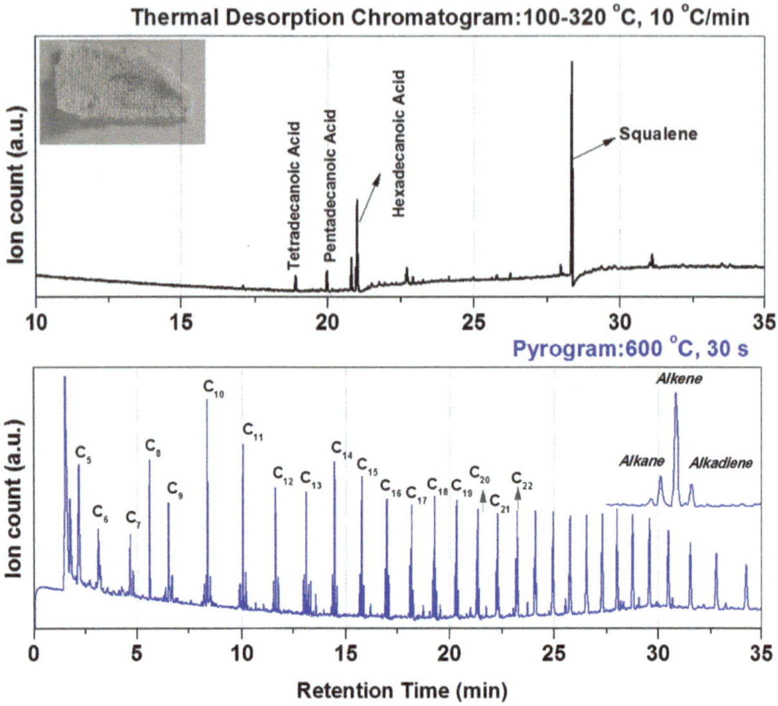

Fig. 6 Thermal desorption and fast pyrolysis GC/MS total ion chromatogram obtained from pyrolysis of waste PE sample collected from a beach in Chennai (12.999529°N 80.272411°E). The typical triplet of alkane, alkene and alkadiene for different hydrocarbon chain lengths are given in the inset of the pyrogram. The experiment was performed in a Shimadzu GC-2010, QP2010 GC/MS system at the following conditions. Ion source temperature = 250 °C, ionization voltage = 70 eV, column flow = 1.5 mL min^{-1} He (99.9995% pure), split ratio = 100, injector temperature = 300 °C, GC column UA-5 (30 m × 0.25 mm i.d. × 0.25 μm), column oven program: 40 °C (2 min)—10 °C min^{-1}—300 °C (10 min), *m/z*: 19–500 *Da*.

pressures of H_2 using this set-up (Kumagai and Yoshioka, 2021). Kumagai and co-workers (Kumagai et al., 2020; Kumagai and Yoshioka, 2021) developed a pyrolysis-gas phase derivatization-GC/MS technique using this pyrolyzer to analyze the pyrolysates such as amines and isocyanates from polyurethanes, which would otherwise react and polymerize in the GC column during analysis. Reaction progress, especially in catalytic upgradation, can be monitored in this system by selectively heating the catalytic reactor at a specific rate and monitoring the upgraded products. Kumagai et al. (2020) adopted the tandem microreactor for upgrading the pyrolysates from PET through a catalyst bed composed of different metal oxides like MgO, ZnO, TiO_2 and ZrO_2. The pyrolysates produced from PET at 450 °C were upgraded at 700 °C using the catalysts in the second reactor. By employing a deactivated column and direct MS injection, the evolution of compounds from the catalyst bed when the temperature of the second reactor was ramped at $10\,°C\,min^{-1}$ to the final temperature was tracked. It was shown that catalytic deoxygenation of benzoic acid and terephthalic acid generated from PET to benzene was promoted by high base strength catalyst like ZnO.

Besides using this system to study the reactions, Py-GC/MS has also been employed to understand the aging mechanism of insulation materials. Gautam et al. (2016) used micropyrolyzer-GC/MS system to probe the mechanism of treeing activity in crosslinked PE (XLPE) insulation when subjected to different dose of electron beam irradiation. The generation of oxygenated compounds like carbonyl and alcohols from XLPE dosed at high radiation levels was correlated with the weakening of the peroxy crosslinks in the XLPE sample. Recently, the micropyrolyzer is gaining popularity for identifying and quantifying the microplastics from complex solid phase matrices like biosolids, soil and solids from waste treatment units (Hermabessiere et al., 2018; Okoffo et al., 2020). The unique F-search software developed by Frontier Lab aids in identifying the polymer based on the signature peaks in the pyrogram, and importantly, the microplastics library can be used to identify the polymer type once the matrix interference is identified and understood. The microplastic in a complex matrix can be identified by monitoring the characteristic compound that is unique to a specific polymer. For example, 2,4-dimethyl-1-heptene, styrene, methyl methacrylate and ε-caprolactam are unique indicators of PP, PS, PMMA and nylon-6 (Hermabessiere et al., 2018). Other polymers like PE, PET, PVC contain multiple marker compounds, which makes the identification task difficult.

One of the unique features of micropyrolyzer is the ability to perform evolved gas analysis (EGA) to monitor the kinetics of gas evolution during non-isothermal pyrolysis of polymers. EGA profile is essentially the evolution of total ions generated in the MS when all the pyrolysates are introduced. In fact, the EGA profile can be matched very well with the DTG obtained from the TGA for kinetic measurements. Fig. 7A–E depicts the comparison of EGA profiles of various polymers with their DTGs. EGA profiles are usually obtained by bypassing the GC column and allowing the pyrolysis vapors from the pyrolyzer directly to the MS through a deactivated transfer line and vent-free adapter. Even though EGA employs a much lower sample mass as compared to that used in a conventional TGA, the temperature corresponding to maximum mass loss rate/gas evolution rate (peak temperature) from EGA and DTG match well (Fig. 7F)

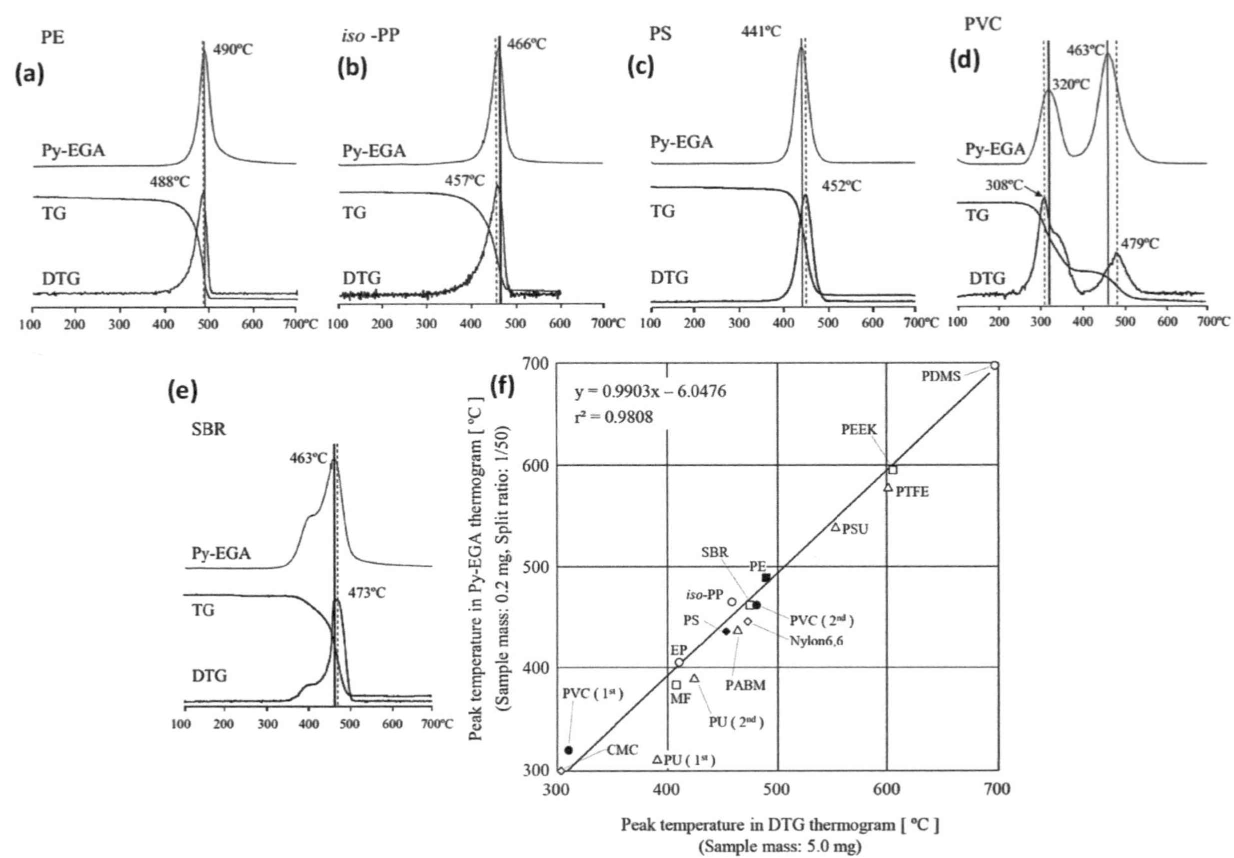

Fig. 7 (A–E) Pyrolysis-evolved gas analysis (EGA) chromatograms of different polymers using 0.2 mg at split ratio 50 and TGA mass loss and differential mass loss profiles using 5 mg of sample at 20 °C min^{-1} heating rate. (F) Comparison of temperatures corresponding to maximum mass loss rate from EGA and TGA (Shiono et al., 2015).

(Shiono et al., 2015). Recently, Shiono et al. (2021) showed that EGA can also be performed in oxidative environment to study the combustion and gasification kinetics of polymer samples by diluting the oxygen gas with helium before admitting into the MS detector.

3.5 Curie-point pyrolyzer

The sample heating mechanism in a Curie point pyrolyzer is much more unique than that in the other pyrolyzers. The pyrolyzer heats the sample at the Curie point of the sample holder, which is called as the Pyrofoil. Generally, the Curie point of a ferromagnetic material is the temperature at which the material loses its magnetic property. Pyrofoils are composed of ferromagnetic materials like Fe, Co, and Ni, which have different Curie points depending on their composition. In the pyrolyzer, the sample, which is in intimate contact with the pyrofoil, is subjected to high frequency radio waves, which induce a current in the pyrofoil. This raises the temperature of the pyrofoil until its Curie point. As the foil loses its magnetic property, the current is cut off from the pyrofoil, which decreases the foil temperature. The pyrofoil regains its magnetic property as the temperature decreases, thus maintaining the Curie temperature for the pre-determined residence time. The Curie-point pyrolyzer is shown to attain the pyrolysis temperature within 0.2 s, which is independent of the magnitude of the temperature (Oguri and Kirn, 2005). Therefore, it is mostly used to perform Py-GC/MS experiments to identify the pyrolysates from a particular sample under fast heating rate conditions. Slow heating experiments cannot be performed using this pyrolyzer. As the pyrolysis temperature is determined by the Curie-point of the ferromagnetic alloy material, the temperature cannot be set at will by the user. The typical pyrofoils available in the market correspond to Curie-point temperatures 386, 423, 445, 485, 500, 590, 650, 670, 740, 764, 920, 1040 °C (Japan Analytical Industry Co. Ltd, 2022).

Early studies on the use of Curie-point pyrolyzer were focused mostly on cellulose pyrolysis to unravel the pyrolysates such as anhydrosugars and furan derivatives (Ohnishi et al., 1975; Pouwels et al., 1989). Although there are some studies in the past decades on the use of Curie-point pyrolyzer-GC/MS to characterize the pyrolysates from ammonium perchlorate (Barnett and Montoya, 2002), oil shale (Huang et al., 2015), sewage sludge (Fan and He, 2016) and jet fuels (Priyadarshi et al., 2022; Priyadarshi and Vinu, 2022), the studies have been relatively less as compared to the Pyroprobe and micropyrolyzer. Catalytic fast pyrolysis can be performed in in-situ mode by packing the feedstock-catalyst mixture within the pyrofoil as demonstrated by Priyadarshi and Vinu (2022).

In an attempt to compare the mass loss characteristic of plastic pyrolysis in Curie-point pyrolyzer with the Pyroprobe, isothermal mass loss experiments were performed on PP at 500 °C (Fig. 8). This is also the first ever mass loss data generated using the Curie-point pyrolyzer. The PP pellets were first cryogenically ground using liquid nitrogen and sieved to obtain particle size of less than or equal to 200 μm. The sample of about 2 mg was then taken in the respective sample holders, i.e., quartz tube for Pyroprobe and Pyrofoil for Curie-point pyrolyzer. Carrier gas (helium) flow rate of 120 mL min^{-1} was maintained in the experiments to ensure an inert atmosphere, and to evacuate the pyrolysis products as soon as they are formed to avoid secondary reaction. The sample in the quartz tube was plugged by glass wool at both ends to prevent loss of particles. The Pyroprobe was initially maintained at 40 °C, which was ramped at 20,000 °C s^{-1} to 500 °C. The interface, which encases the probe, was maintained at 250 °C throughout the experiment to prevent condensation of pyrolysis vapors. In the case of the Curie-point pyrolyzer, a Pyrofoil with Curie point of 500 °C was used to pyrolyze the PP sample. The sample was pyrolyzed at residence times between 1 s and 50 s.

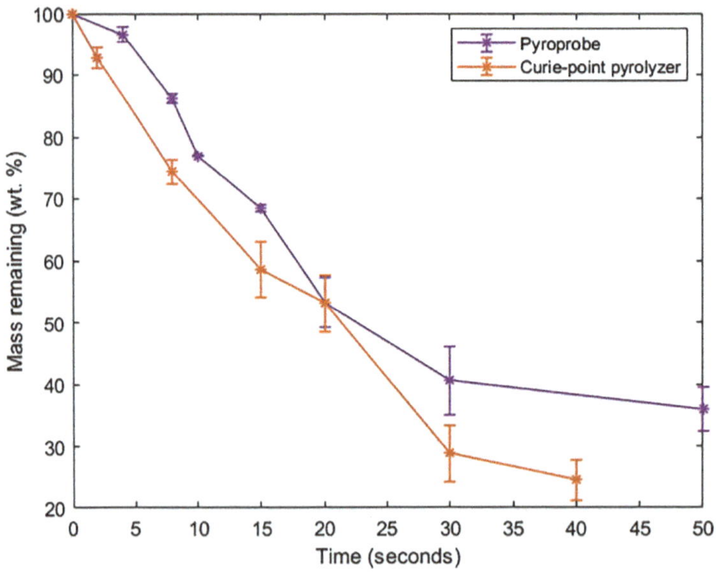

Fig. 8 Isothermal mass loss profiles of PP when pyrolyzed at 500 °C in a Pyroprobe reactor (Pyroprobe 5200 system, C.D.S. Analytical, U.S.A.) (sample mass = 2 mg) and Curie-point pyrolyzer (JHI-07 hybrid pyrolyzer, Japan Analytical Industry *Co.* Ltd., Japan) (sample mass = 2 mg).

As mentioned in Section 3.3, each data point on the graph is a separate experiment performed with Pyroprobe and Curie-point pyroyzer at different sample residence times. The conversion was calculated by accurately measuring the sample mass before and after pyrolysis using a microbalance.

The mass loss profiles show that, at the same residence time, the polymer conversion in the Curie-point pyrolyzer is more than that in the Pyroprobe. This can be attributed to the relatively slow heating of the sample in the Pyroprobe, even though the claimed heating rate of the sample in the Curie-point pyrolyzer is yet to be validated through experiments or models. Nevertheless, the conversion of PP sample is not remarkably different from both the pyrolyzers as the pyrolysis time scale in both the cases is of the order of few tens of seconds. One possible explanation is that PP pyrolysis is kinetically controlled for the size of particles considered in the experiments, which was also observed in the *Py-Bi* plot of PP. More detailed experimental and modeling studies are required to understand the heating behavior of the sample and the exact heating rate in a Curie-point pyrolyzer.

3.6 Pyrola

Pyrola is a pyrolyzer manufactured by Pyrolab (Sweden), which has been recently employed in plastic pyrolysis studies. In Pyrola 2000 system, the sample is placed directly onto the platinum filament. Thus, the sample is in direct contact with the platinum filament to ensure conductive heat transfer to the sample. Studies have claimed that the sample reaches the pyrolysis temperature within 8 ms (Evangelopoulos et al., 2015; Sophonrat and Yang, 2017; Supriyanto et al., 2021; Usino et al., 2020). In the previous works, this pyrolyzer has been hyphenated to a GC with a thermal conductivity or flame ionization detector (TCD/FID). Supriyanto et al. (2021) used the pyrolyzer in the Py-GC/MS/FID configuration to quantify the primary pyrolysis products from PE, PP and PS. Nearly 75 wt% of C_4-C_{12} hydrocarbons (alkenes, alkanes and alkadienes) from LDPE, 80 wt% of C_3-C_{12} hydrocarbons (alkyl-alkenes, alkenes and alkanes) from PP, and 95 wt% monoaromatics from PS were obtained using this set-up. Not many pyrolysis studies are available in the literature using this equipment.

3.7 Py-GC × GC

Due to its capability to be interfaced with the GC/MS system, the aforementioned analytical pyrolysis systems are extensively used to obtain qualitative and quantitative information on pyrolysis products. However, pyrolysis oil from waste plastics can contain oxygen, nitrogen, and sulfur-based organics, which can hinder the analysis using a conventional one-dimensional GC (Miskolczi et al., 2004). Moreover, many waste oil products either belong to the same homologous series or only differ in the degree of unsaturation. Thus, the analysis involves compounds with very close boiling points. For example, alkenes and alkadienes commonly found in the pyrolysis oil from PE have similar boiling points. The relatively smaller amount of alkadienes makes quantification of the co-eluted fraction (alkene + alkadiene) difficult using the one-dimensional GC/MS. The similarities in the mass spectra of alkadienes and mono-unsaturated naphthenes can lead to erroneous identification. Thus, two-dimensional GC × GC integrated with multiple detectors like MS, FID and chemiluminescence detector are recently being used to avail accurate quantification of compounds (Thi et al., 2021; Toraman et al., 2014). In a typical GC × GC, the compounds are separated in two different columns containing polar and non-polar stationary phases. Fig. 9A depicts the GC × GC chromatogram of the plastic waste pyrolysis oil containing the heavy diesel-range fractions analyzed using a polar column (Stabilwax, 60 m × 0.25 mm × 0.25 μm) followed by a non-polar column (Rxi-5 ms, 2 m × 0.1 mm × 0.1 μm) combination. 3-Chlorothiophene was used an internal standard. The analysis presented in Fig. 9B demonstrates that the GC × GC system with this specific reverse-phase column combination demonstrated an excellent separation of n-paraffins, α-olefins, diolefins, iso-olefins, isoparaffins and monoaromatics. Studies on analytical pyrolyzer hyphenated to a GC × GC system are scarce in the literature, and it is highly recommended to obtain high quality data with accurate mass balance closure using such advanced instrumentation.

3.8 PHASR

PHASR (Pulse-Heated Analysis of Solid Reactions) is another pyrolyzer recently used in studying plastic pyrolysis. The pyrolyzer was designed by Paul Dauenhauer group at the University of Minnesota to understand kinetics of cellulose pyrolysis. In this pyrolyzer, a thin film of the sample is placed in direct contact with the heating element, which pyrolyzes the sample by resistive heating. The capability of performing pyrolysis at shorter residence times (millisecond timescale) and the ability of conducting thin-film pyrolysis (film thickness ~ 10 μm) makes PHASR a potential analytical system to derive kinetic parameters free from transport effects. The details of the construction of the PHASR system for plastics pyrolysis are given in the study by Zolghadr et al. (2021). The major components of the PHASR reactor that

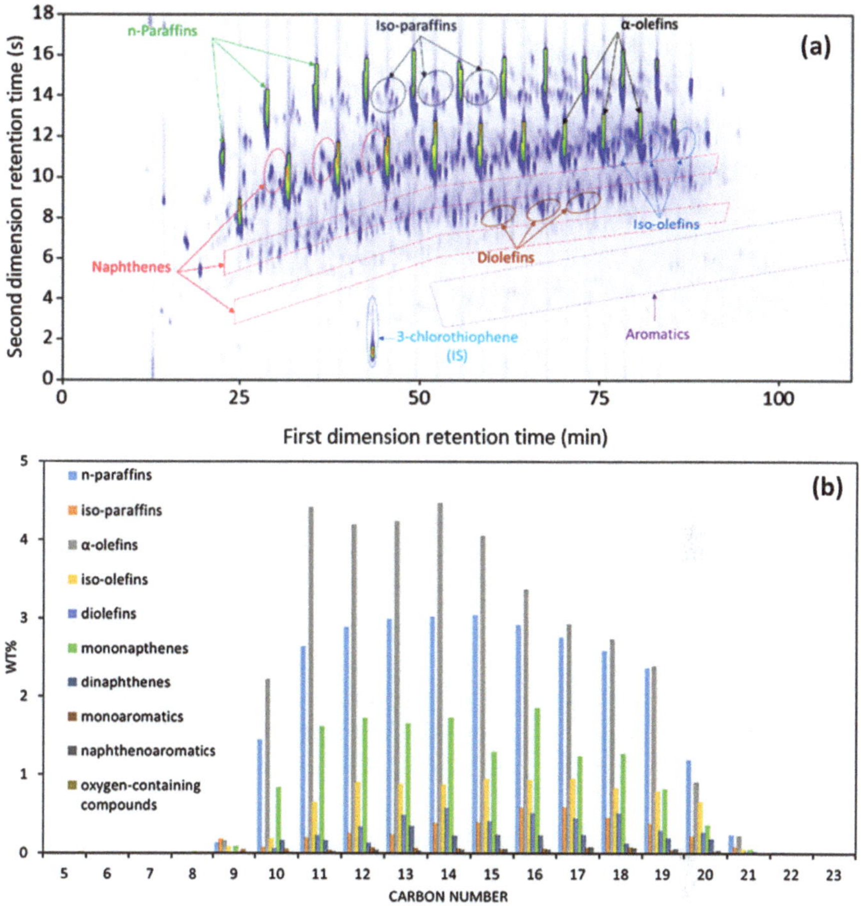

Fig. 9 (A) GC × GC/MS of the diesel fraction from plastic pyrolysis oil using Stabilwax × Rxi-5 ms column combination, (B) PIONA distribution (Thi et al., 2021).

work in tandem include (a) heating loop with accurate PID control, (b) cooling loop, and (c) hyphenation with the GC system. The sample is heated resistively to the desired pyrolysis temperature using electrical pulses whose duration can be changed from 20 ms to 2 s (Maduskar et al., 2018; Zolghadr et al., 2021). Another unique feature of this system over the existing analytical pyrolyzers is the ability to visualize the pyrolyzing sample using high speed imaging at a resolution of 10 ms. Zolghadr et al. (2021) revealed the key melt-phase phenomena during LDPE pyrolysis like bubbling, foaming, coalescing of bubbles, bursting of bubbles and evaporation into the vapor cloud. Arresting the secondary decomposition of primary pyrolysates is ensured in the PHASR via rapid sample quenching, dilution of pyrolysate vapors with inert carrier gas, and maintaining ideal plug flow of the gases in the reactor. The results of LDPE pyrolysis in PHASR reactor confirm that pyrolysis is completed within 2 s at temperatures above 600 °C with the evolution of hydrocarbons of chain lengths < C20. Thus, PHASR may potentially open up new avenues in understanding plastic pyrolysis in pure kinetic regimes.

4. Conclusions

This chapter has showcased the operating principles and the potential of different analytical devices to generate kinetic data for polymer pyrolysis. It is evident that each system has its own merits which can be a combination of operational simplicity, faster sample heating rate, capability to interface with another equipment (GC or FTIR), with a catalytic unit in tandem or operate at higher reaction pressure, and analysis of reaction intermediates by effective and fast quenching. In order to obtain a holistic understanding of the pyrolysis process, insights from two or more techniques may be required. TGA has proved as an indispensable technique both for the understanding of the thermal behavior/stability of new materials and mixtures of real solid plastic waste streams, and to evaluate the key reaction steps occurring under slow heating conditions through TG-FTIR and TG-MS. Even though the rate parameters derived from TGA mass loss data are global in nature, they can be applied in reactor-level models where only the conversion of the plastic sample during pyrolysis is required. Kinetic models that aim to predict the product distribution from pyrolysis of polymer wastes require experimental data obtained in transport-free, kinetic regime. Various studies also emphasize the need to attain the required pyrolysis temperature instantaneously, and to prevent the secondary cracking reactions of the primary pyrolysates either through faster removal of vapors or quenching of the reaction.

Analytical Py-GC/MS conducted at a microscale has become a norm in pyrolysis research as it provides valuable information on pyrolysate composition. The actual sample temperature is not usually measured in these set-ups, and it is assumed to be the temperature of the furnace or platinum coil or pyrofoil. Different studies have also proved through experiments and simulation that there is a lag of few seconds until the sample reaches the set temperature. Therefore, an online temperature measurement capability in the commercially available analytical pyrolyzers will help resolve the doubt-factor about the exact sample temperature during pyrolysis. In this regard, the emerging PHASR reactor has a greater edge as it involves accurate sample temperature measurement, faster heating and cooling of the reaction sample along with imaging of the pyrolyzing polymeric melt phase.

Neat polyolefins like PP, PE, PS and PMMA have been the polymers of choice in a large number of solid plastic waste pyrolysis works in the literature. The pyrolysis chemistry, mechanism and detailed kinetics of these polymers are well known today (Marongiu et al., 2007; Vinu and Broadbelt, 2012). While these polymers serve as benchmarks to evaluate the efficacy of a new pyrolyzer or catalyst, due focus should be given to other polymers like PET, PC, polyurethanes, multilayered packaging, condensation polymers, rubbers, and other polymers belonging to SPI #7 category. These are difficult polymers to crack and generate valuable products in selective amounts. The kinetics of product evolution is also complicated as these polymers contain hetero atoms like oxygen, nitrogen and sulfur. In the study of real plastic waste mixtures obtained from landfills/dumpsites using analytical pyrolysis experiments, the major challenge lies in obtaining a homogeneous sample, especially because of the sub-milligram to milligram levels of sample that is allowed in these techniques. This requires proper coning and quartering techniques to obtain a representative sample, and multiple repetitions to obtain reasonable standard deviation in the values of the parameters. One of the limitations of a large number of studies in the literature reporting pyrolysate composition is the use of non-calibrated GC/MS peak area% as a metric to quantify the pyrolysates. While this is accepted only as a relative metric to compare the organics in the pyrolysis products, it does not aid in providing a mass balance closure of the products. Moreover, the non-condensable gases are not quantified in many Py-GC/MS studies as FID and TCD would be required to accurately quantify the low molecular weight gases. Reporting the mass balance closure is an important task that needs to be standardized across various studies so that the results can be reasonably compared. The authors are confident that the current advancements in analytical pyrolysis will pave the way for a better understanding of the kinetics of the process in order to make pyrolysis a promising resource recovery strategy for a circular plastics economy.

Acknowledgments

R.V. thanks National Center for Combustion Research and Development, a Department of Science and Technology (DST), India, funded center and the Science and Engineering Research Board (SERB) of DST, India, for funding analytical pyrolysis research over the past decade. Frontier Laboratories, Japan, is also thanked for gifting a double-shot micropyrolyzer for analytical pyrolysis experiments and evolved gas analysis.

References

Aboulkas A, El Harfi K, El Bouadili A: Thermal degradation behaviors of polyethylene and polypropylene. Part I: pyrolysis kinetics and mechanisms, *Energy Convers Manag* 51:1363–1369, 2010. https://doi.org/10.1016/j.enconman.2009.12.017.

Aguado R, Olazar M, Vélez D, Arabiourrutia M, Bilbao J: Kinetics of scrap tyre pyrolysis under fast heating conditions, *J Anal Appl Pyrolysis* 73:290–298, 2005. https://doi.org/10.1016/j.jaap.2005.02.006.

Al-Salem SM, Lettieri P: Kinetic study of high density polyethylene (HDPE) pyrolysis, *Chem Eng Res Des* 88:1599–1606, 2010. https://doi.org/10.1016/j.cherd.2010.03.012.

Anderson DA, Freeman ES: The kinetics of the thermal degradation of polystyrene and polyethylene, *J Polym Sci* 54:253–260, 1961. https://doi.org/10.1002/pol.1961.1205415920.

Anene AF, Fredriksen SB, Sætre KA, Tokheim L-A: Experimental study of thermal and catalytic pyrolysis of plastic waste components, *Sustainability* 10:3979, 2018. https://doi.org/10.3390/su10113979.

Barnett J, Montoya BM: *Viability of applying curie point pyrolysis/gas chromatography techniques for characterization of ammonium perchlorate based propellents*, 2002 https://www.osti.gov/servlets/purl/801003-AEoYDT/native/. Accessed Jul. 2022.

Benson SW, Cruickshank FR, Golden DM, et al.: Additivity rules for the estimation of thermochemical properties, *Chem Rev* 69:279–324, 1969.

Blazsó M: Chlorinated products of plastic pyrolysis, *Macromol Symp* 135:237–245, 1998. https://doi.org/10.1002/masy.19981350126.

Bockhorn H, Hornung A, Hornung U, Schawaller D: Kinetic study on the thermal degradation of polypropylene and polyethylene, *J Anal Appl Pyrolysis* 48:93–109, 1999. https://doi.org/10.1016/S0165-2370(98)00131-4.

Ceamanos J, Mastral JF, Millera A, Aldea ME: Kinetics of pyrolysis of high density polyethylene. Comparison of isothermal and dynamic experiments, *J Anal Appl Pyrolysis* 65:93–110, 2002. https://doi.org/10.1016/S0165-2370(01)00183-8.

Chowlu ACK, Reddy PK, Ghoshal AK: Pyrolytic decomposition and model-free kinetics analysis of mixture of polypropylene (PP) and low-density polyethylene (LDPE), *Thermochim Acta* 485:20–25, 2009. https://doi.org/10.1016/j.tca.2008.12.004.

Conesa JA, Font R, Marcilla A: Comparison between the pyrolysis of two types of polyethylenes in a fluidized bed reactor, *Energy Fuel* 11:126–136, 1997. https://doi.org/10.1021/ef960098w.

Costa P, Pinto F, Ramos AM, Gulyurtlu I, Cabrita I, Bernardo MS: Study of the pyrolysis kinetics of a mixture of polyethylene, polypropylene, and polystyrene, *Energy Fuel* 24:6239–6247, 2010. https://doi.org/10.1021/ef101010n.

Czégény Z, Jakab E, Blazsó M, Bhaskar T, Sakata Y: Thermal decomposition of polymer mixtures of PVC, PET and ABS containing brominated flame retardant: formation of chlorinated and brominated organic compounds, *J Anal Appl Pyrolysis* 96:69–77, 2012. https://doi.org/10.1016/j.jaap.2012.03.006.

del Hernández MR, García ÁN, Gómez A, Agulló J, Marcilla A: Effect of residence time on volatile products obtained in the HDPE pyrolysis in the presence and absence of HZSM-5, *Ind Eng Chem Res* 45:8770–8778, 2006. https://doi.org/10.1021/ie060728.

Dubdub I, Al-Yaari M: Pyrolysis of mixed plastic waste: I. Kinetic study, *Materials* 13:4912, 2020. https://doi.org/10.3390/ma13214912.

Eigenmann HK, Golden DM, Benson SW: Revised group additivity parameters for the enthalpies of formation of oxygen-containing organic compounds, *J Phys Chem* 77:1687–1691, 1973. https://doi.org/10.1021/j100632a019.

Encinar JM, González JF: Pyrolysis of synthetic polymers and plastic wastes. Kinetic study, *Fuel Process Technol* 89:678–686, 2008. https://doi.org/10.1016/j.fuproc.2007.12.011.

Evangelopoulos P, Kantarelis E, Yang W: Investigation of the thermal decomposition of printed circuit boards (PCBs) via thermogravimetric analysis (TGA) and analytical pyrolysis (Py–GC/MS), *J Anal Appl Pyrolysis* 115:337–343, 2015. https://doi.org/10.1016/j.jaap.2015.08.012.

Fabbri D, Trombini C, Vassura I: Analysis of polystyrene in polluted sediments by pyrolysis-gas chromatography-mass spectrometry, *J Chromatogr Sci* 36:600–604, 1998. https://doi.org/10.1093/chromsci/36.12.600.

Fabbri D, Tartari D, Trombini C: Analysis of poly(vinyl chloride) and other polymers in sediments and suspended matter of a coastal lagoon by pyrolysis-gas chromatography-mass spectrometry, *Anal Chim Acta* 413:3–11, 2000. https://doi.org/10.1016/S0003-2670(00)00766-2.

Fan H, He K: Fast pyrolysis of sewage sludge in a curie-point pyrolyzer: the case of sludge in the city of shangai, China, *Energy Fuel* 30:1020–1026, 2016. https://doi.org/10.1021/acs.energyfuels.5b02235.

Faravelli T, Bozzano G, Colombo M, Ranzi E, Dente M: Kinetic modeling of the thermal degradation of polyethylene and polystyrene mixtures, *J Anal Appl Pyrolysis* 70:761–777, 2003. https://doi.org/10.1016/S0165-2370(03)00058-5.

Gao Z, Amasaki I, Nakada M: A thermogravimetric study on thermal degradation of polyethylene, *J Anal Appl Pyrolysis* 67:1–9, 2003. https://doi.org/10.1016/S0165-2370(02)00010-4.

García AN, Font R, Marcilla A: Kinetic studies of the primary pyrolysis of municipal solid waste in a Pyroprobe 1000, *J Anal Appl Pyrolysis* 23:99–119, 1992. https://doi.org/10.1016/0165-2370(92)80016-F.

García AN, Font R, Marcilla A: Flash pyrolysis of municipal solid waste in a pyroprobe 1000, In Bridgwater AV, editor: *Advances in thermochemical biomass conversion*, Dordrecht, 1993, Springer Netherlands, pp 793–805. https://doi.org/10.1007/978-94-011-1336-6_62.

Gautam R, Vinu R, Sarathi R, Acharya S, Kumar M, Sharma A: Understanding electrical treeing activity in electron beam irradiated XLPE cable insulation, *IEEE Trans Dielectr Electr Insul* 23:1652–1662, 2016. https://doi.org/10.1109/TDEI.2016.005621.

Gautam R, Varma A, Vinu R: Apparent kinetics of fast pyrolysis of four different microalgae and product analysis using Py-FTIR and Py-GC/MS, *Energy Fuel* 31:12339–12349, 2017. https://doi.org/10.1021/acs.energyfuels.7b02520.

Hermabessiere L, Himber C, Boricaud B, et al.: Optimization, performance, and application of a pyrolysis-GC/MS method for the identification of microplastics, *Anal Bioanal Chem* 410:6663–6676, 2018. https://doi.org/10.1007/s00216-018-1279-0.

Huang Y, Han X, Jiang X: Charcaterization of dachengzi oil shale fast pyrolysis by Curie-point pyrolysis-GC-MS, *Oil Shale* 32:134–150, 2015. https://doi.org/10.3176/oil.2015.2.04.

Japan Analytical Industry Co. Ltd:, 2022 http://www.jai.co.jp/english/products/py/pyrofoil.html. Accessed Sep 2022.

Jin Z, Chen D, Yin L, Hu Y, Zhu H, Hong L: Molten waste plastic pyrolysis in a vertical falling film reactor and the influence of temperature on the pyrolysis products, *Chin J Chem Eng* 26:400–406, 2018. https://doi.org/10.1016/j.cjche.2017.08.001.

Kannan P, Shoaibi AA, Srinivasakannan C: Temperature effects on the yield of gaseous olefins from waste polyethylene via flash pyrolysis, *Energy Fuel* 28:3363–3366, 2014. https://doi.org/10.1021/ef500516n.

Kayacan İ, Doğan ÖM: Pyrolysis of low and high density polyethylene. Part I: non-isothermal pyrolysis kinetics, *Energy Sources A: Recovery Util Environ Eff* 30:385–391, 2008. https://doi.org/10.1080/15567030701457079.

Korobeinichev OP, Paletsky AA, Gonchikzhapov MP, et al.: Kinetics of thermal decomposition of PMMA at different heating rates and in a wide temperature range, *Thermochim Acta* 671:17–25, 2019. https://doi.org/10.1016/j.tca.2018.10.019.

Krishna JVJ, Korobeinichev OP, Vinu R: Isothermal fast pyrolysis kinetics of synthetic polymers using analytical pyroprobe, *J Anal Appl Pyrolysis* 139:48–58, 2019. https://doi.org/10.1016/j.jaap.2019.01.008.

Krishna JVJ, Kumar SS, Korobeinichev OP, Vinu R: Detailed kinetic analysis of slow and fast pyrolysis of poly(methyl methacrylate)-flame retardat mixtures, *Thermochim Acta* 687, 2020, 178545. https://doi.org/10.1016/j.tca.2020.178545.

Kruse TM, Woo OS, Broadbelt LJ: Detailed mechanistic modeling of polymer degradation: application to polystyrene, *Chem Eng Sci* 56:971–979, 2001. https://doi.org/10.1016/S0009-2509(00)00312-2.

Kruse TM, Wong H-W, Broadbelt LJ: Mechanistic modeling of polymer pyrolysis: polypropylene, *Macromolecules* 36:9594–9607, 2003. https://doi.org/10.1021/ma030322y.

Kruse TM, Levine SE, Wong H-W, et al.: Binary mixture pyrolysis of polypropylene and polystyrene: a modeling ad experimental study, *J Anal Appl Pyrolysis* 73:342–354, 2005. https://doi.org/10.1016/j.jaap.2005.03.006.

Kumagai S, Yoshioka T: Chemical feedstock recovery from hard-to-recycle plastics through pyrolysis-based approaches and pyrolysis-gas chromatography, *Bull Chem Soc Jpn* 94:2370–2380, 2021. https://doi.org/10.1246/bcsj.20210219.

Kumagai S, Yamasaki R, Kameda T, et al.: Catalytic pyrolysis of poly(ethylene terephthalate) in the presence of metal oxides for aromatic hydrocarbon recovery using Tandem μ-reactor-GC/MS, *Energy Fuel* 34:2492–2500, 2020. https://doi.org/10.1021/acs.energyfuels.9b02915.

Lédé J, Authier O: Temperature and heating rate of solid particles undergoing a thermal decomposition. Which criteria for characterizing fast pyrolysis? *J Anal Appl Pyrolysis* 113:1–14, 2015. https://doi.org/10.1016/j.jaap.2014.11.013.

Levine SE, Broadbelt LJ: Detailed mechanistic modeling of high-density polyethylene pyrolysis: low molecular weight product evolution, *Polym Degrad Stab* 94:810–822, 2009. https://doi.org/10.1016/j.polymdegradstab.2009.01.031.

Liebman SA, Ahlstrom DH, Foltz CR: Thermal degradation studies of PVC with time-resolved pyrolysis GC and derivative TGA, *J Polym Sci Polym Chem Ed* 16:3139–3150, 1978. https://doi.org/10.1002/pol.1978.170161208.

Linstrom P: *NIST chemistry WebBook—SRD 69*, 2017, National Institute of Standards and Technology. https://doi.org/10.18434/T4D30. Date Accessed: June 2022.

Liu X, Li X, Liu J, et al.: Study of high density polyethylene (HDPE) pyrolysis with reactive molecular dynamics, *Polym Degrad Stab* 104:62–70, 2014. https://doi.org/10.1016/j.polymdegradstab.2014.03.022.

Liu H, Wang C, Zhang J, Zhao W, Fan M: Pyrolysis kinetics and thermodynamics of typical plastic waste, *Energy Fuel* 34:2385–2390, 2020. https://doi.org/10.1021/acs.energyfuels.9b04152.

Lopez G, Artetxe M, Amutio M, Alvarez J, Bilbao J, Olazar M: Recent advances in the gasification of waste plastics. A critical overview, *Renew Sustain Energy Rev* 82:576–596, 2018. https://doi.org/10.1016/j.rser.2017.09.032.

Maduskar S, Facas GG, Papageorgiou C, Williams CL, Dauenhauer PJ: Five rules for measuring biomass pyrolysis rates: pulse-heated analysis of solid reaction kinetics of lignocellulosic biomass, *ACS Sustain Chem Eng* 6:1387–1399, 2018. https://doi.org/10.1021/acssuschemeng.7b03785.

Marongiu A, Faravelli T, Bozzano G, Dente M, Ranzi E: Thermal degradation of poly(vinyl chloride), *J Anal Appl Pyrolysis* 70:519–553, 2003. https://doi.org/10.1016/S0165-2370(03)00024-X.

Marongiu A, Faravelli T, Ranzi E: Detailed kinetic modeling of the thermal degradation of vinyl polymers, *J Anal Appl Pyrolysis* 78:343–362, 2007. https://doi.org/10.1016/j.jaap.2006.09.008.

Marrero-Morejón J, Pardillo-Fontdevila E: Estimation of pure compound properties using group-interaction contributions, *AICHE J* 45:615–621, 1999. https://doi.org/10.1002/aic.690450318.

Miranda R, Yang J, Roy C, Vasile C: Vacuum pyrolysis of commingled plastics containing PVC I. Kinetic study, *Polym Degrad Stab* 72:469–491, 2001. https://doi.org/10.1016/S0141-3910(01)00048-9.

Miskolczi N, Bartha L, Deák G, Jóver B: Thermal degradation of municipal plastic waste for production of fuel-like hydrocarbons, *Polym Degrad Stab* 86:357–366, 2004. https://doi.org/10.1016/j.polymdegradstab.2004.04.025.

Muravyev NV, Monogarov KA, Asachenko AF, et al.: Pursuing reliable thermal analysis techniques for energetic materials: decomposition kinetics and thermal stability of dihydroxylammonium 5,5′-bistetrazole-1,1′-diolate (TKX-50), *Phys Chem Chem Phys* 19:436–449, 2017. https://doi.org/10.1039/C6CP06498A.

Nair V, Vinu R: Production of guaiacols via catalytic fast pyrolysis of alkali lignin using titania, zirconia and ceria, *J Anal Appl Pyrolysis* 119:31–39, 2016. https://doi.org/10.1016/j.jaap.2016.03.020.

O'Neal HE, Benson SW: Entropies and heat capacities of cyclic and polycyclic compounds, *J Chem Eng Data* 15:266–276, 1970.

Oguri N, Kirn P: *Design and applications of a curie point pyrolyzer*, 2005 http://www.jai.co.jp/english/tech_data/py/PY011.pdf. Accessed on Aug. 2022.

Ohnishi A, Kato K, Takagi E: Curie-point pyrolysis of cellulose, *Polym J* 7:431–437, 1975.

Ojha DK, Vinu R: Fast co-pyrolysis of cellulose and polypropylene using Py-GC/MS and Py-FT-IR, *RSC Adv* 5:66861–66870, 2015a. https://doi.org/10.1039/C5RA10820A.

Ojha DK, Vinu R: Resource recovery via catalytic fast pyrolysis of polystyrene using zeolites, *J Anal Appl Pyrolysis* 113:349–359, 2015b. https://doi.org/10.1016/j.jaap.2015.02.024.

Ojha DK, Vinu R, Pandey A, et al.: Co-pyrolysis of lignocellulosic biomass with waste plastics for resource and energy recovery, In *Waste biorefinery: potential and perspectives*, 2019, Elsevier. https://doi.org/10.1016/B978-0-444-63992-9.00012-4.

Ojha DK, Viju D, Vinu R: Fast pyrolysis kinetics of alkali lignin: evaluation of apparent rate parameters and product time evolution, *Bioresour Technol* 241:142–151, 2017. https://doi.org/10.1016/j.biortech.2017.05.084.

Ojha DK, Viju D, Vinu R: Fast pyrolysis kinetics of lignocellulosic biomass of varying compositions, *Energy Conv Manage* X 10, 2021, 100071. https://doi.org/10.1016/j.ecmx.2020.100071.

Okoffo ED, Ribiero F, O'Brien JW, et al.: Identification and quantification of selected plastics in biosolids by pressurized liquid extraction combined with double-shot pyrolysis gas chromatography-mass spectrometry, *Sci Total Environ* 715, 2020, 136924. https://doi.org/10.1016/j.scitotenv.2020.136924.

Okonsky ST, Krishna JVJ, Toraman HE: Catalytic co-pyrolysis of LDPE and PET with HZSM-5, H-beta, and HY: experiments and kinetic modelling, *React Chem Eng* 7:2175–2191, 2022. https://doi.org/10.1039/D2RE00144F.

Osman AI, Farrell C, Al-Muhtaseb AH, Al-Fatesh AS, Harrison J, Rooney DW: Pyrolysis kinetic modelling of abundant plastic waste (PET) and in-situ emission monitoring, *Environ Sci Eur* 32:112, 2020. https://doi.org/10.1186/s12302-020-00390-x.

Özsin G, Pütün AE: TGA/MS/FT-IR study for kinetic evaluation and evolved gas analysis of a biomass/PVC co-pyrolysis process, *Energy Convers Manag* 182:143–153, 2019. https://doi.org/10.1016/j.enconman.2018.12.060.

Pan R, Duque JVF, Debenest G: Investigating waste plastic pyrolysis kinetic parameters by genetic algorithm coupled with thermogravimetric analysis, *Waste Biomass Valor* 12:2623–2637, 2021. https://doi.org/10.1007/s12649-020-01181-4.

Paulsen AD, Mettler MS, Dauenhauer PJ: The role of sample dimension and temperature in cellulose pyrolysis, *Energy Fuel* 27:2126–2134, 2013. https://doi.org/10.1021/ef302117j.

Physical and Chemical Property Prediction: *Experimental properties & databases*, 2010, Cheméo. https://www.chemeo.com/. Date accessed June 2022.

Pouwels AD, Eijkel GB, Boon JJ: Curie-point pyrolysis-capillary gas chromatography-high resolution mass spectrometry of microcrystalline cellulose, *J Anal Appl Pyrolysis* 14:237–280, 1989.

Priyadarshi S, Vinu R: Catalytic fast pyrolysis of JP-10 and 3-carene using analytical Py-GC/MS for the production of low molecular weight hydrocarbons, *Energy Fuel* 36:12144–12159, 2022. https://doi.org/10.1021/acs.energyfuels.2c01436.

Priyadarshi S, Kishore MSN, Vinu R: Analytical pyrolysis of jet fuel using different free radical initiators to produce low molecular weight hydrocarbons, *J Anal Appl Pyrolysis* 162, 2022, 105430. https://doi.org/10.1016/j.jaap.2021.105430.

Proano-Aviles J, Lindstrom JK, Johnston PA, Brown RC: Heat and mass transfer effects in a furnace-based micropyrolyzer, *Energy Technol* 5:189–195, 2017. https://doi.org/10.1002/ente.201600279.

Roy D, Thodos G: Thermal conductivity of gases. Hydrocarbons at normal pressures, *Ind Eng Chem Fundam* 7:529–534, 1968.

Roy D, Thodos G: Thermal conductivity of gases. Organic compounds at atmospheric pressure, *Ind Eng Chem Fundam* 9:71–79, 1970.

Seshasayee MS, Savage PE: Oil from plastic via hydrothermal liquefaction: production and characterization, *Appl Energy* 278, 2020, 115673. https://doi.org/10.1016/j.apenergy.2020.115673.

Sezgi NA, Cha WS, Smith JM, McCoy BJ: Polyethylene pyrolysis: theory and experiments for molecular-weight-distribution kinetics, *Ind Eng Chem Res* 37:2582–2591, 1998. https://doi.org/10.1021/ie980106r.

Sharma V, Hossain AK, Griffiths G, et al.: Plastic waste to liquid fuel: a review of technologies, application, and challenges, *Sustain Energy Technol Assess* 53, 2022, 102651. https://doi.org/10.1016/j.seta.2022.102651.

Shiono A, Watanabe A, Watanabe C, Teramae N, Nemoto N, Ohtani H: Thermoanalytical characterization of polymers: a comparative study between thermogravimetry and evolved gas analysis using a temperature-programmable pyrolyzer, *Polym Test* 42:54–61, 2015. https://doi.org/10.1016/j.polymertesting.2014.12.011.

Shiono A, Watanabe A, Watanabe C, et al.: Evolved gas analysis-mass spectrometry in an oxidative atmosphere using a temperature-programmable furnace-type pyrolyzer, *J Anal Appl Pyrolysis* 156, 2021, 105122. https://doi.org/10.1016/j.jaap.2021.105122.

Sichina WJ: *Characterization of polymers*, 2000 https://www.eng.uc.edu/~beaucag/Classes/Characterization/DSCParts/TGA_Notes.pdf. Accessed June 2022.

Sophonrat N, Yang W: Effect of mixing methods of polyethylene and cellulose on volatile products from its co-pyrolysis, *Energy Procedia* 142:315–320, 2017. https://doi.org/10.1016/j.egypro.2017.12.050.

Stein SE, Golden DM, Benson SW: Predictive scheme for thermochemical properties of polycyclic aromatic hydrocarbons, *J Phys Chem* 81:314–317, 1977.

Supriyanto, Ylitervo P, Richards T: Gaseous products from primary reactions of fast plastic pyrolysis, *J Anal Appl Pyrolysis* 158, 2021, 105248. https://doi.org/10.1016/j.jaap.2021.105248.

Tang Y, Ma X, Wang Z, Wu Z, Yu Q: A study of the thermal degradation of six typical municipal waste components in CO_2 and N_2 atmospheres using TGA-FTIR, *Thermochim Acta* 657:12–19, 2017. https://doi.org/10.1016/j.tca.2017.09.009.

Teixeira AR, Gantt R, Joseph KE, et al.: Spontaneous aerosol ejection: origin of inorganic particles in biomass pyrolysis, *ChemSusChem* 9:1322–1328, 2016. https://doi.org/10.1002/cssc.201600112.

The New Plastics Economy: Rethinking the Future of Plastics: *World economic forum*, 2016 https://www3.weforum.org/docs/WEF_The_New_Plastics_Economy.pdf. Accessed Sep. 2022.

Thi HD, Djokic MR, Van Geem KM: Detailed group-type characterization of plastic-waste pyrolysis oils: by comprehensive two-dimensional gas chromatography including linear, branched, and di-olefins, *Separations* 8:103, 2021. https://doi.org/10.3390/separations8070103.

Toraman HE, Dijkmans T, Djokic MR, Van Geem KM, Marin GB: Detailed compositional characterization of plastic waste pyrolysis oil by comprehensive two-dimensional gas-chromatography coupled to multiple detectors, *J Chromatogr A* 1359:237–246, 2014. https://doi.org/10.1016/j.chroma.2014.07.017.

Tyn MT, Calus WF: Estimating liquid molal volume, *Processing* 21:16–17, 1975.

United Nations Environment Programme: *Single use plastics: a roadmap for sustainability*, 2018, ISBN:978-92-807-3705-9.

Usino DO, Supriyanto, Ylitervo P, Pettersson A, Richards T: Influence of temperature and time on initial pyrolysis of cellulose and xylan, *J Anal Appl Pyrolysis* 147, 2020, 104782. https://doi.org/10.1016/j.jaap.2020.104782.

Venkatesan K, Prashanth F, Kaushik V, Choudhari H, Mehta D, Vinu R: Evaluation of pressure and temperature effects on hydropyrolysis of pine sawdust: pyrolysate composition and kinetics studies, *React Chem Eng* 5:1484–1500, 2020. https://doi.org/10.1039/D0RE00121J.

Viju D, Gautam R, Vinu R: Application of the distributed action energy model to the kinetic study of pyrolysis of*Nannochloropsis oculata*, *Algal Res* 35:168–177, 2018. https://doi.org/10.1016/j.algal.2018.08.026.

Vinu R, Broadbelt LJ: Unraveling reaction pathways and specifying reaction kinetics for complex systems, *Annu Rev Chem Biomol Eng* 3:29–54, 2012. https://doi.org/10.1146/annurev-chembioeng-062011-081108.

Vinu R, Ojha DK, Nair V: Polymer pyrolysis for resource recovery, In Reedijk J, editor: *Elsevier reference module in chemistry, molecular sciences and chemical engineering*, Waltham, MA, 2016, Elsevier. https://doi.org/10.1016/B978-0-12-409547-2.11641-5.

Vyazovkin S, Wight CA: Isothermal and nonisothermal reaction kinetics in solids: in search of ways toward consensus, *J Phys Chem A* 101:8279–8284, 1997. https://doi.org/10.1021/jp971889h.

Vyazovkin S, Burnham AK, Criado JM, Pérez-Maqueda LA, Popescu C, Sbirrazzuoli N: ICTAC Kinetics Committee recommendations for performing kinetic computations on thermal analysis data, *Thermochim Acta* 520:1–19, 2011. https://doi.org/10.1016/j.tca.2011.03.034.

Williams EA, Williams PT: Analysis of products derived from the fast pyrolysis of plastic waste, *J Anal Appl Pyrolysis* 40–41:347–363, 1997. https://doi.org/10.1016/S0165-2370(97)00048-X.

Wu C-H, Chang C-Y, Hor J-L, Shih A-M, Chen L-W, Chang F-W: On the thermal treatment of plastic mixtures of MSW: pyrolysis kinetics, *Waste Manag* 13:221–235, 1993. https://doi.org/10.1016/0956-053X(93)90046-Y.

Wu J, Chen T, Luo X, Han D, Wang Z, Wu J: TG/FTIR analysis on co-pyrolysis behavior of PE, PVC and PS, *Waste Manag* 34:676–682, 2014. https://doi.org/10.1016/j.jaap.2021.105122.

Yang B, Chen M: Py-FTIR-GC/MS analysis of volatile products of automobile shredder pyrolysis residue pyrolysis, *Polymers* 12:2734, 2020. https://doi.org/10.3390/polym12112734.

Yousef S, Eimontas J, Striūgas N, Zakarauskas K, Praspaliauskas M, Abdelnaby MA: Pyrolysis kinetic behavior and TG-FTIR-GC-MS analysis of metallised food packaging plastics, *Fuel* 282, 2020, 118737. https://doi.org/10.1016/j.fuel.2020.118737.

Yuan Z, Zhang J, Zhao P, et al.: Synergistic effect and chlorine-release behaviors during co-pyrolysis of LLDPE, PP, and PVC, *ACS Omega* 5:11291–11298, 2020. https://doi.org/10.1021/acsomega.9b04116.

Zhang W, Zhang J, Ding Y, He Q, Lu K, Chen H: Pyrolysis kinetics and reaction mechanism of expandable polystyrene by multiple kinetics methods, *J Clean Prod* 285, 2021, 125042. https://doi.org/10.1016/j.jclepro.2020.125042.

Zhou H, Meng AH, Long YQ, Li QH, Zhang YG: Interactions of municipal solid waste components during pyrolysis: a TG-FTIR study, *J Anal Appl Pyrolysis* 108:19–25, 2014. https://doi.org/10.1016/j.jaap.2014.05.024.

Zhou X, Broadbelt LJ, Vinu R: Mechanistic understanding of thermochemical conversion of polymers and lignocellulosic biomass, In Van Geem K, editor: *Advances in chemical engineering: thermochemical process engineering*, vol. 49, U.K, 2016, Elsevier, pp 95–198. https://doi.org/10.1016/bs.ache.2016.09.002.

Zhou J, Liu G, Wang S, Zhang H, Xu F: TG-FTIR and Py-GC/MS study of the pyrolysis mechanism and composition of volatiles from flash pyrolysis of PVC, *J Energy Inst* 93:2362–2370, 2020. https://doi.org/10.1016/j.joei.2020.07.009.

Zolghadr A, Sidhu N, Mastalski I, et al.: On the Method of pulse-heated analysis of solid reactions (PHASR) for polyolefin pyrolysis, *ChemSusChem* 14:4214–4227, 2021. https://doi.org/10.1002/cssc.202002667.

CHAPTER FOUR

Catalytic conversion of SPW and products upgrading

Ozge Deniz Bozkurt[a], Sean Timothy Okonsky[a], Konstantinos Alexopoulos[a,*], and Hilal Ezgi Toraman[a,b,c,*]
[a]Department of Chemical Engineering, The Pennsylvania State University, University Park, PA, United States
[b]Department of Energy and Mineral Engineering, The Pennsylvania State University, University Park, PA, United States
[c]Institutes of Energy and the Environment, The Pennsylvania State University, University Park, PA, United States
*Corresponding authors: e-mail address: kxa5325@psu.edu; hzt5148@psu.edu

Contents

Abstract

Chemical recycling is the emerging alternative to the well-established mechanical recycling toward the circular economy of solid plastic waste. Incorporation of catalysts into chemical recycling processes such as solvolysis, hydrothermal liquefaction, pyrolysis and gasification enables lower temperature operation and enhanced target product selectivity. Plastic waste, and/or thermochemically produced gaseous/liquid products, can be integrated into existing heterogeneous catalytic processes such as hydrogenolysis, Fischer Tropsch synthesis, methanol routes, catalytic cracking and hydroreforming (hydrocracking) by the action of metal and/or acid active sites to produce plastic monomers (ethylene and propylene), chemicals (benzene, toluene,

Advances in Chemical Engineering, Volume 60
ISSN 0065-2377
https://doi.org/10.1016/bs.ache.2022.09.006

ethylbenzene, xylene and methanol) and fuels (gasoline, jet fuel and diesel). In this review, catalytic processes directly or indirectly related with chemical plastic recycling are elaborated by discussing structure-performance relationships from the recent experimental and computational research performed over the last 10 years, assessing proposed catalytic reaction mechanisms and overviewing pilot to early commercial scale catalytic chemical recycling applications with projected future directions.

1. Introduction

1.1 Chemical recycling process pathways

Mechanical recycling is a commercially exploited technology for converting plastic waste back into consumer plastics with common methods of sorting, washing, grinding, washing, melting and extrusion. Chemical recycling includes breaking the bonds of the plastic polymers toward plastic monomers, chemicals, and fuels (Ragaert et al., 2017; Schweighuber et al., 2021). Solvolysis as a low temperature chemical recycling method is applicable to heteroatom containing plastic polymers such as PET and PU due to the ease of breaking of the C–O and C–N bonds, whereas higher temperature operations such as hydrogenolysis, hydrothermal liquefaction, pyrolysis and gasification are required to break the C–C bonds in plastic polymers (Vollmer et al., 2020). Chemical recycling has advantages over mechanical recycling such as the utilization of mixed plastic waste as feedstock (Dogu et al., 2021), depolymerization of polyolefins with strong C–C backbones (Huang et al., 2022), and incorporation of catalysts for the selective production of hydrocarbons for a circular economy of plastics (Mark et al., 2020).

The processes for the primary catalytic conversion of plastic waste are solvolysis, hydrogenolysis, hydrothermal liquefaction, pyrolysis and gasification, which are schematized in Fig. 1. The figure also embodies the secondary catalytic processes for upgrading the primary conversion products, including catalytic cracking, hydroreforming, Fischer Tropsch synthesis, methanol synthesis and methanol to olefins/gasoline. "Hydro" term refers to water in hydrothermal treatment, and to hydrogen in hydrogenolysis and in hydroreforming. These primary and secondary catalytic processes yield in plastic monomers, chemicals and fuels, contributing to plastic waste recycling and upcycling. The pressure and temperature intervals of the mentioned catalytic conversion processes together with main group of catalysts are presented in Fig. 2. The action of the acid and metal catalysts in breaking bonds and forming new bonds will be exemplified throughout the chapter.

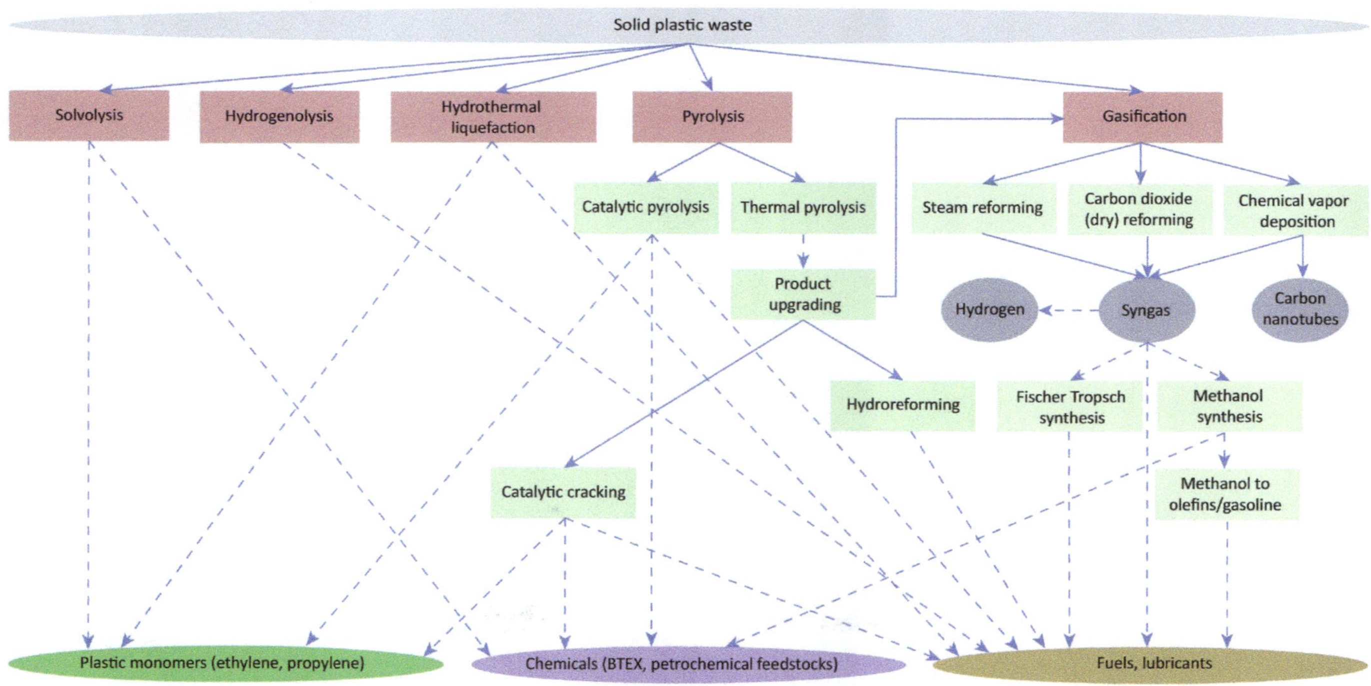

Fig. 1 Process pathways and end-products in chemical plastic recycling of solid plastic waste.

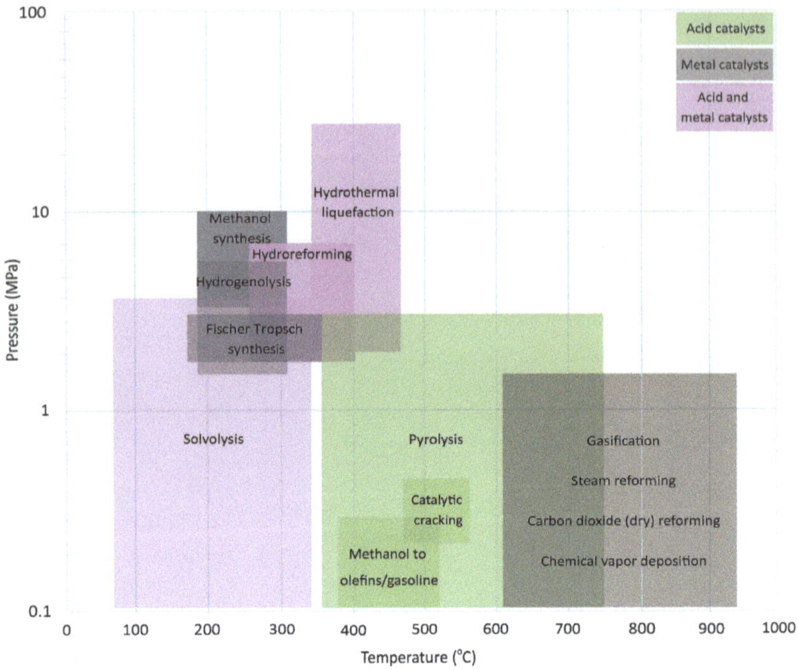

Fig. 2 Pressure and temperature ranges of catalytic processes directly and indirectly related to the chemical recycling of solid plastic waste. The processes are notated with the type of catalysts employed.

Solvolysis can be employed non-catalytically or catalytically in nucleophilic solvents using Lewis acids such as Zn^{+2} (Ellis et al., 2021a; Vollmer et al., 2020). Hydrogenolysis occurs in hydrogen environment over noble metal catalysts (Pt, Ru) with dehydrogenation/hydrogenation functions (Ellis et al., 2021a, 2021b; Jaydev et al., 2021; Kots et al., 2021; Rorrer et al., 2021a, 2021b). Hydrothermal liquefaction is conducted under supercritical water, mostly without a catalyst (Chand et al., 2022; Chen et al., 2019; Dimitriadis and Bezergianni, 2017; Dogu et al., 2021; Helmer Pedersen and Conti, 2017; Hongthong et al., 2020; Seshasayee and Savage, 2021; Wang et al., 2017; Yang et al., 2022). Pyrolysis occurs under inert atmosphere either non-catalytically or catalytically (mostly acid catalysts such as zeolites with cracking activity). Gasification operates under steam, carbon dioxide, air, or nitrogen atmosphere usually with Ni-based reforming catalysts to maximize hydrogen production (Dogu et al., 2021; Huang et al., 2022; Lopez et al., 2018; Soni et al., 2021; Weiland et al., 2021; Yang et al., 2020, 2022). Pyrolysis products can be further upgraded with gasification in steam, carbon dioxide, air or nitrogen environment (Acomb et al., 2016; Aminu et al., 2020, 2022; Chai et al., 2020a, 2020b; Huang et al., 2021; Saad and Williams, 2016a, 2016b, 2017a, 2017b; Wu and Williams, 2009; Yang et al., 2015; Yao and Wang, 2020; Yao et al., 2018). The main product of gasification is syngas, a mixture of carbon monoxide and hydrogen (Ali et al., 2015; de Klerk, 2020). Syngas can be converted into a crude by Fischer-Tropsch synthesis on Fe or Co catalysts with CO dissociation function, or into methanol over Cu based with CO hydrogenation function (Cheng et al., 2017; Laudenschleger et al., 2020; Liu et al., 2019; Ojeda et al., 2010). Methanol can be utilized to produce olefins and aromatics on zeolite catalysts with dehydration, alkylation, and cyclization functions (Cheng et al., 2017; de Klerk, 2020; Huang et al., 2022; Ragaert et al., 2017; Stöcker, 2010). The crudes from the thermochemical processes can be used as refinery feedstocks in non-catalytic (steam crackers) or catalytic (fluid catalytic cracking on acid catalysts, hydrocracking on bifunctional metal-acid catalysts) processes toward the production of fuels (Cheng et al., 2017; Dogu et al., 2021; Escola et al., 2015; Huang et al., 2022; Liu et al., 2021; Petri, 2009; Ragaert et al., 2017; Ruppert et al., 2012; Serrano et al., 2015; Thybaut and Marin, 2016; Tomasek et al., 2020; Weitkamp, 2012).

1.2 Definition of solid plastic waste (SPW)

Plastics are contributors to municipal solid waste with the subcategories of containers & packaging goods such as PET and HDPE bottles, durable goods such as appliances and furniture, and non-durable goods such as diapers and utensils. US EPA estimated the recycling rate of PET bottles & jars as 29.1% and of HDPE bottles as 29.3% as the most recycled items, yet the overall plastic recycling rate was 8.7%, in 2018 (US EPA, 2018). While the global production volumes of major plastic plastics in 2019 were PP (69 Mt) $\sim$ LDPE (68 Mt) $>$ HDPE (57 Mt) $>$ PVC (39 Mt) $\sim$ PET (38 Mt) $>$ PS (29 Mt), the plastics such as PP, LDPE, HDPE, and PET used in food packaging, cosmetics packaging and single use bags were associated with high amounts of waste generation almost equal to their production volumes (e.g., 57 Mt for PP and 63 Mt for LDPE waste production) (Huang et al., 2022; Martín et al., 2021).

Plastic waste can be categorized into pre-consumer (post-industrial) and post-consumer groups. The former group enters the sorting step or directly to the recycling step, while the latter group follows the collection, sorting and recycling steps (EC, 2021). The widely used notation of "solid plastic waste (SPW)" in industry refers to **post-consumer waste** obtained at the end of the sorting step. The collected plastic waste can be a mixture of PET, PE, PP, PS, PVC materials as well as multilayer films consisting of several plastics. Sorting process covers physical separation of different plastics with manual, near-infrared spectroscopy (NIR), Eddy current (to remove nonferrous metals such as aluminum), density, and float/sink techniques (Antonopoulos et al., 2021; Ragaert et al., 2017). The pre- and post-consumer plastics are heterogeneous materials of plastic polymers and additives, wherein the latter group is used for improving the performance, functionality and aging properties of the polymer. The common additives are plasticizers, antioxidants, heat stabilizers, slip agents, pigments, fillers and reinforcement fibers (Hahladakis et al., 2018). A post-consumer plastic packaging waste from the sink fraction consists not only of PET, PVC, PE, PP, PS, PA type plastics, but also includes rubber, paper, metals, fibers, fillers and additives (Bonifazi et al., 2016).

The term SPW used in plastic recycling research may represent (i) a plastic polymer (e.g., isotactic PP wherein the methyl groups are at the same side of the chain), (ii) a post-consumer single stream plastic (e.g., an HDPE bottle), (iii) a physical mixture of two different plastic polymers (widely used in co-pyrolysis), (iv) a model mixture of several plastic polymers, or (v) a sample from post-consumer mixed plastic waste. The catalytic research mostly focuses on pure or single stream plastics, while avoiding additives (thus Br, Cl, P and metals) and contaminants (e.g., food, chemicals). Widely studied plastic polymers in catalytic plastic recycling are shown in Fig. 3.

1.3 Advantages of catalysis in SPW conversion

Employing catalysts in chemical recycling of plastics provides benefits covering energy efficiency and product selectivity. Catalysts reduce the reaction temperature of hydrothermal processes by maintaining lower activation barriers for polymer conversion reactions such as the action of acid sites in catalytic cracking, metal sites in hydrogenolysis, or bifunctional acid-metal sites in hydrocracking (Ellis et al., 2021b; Mark et al., 2020; Vollmer et al., 2020). Molecular structures of widely studied plastic polymers, not surprisingly as the plastics with highest waste output, are shown in Fig. 3. High energy C–C bond between sp^3 hybridized carbons in PE, PP and PS are more difficult to break compared to the C–O (ester) bond in PET (Ellis et al., 2021b; Vollmer

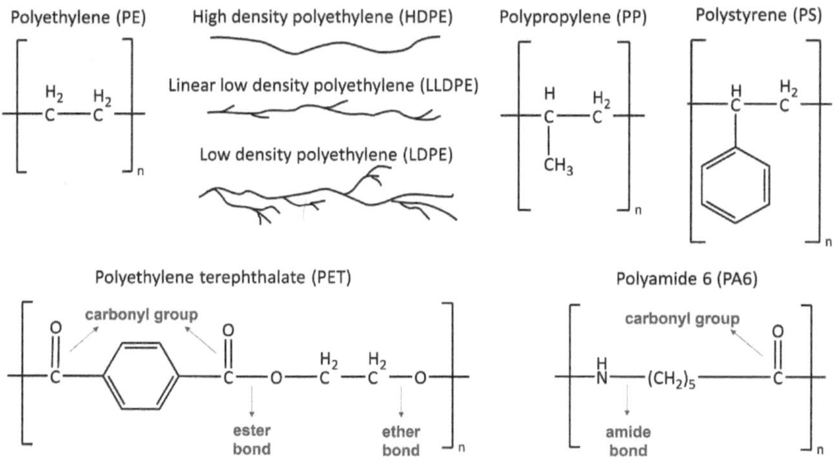

Fig. 3 The structures of the most common plastic polymers studied in chemical plastic recycling.

et al., 2020). The C–C bonds can be activated via protonation over Brønsted acid sites (carbocation mechanism) or via dehydrogenation (C–H activation) over metal sites on catalysts (Ellis et al., 2021a; Escola et al., 2015; Fu et al., 2018; Huang et al., 2021; Kots et al., 2021; Liu et al., 2016, 2021; Mangesh et al., 2020; Munir et al., 2018; Petri, 2009; Rorrer et al., 2021b; Ruppert et al., 2012; Soni et al., 2021; Thybaut and Marin, 2016; van Santen et al., 2011; Vollmer et al., 2020; Weitkamp, 2012; Yao et al., 2021). Tunability of catalyst structural properties including pore size (hierarchical porosity), number of active sites (Brønsted-Lewis acid sites, metal dispersion) and thermal stability (zeolite steaming) allow control on the catalytic performance such as narrower product distribution, coking resistance, and enhanced lifetime (Cheng et al., 2017; Christopher et al., 2022; Escola et al., 2011, 2015; Huang et al., 2022; Mark et al., 2020; Serrano et al., 2015; Soni et al., 2021). Heterogeneous catalysts widely applied in thermochemical processes can be easily separated from gaseous or liquid products (Al-Sabagh et al., 2016; Ellis et al., 2021a; Mark et al., 2020; Niu et al., 2021; Rorrer et al., 2021b; Sánchez-Rivera and Huber, 2021; Soni et al., 2021; Wang et al., 2017; Yang et al., 2022). Some spent catalysts are also eligible to in-situ or ex-situ regeneration by air at elevated temperatures to remove the coke deposit, as performed on fluid catalytic cracking (FCC) and hydrocracking catalysts (Chai et al., 2020b; Huang et al., 2022; Ragaert et al., 2017; Stöcker, 2010; Yang et al., 2022).

2. Catalytic conversion and product upgrading processes of SPW

2.1 Solvolysis

Solvolysis, also known as chemolysis, is the depolymerization of condensation polymers including polyesters (PET, PLA, PHB), polyamides (PA6, PA12, PA66), polyurethanes (PU) and polycarbonates (PC) into their monomers (Ellis et al., 2021a; Mark et al., 2020; Vollmer et al., 2020; Yang et al., 2022). Obtained monomers are then polycondensed into virgin grade polymers. Solvolysis is operated at 70–340 °C and 0.1–4 MPa (Fig. 2) (Yang et al., 2022). Solvolysis methods are named based on the nucleophile type; for instance, water is the nucleophile in hydrolysis, ethylene glycol in glycolysis, ethanolamine in aminolysis, and methanol in methanolysis. The reaction initiates by the attack of nucleophile's heteroatom (O, N) to the carbonyl groups of the polymer, followed by the cleavage of ether, ester or amide bonds (Horn et al., 2012).

Glycolysis of **PET** has been widely studied in literature to obtain the monomer bis(2-hydroxyethyl) terephthalate (BHET). Reduction of the temperature is possible by employing catalysts (Jehanno et al., 2018; Payne and Jones, 2021), while reaction time can be shortened by microwave heating (Vollmer et al., 2020; Yang et al., 2022). Homogeneous catalysts zinc acetate, organic acid-base catalysts (urea, triazabicyclodecene-TBD, methane sulfonic acid) and urea based ionic liquids achieved improved monomer yields at temperatures lower than 200 °C (Ellis et al., 2021a; Jehanno et al., 2018; Payne and Jones, 2021). Heterogeneous catalysis in PET depolymerization aims to overcome catalyst-product separation issues. BHET yields over 80 wt% achieved using novel catalysts such as Fe-loaded carbon nanotubes and metal organic frameworks (Al-Sabagh et al., 2016; Yang et al., 2021). Silica coated magnetic Fe_3O_4 nanoparticles, further functionalized with a Lewis acidic ionic liquid [methylimidazolium] [$FeCl_4$] provided over 90 wt% BHET yield at 180 °C for 10 consecutive catalytic cycles (Cano et al., 2020).

Enzyme based hydrolysis of **PET** is a promising research area toward depolymerizing PET to its monomers terephthalic acid (TPA) and ethylene glycol (EG) at low temperatures. Cutinase (polyester hydrolase) enzymes of bacteria and fungi in nature depolymerize the polyester structured cutin polymer in plants. Leaf-branch compost cutinase enzyme has been modified by protein engineering techniques toward higher PET depolymerization activity and higher thermal stability, providing 90 wt% PET conversion at 72 °C with almost doubled activity compared to the wild type enzyme (3 mg_{enzyme}/g_{PET}, in 10 h) (Tournier et al., 2020). A polyester hydrolase (PETase) isolated from a mixture of compost microorganisms, demonstrated extra-ordinarily high PET depolymerization activity, with complete conversion of PET into TPA and EG at 70 °C (0.6 mg_{enzyme}/g_{PET}, in 24 h) (Sonnendecker et al., 2022). Active site of PETase is postulated as the serine aminoacid residue, attacking as a nucleophile to the ester bond in PET (Joo et al., 2018).

2.2 Hydrogenolysis

Hydrogenolysis is the cleavage of a C–X single bond with hydrogen ($R-X + H_2 \rightarrow R-H + H-X$), wherein X can be an alkyl chain R, or a functional group including a heteroatom such as S, N or O. Very recently, hydrogenolysis studies can be found in plastic waste upcycling research toward depolymerization of polyolefins (Kots et al., 2021; Martín et al., 2021; Rorrer et al., 2021a, 2021b; Yao et al., 2021) and heteroatom containing polymers such as PET and polycarbonates (Kratish et al., 2020; Westhues et al., 2018). The process has been suggested as a milder alternative to hydroreforming (Huang et al., 2022), and to pyrolysis (Sánchez-Rivera and Huber, 2021) in terms of lower temperature (<300 °C) operation (Fig. 2).

Group VIII metals, known as transition metals including Fe, Co, Ni and noble metals are prominent metals used in catalysis with their outstanding ability of chemisorbing hydrocarbons compared to other group metals (Sinfelt, 1991). Noble metals such as **Ru, Pd** and **Pt** catalyze the dehydrogenation reaction of an adsorbed aliphatic hydrocarbon into an adsorbed reactive intermediate, named "C–H activation." The hydrogen deficient reactive intermediate, such as an olefin intermediate then undergoes C–C bond cleavage to form olefin products. Hydrogenation of the product olefin reveals paraffins (Ellis et al., 2021a; Sinfelt, 1991), or can proceed with olefin metathesis in which the groups attached to the double bonds are exchanged, resulting in new product combinations (Ellis et al., 2021b; Grubbs, 2004).

Various polyolefin feedstock including **PE** (molecular weight (MW): 4 kDa), commercial **LDPE**, post-consumer **LDPE** plastic bottle, **PP** (isotactic, MW: 12 and 340 kDa), and **PP-HDPE** model mixed plastic waste (with PP of 340 kDa) have recently been subjected to hydrogenolysis at 200–250 °C temperature, 2–6 MPa H_2 pressure, 4–24 h reaction time and

4:1–2:1 plastic:catalyst ratio over Ru nanoparticles on carbon support (5 wt% **Ru/C**). The polyolefins were depolymerized into gaseous (C_1–C_6) and liquid (C_7–C_{45}) products. Product selectivity was influenced by temperature, H_2 pressure, reaction time, feedstock type (PP yielding in iso-paraffinic products), and catalyst lifetime (slight sintering on the spent catalyst yielding higher molecular weight products) (Rorrer et al., 2021b, 2021a). The findings are promising for technological deployment not only because the obtained products can be further processed into fuels, lubricants or steam cracked to produce olefins; but also the Ru/C catalyst with commercial availability showed resistance to deactivation (Sánchez-Rivera and Huber, 2021). A catalyst screening study of isotactic **PP** (MW: 250 kDa) hydrogenolysis at 250 °C, 3 MPa H_2, 16 h reaction time and 20:1–30:1 plastic:catalyst ratio using supported metal catalysts prepared with wetness impregnation method (5 ± 0.6 wt% metal loading, **Pd, Rh, Ir, Ni, Pt and Ru** metals on **TiO_2, SiO_2, C, Al_2O_3 and CeO_2** supports) demonstrated liquid hydrocarbon formation in 7–66 wt% yield range over Ru metal sites. The maximized liquid product yield of 66 wt% over the Ru/TiO_2 (anatase) catalyst was related to several factors including the small nanoparticle size (1.2 nm, determined by STEM) preference of internal C–C bond breaking to terminal C–C, and preferred desorption of high molecular weight products. The average molecular weight of the liquid products, determined by gel permeation chromatography were in the range of 0.3–1.4 kDa (1 kDa = 1 kg/mol) (Kots et al., 2021). This molecular weight interval corresponds to the C_{30+} vacuum distillation products in a petroleum refinery such as vacuum gas oil (VGO) and lubrication oils with boiling points between 350 and 600 °C. VGO can be converted into fuels and petrochemical feedstock via FCC or hydrocracking processes, while lubricants can be produced via solvent dewaxing or catalytic dewaxing of lubrication oils (Boduszynski, 1987; Cheremisinoff and Rosenfeld, 2009; Marafi et al., 2010; Speight, 2011).

As alternative to Ru, Pt has been investigated in the hydrogenolysis of plastic polymers. Carbon supported platinum catalyst, **5 wt% Pt/C**, achieved complete conversion of **PP** into liquid products including motor oil (C_{21}–C_{45}), diesel (C_{13}–C_{20}) and gasoline (C_6–C_{12}) range hydrocarbons with wt% selectivity of 77%, 20% and 3% respectively, at 300 °C and 1.5 MPa H_2 pressure in 24 h. Product selectivity was significantly influenced by Pt nanoparticle size determined by high-angle annular dark-field scanning transmission electron microscopy (HAADF-STEM), with higher selectivity at smaller nanoparticle size (2.4 nm being the optimum). Temperature programmed desorption experiments with heptane, hexane and pentane revealed higher adsorption strength of the carbon support for longer chains, explaining the bias toward liquid product formation, rather than gaseous product (Jaydev et al., 2021).

In addition to metal nanoparticle size and polymer-support interaction, pore size in a porous support act influences on product selectivity. In a novel core-shell catalyst design inspired by processive enzymatic polymer degradation, the **Pt/SiO_2** as the core was covered by mesoporous silica **$mSiO_2$** as the shell whose pore size can be controlled during catalyst synthesis. At 300 °C and 1.4 MPa H_2, **$mSiO_2/Pt/SiO_2$** catalyst with 0.004 wt% Pt completely converted **HDPE** (MW: 130 kDa) in 24 h into liquid and gas hydrocarbons. The average carbon number of the liquid products were C_{14}, C_{16} and C_{18} for the 1.7, 2.4 and 3.5 nm mesopore size, respectively, indicating diesel range paraffins (Tennakoon et al., 2020).

Exothermic hydrogenolysis can be coupled with endothermic aromatization for a thermodynamically favored reaction cascade on the metal and acid active sites. For instance, 1.5 wt% Pt dispersed as ~1 nm nanoparticles on alumina, **Pt/Al_2O_3**, enabled the conversion of **PE** (MW: 3.5 kDa**)** into long chain alkyl aromatics with up to 80 wt% yields at a moderate temperature of 280 °C in 24 h without an external H_2 source (Zhang et al., 2020). In the proposed reaction mechanism, **PE** (C_{132}) forms alkyl aromatics (~C_{30}) either by 1,6 ring closure → hydrogenolysis → dehydroaromatization, or by hydrogenolysis → 1,6 ring closure → dehydroaromatization routes. Hydrogenolysis decreases the molecular weight of the paraffin and aromatic reactants, while the hydrogen balance is preserved with the coupling of H_2 generating aromatization with H_2 consuming hydrogenolysis. Plasma assisted hydrogenolysis at ambient temperature and pressure has also been investigated in polyolefin depolymerization. **PE** was converted into C_1–C_3 paraffins at >95 wt% selectivity with CH_4 being the major product with 70 wt% selectivity. Introduction of catalysts to the high energy hydrogen plasma influenced product selectivity, for instance acidic **SAPO-34** zeolite with C–C coupling capability led to the production of more olefins, while **Pt/C** metallic catalyst with H_2 dissociative adsorption function provide more paraffins (Yao et al., 2021).

2.3 Hydrothermal liquefaction

Hydrothermal liquefaction (HTL) is the decomposition of carbonaceous material under oxygen -free conditions into oil, gas and solid (char) with subcritical or supercritical water, or organic solvents such as ethanol and methanol. HTL operation is carried out at moderate temperatures of 280–450 °C, and pressure range of 4–25 MPa (Fig. 2). Fluids above their supercritical temperature (T_c) and pressure (p_c) are called supercritical fluids with altered dielectric constant, density, and ionic strength. Water transforms into supercritical form above 374 °C and 22.1 MPa, resulting in lowered dielectric constant (from 80 to 10), becoming a less-polar solvent to dissolve hydrocarbons (Dogu et al., 2021; Huang et al., 2022; Yang et al., 2022). HTL process parameters include temperature, pressure, feed/water ratio, residence time and the existence of a catalyst. So far batch, plug flow and CSTR type reactors have been investigated in HTL research (Yang et al., 2022).

The dilution effect by dissolution of the reactant in water phase limits the formation of char and condensed polymers. Production of lower molecular weight products in HTL, compared to pyrolysis is due to hydrogen donation from water to the reaction medium, which favors the chain termination reactions of radicals with hydrogen instead of combining with larger radicals (Akiya and Savage, 2002). Branching structure of polyolefins may result in olefin rich or paraffin rich products. For instance, branched structure of PP with the lower energy C–CH_3 bonds (compared to the primary C–C bonds) promotes thermal cleavage of the C–CH_3 bonds forming radicals. The subsequent β-scission of the radicals forms olefins. Furthermore, olefin cyclization and dehydrogenation yields in aromatics (Chen et al., 2019). Synergistic interactions between mixed plastic waste and biomass toward higher oil yield at lower temperatures in supercritical HTL are speculated to occur with hydrogen transfer from polyolefins to biomass components such as cellulose, starch and lignin (Seshasayee and Savage, 2021).

HTL of polyolefins and heteroatom containing polymers as well as post-consumer plastic waste have been studied, achieving 60–90 wt% liquefaction yields. Advantages of the HTL process are the handling of mixed, unsorted, moisturized plastic waste and reduced coke formation. Although water is regarded as a green solvent, substantial water usage in HTL, combined with high pressures necessitate high energy use and specialized equipment (Dogu et al., 2021). Supercritical HTL of several single-stream plastics including **PET**, **polycarbonate (PC)**, **styrene butadiene (SB)** and **polybutylene terephthalate (PBT)** at 400 °C and 25 MPa revealed different oil, aqueous, solid and gas yields. For PC the yield of oil phase was 99.8 wt% and the product mixture consisted of bisphenol A monomer (BPA), as well as the phenolic cracking products of BPA. However, in the case of PET, the solid phase yield was almost 70 wt%

consisting of terephthalic acid (TPA) monomers (Helmer Pedersen and Conti, 2017). Supercritical HTL of **PP** at 23 MPa operated at a temperature range of 380–450 °C demonstrated the effect of temperature and reaction time on product yields. PP to oil yield was significantly improved by 425 °C, while the olefins were transformed into cyclics and aromatics at extended reaction times from 1 to 4 h (Chen et al., 2019).

Homogenous acid and alkali catalysts **(H_2SO_4, K_2CO_3, Na_2CO_3)** as well as heterogeneous catalysts including zeolites **(HY, HBeta, HZSM-5)** and mesoporous silica **(MCM-41)** have been demonstrated to improve liquid yields in biomass HTL (Dimitriadis and Bezergianni, 2017). Slight improvements in bio-oil quality such as lower oxygen, sulfur and nitrogen contents were achieved over metal based catalysts **Ni**, **Pd**, **Pt**, **Ru**, **Co-Mo** with hydrogenation activity in microalgae HTL (Duan and Savage, 2011). Catalytic HTL studies with plastics are rare in literature. Supercritical water gasification of polyolefins **(LDPE, HDPE, PP** and **PS**) at 450 °C on **$RuO_2/\gamma\text{-}Al_2O_3$** catalyst provided methane rich gas products with up to 60 vol% CH_4 (Onwudili and Williams, 2016). Subcritical HTL of polyamide-nylon at 2 MPa and 345 °C was conducted with and without catalyst **(HBeta)**. While the existence of HBeta did not vary the product distribution, decomposition of polyamide-nylon was accelerated by hydrolysis of the amide (–CONH–) group upon protonation of the carbonyl oxygen, followed by the water attack on the carbonyl group (Wang et al., 2017). Subcritical catalytic co-HTL of **PP** with biomass over **aluminosilicate** catalyst at lower temperatures (350 °C) yielded in 30 wt% plastic conversion, while pure PP depolymerization was negligible under the mentioned conditions, pointing out to a synergy between plastic waste and biomass (Hongthong et al., 2020). Sewage sludge with 0.1 g/kg **microplastics** content (mainly PU, PP and PE) was subjected to supercritical HTL at 400 °C and 30 MPa using homogeneous alkali catalysts **(K_2CO_3 and NaOH)** wherein 97 wt% of the microplastic mass consisting mainly of PU, alkyd PET resin, PP, PE, and PVC was converted into bio-crude in HTL. The identification and quantification of microplastics before and after HTL was performed by combined light microscopy and FTIR (Chand et al., 2022).

2.4 Pyrolysis

Pyrolysis is a thermochemical process which heats feedstocks at temperatures ranging from 350 to 750 °C in an inert atmosphere, most often under atmospheric pressure (Fig. 2) (Lopez et al., 2017). It is a known technique for converting solid feedstocks, such as biomass and plastic waste, into a valuable mixture of chemicals which can then be further upgraded via catalysts. Catalysts are also able to be integrated in with the pyrolysis process allowing for polymer degradation at lower temperatures. While some polymers, such as polystyrene, are able to be pyrolyzed thermally to obtain high yields of its monomeric form (Artetxe et al., 2015), other polymers such as polyethylene and polypropylene require catalysts to upgrade their pyrolysis products to their monomeric form of ethylene and propylene (Eschenbacher et al., 2022a). A review of the various modes of catalytic pyrolysis, a discussion of which polymers and catalysts are most ideal for production of various groups of value-added chemicals, and the ways in which the kinetics of catalytic pyrolysis can be determined, are discussed in the following section.

2.4.1 In-situ and ex-situ pyrolysis

Catalysts can be introduced into the pyrolysis process in a number of different ways. In-situ catalytic pyrolysis refers to when plastic feedstock and catalysts are mixed together, and their mixture is pyrolyzed. This is a popular method for catalytic pyrolysis among researchers, as only one reaction vessel is needed, and the reaction can be conducted in commonly used scientific instruments, such as thermogravimetric analyzers, single furnace micro-pyrolysis reactors, and batch reactors. The catalyst and plastic are often reported to be pre-mixed in separate containers, and then added to the reaction vessel (Berčič et al., 2019; Silva et al., 2020), while it is also reported that the polymer and catalyst are added to the reaction vessel, and then mixed inside of the vessel (Marcilla et al., 2007). Ex-situ catalytic pyrolysis refers to when plastic feedstock is first pyrolyzed in the absence of catalyst, commonly referred to as "thermal pyrolysis," and then the thermal pyrolysis vapors are subsequently passed over a catalyst bed located downstream, often in a separate furnace, from the pyrolysis feedstock. The pyrolysis volatiles can be subjected to catalytic processing on cracking, hydroreforming and gasification catalysts towards a variety of products including plastic monomers, aromatics, fuels, and hydrogen (Sections 2.6 and 2.7).

The catalyst to feedstock weight ratios which have been reported to be used for lab-scale fixed bed reactors are in the range of 1:100–1:2 (Du et al., 2016; Lee et al., 2021; Tian et al., 2022; Yu et al., 2020), for thermogravimetric analyzers the ranges reported are between 1:10–10:1

(Durmuş et al., 2005; Marcilla et al., 2007), and for micro-pyrolysis reactors the ranges reported are between 1.5:1–100:1 (Du et al., 2016; Kumagai et al., 2020; Wang et al., 2019; Xue et al., 2017). There is a trend that for reactor systems which pyrolyze larger amounts of polymers (~1 g), such as lab-scale fixed bed reactors, lower catalyst to feedstock ratios are used, and reactor systems which pyrolyze smaller amounts of polymers (0.2–0.5 mg), such as micro-pyrolysis reactors, higher catalyst to feedstock ratios are used. It has been said that higher catalyst to feedstock ratios are used in micro-pyrolysis reactors due to the short retention times of pyrolysis vapors in the catalyst bed due to the short length of the catalyst bed and the high flow rate of carrier gas which are typical for micro-reactors (Xue et al., 2017).

2.4.2 Production of ethylene/propylene, BTEX and fuels

Adding in a catalyst to the pyrolysis process can alter the distribution of pyrolysis products. Depending on the plastic feedstock and catalyst used, the pyrolysis operating conditions such as catalyst to feedstock ratio and pyrolysis temperature, and the catalyst contact mode (in-situ vs. ex-situ), the distribution of products formed can change dramatically. Desired classes of products which can be enhanced in the presence of catalyst are light olefins such as ethylene and propylene which serve as monomers for polyethylene and polypropylene, monoaromatic hydrocarbons such as BTEX (benzene, toluene, ethylbenzene, and xylene) which serve as valuable building blocks for many platform chemicals, and hydrocarbon fuels such as gasoline, diesel, and jet fuel (Fig. 1).

Table 1 shows the wt% yield of ethylene and propylene which have been observed during the catalytic pyrolysis of various polymers using various catalysts and reactors by different research groups. Propylene yield has been observed to reach a maximum value of 46.3% (Eschenbacher et al., 2021) in **LDPE** pyrolysis, which is higher than the maximum yield of ethylene reported which is 25.3% (Xue et al., 2017). Zeolite catalysts with smaller pore size, such as HZSM-5 (pore size 5.1 × 5.5 Å and 5.3 × 5.6 Å) result in high yields of ethylene and propylene in comparison with larger pore size zeolites such as HY (pore size 7.7 × 7.6 Å), as can be seen in Table 1 (Hasan et al., 2022). Incorporation of metals into the zeolite framework such as boron and phosphorous have been shown to increase catalyst stability and decrease the level of coking during polyethylene pyrolysis (Eschenbacher et al., 2021, 2022b). Decreasing the Si/Al ratio of HZSM-5 enhances the catalyst's acidity, thus furthering the cracking of pyrolysis

Table 1 Reported weight% yields of ethylene and propylene observed during in-situ (I) and ex-situ (E) catalytic pyrolysis.

Polymer (feedstock weight)	Catalyst	C/F ratio	Reactor used[a]	T (R1) °C	T (R2) °C	Ethylene wt% yield	Propylene wt% yield	Reference
LDPE (0.4 mg)	Boron Modified HZSM-5 (B/Meso-Z_med Si/Al)	40:1	Tandem μ-reactor (E)	550	600	16.9	46.3	Eschenbacher et al. (2021)
LDPE (0.4 mg)	Phosphorous modified HZSM-5 (P/mesoZ55-ss)	80:1	Tandem μ-reactor (E)	550	700	15.6	40.7	Eschenbacher et al. (2022b)
LDPE (30 g)	HZSM-5	1:10	CSTR (I)	450	–	6.7	24	Hasan et al. (2022)
LDPE (30 g)	HY	1:10	CSTR (I)	450	–	0.6	9.2	Hasan et al. (2022)
HDPE (300 g, 1 g/min)	HZSM-5 (Si/Al 30)	8:1 min^{-1}	Spouted Bed/ Fixed Bed (E)	500	500	9.39	32	Artetxe et al. (2013)
HDPE (300 g, 1 g/min)	HZSM-5 (Si/Al 80)	8:1 min^{-1}	Spouted Bed/ Fixed Bed (E)	500	500	3.6	24.1	Artetxe et al. (2013)
UHMWPE (0.5 mg)	HZSM-5 (Si/Al 23)	20:1	Tandem μ-reactor (E)	600	600	25.3	41.9	Xue et al. (2017)
UHMWPE (0.5 mg)	HZSM-5 (Si/Al 23)	20:1	Tandem μ-reactor (I)	600	600	5.92	9.39	Xue et al. (2017)

[a](E) refers to pyrolysis reactors operated in ex-situ configuration, and (I) corresponds to reactors operated in in-situ configuration.

volatiles and increasing the yield of ethylene and propylene (Artetxe et al., 2013). When polyolefins, such as ultra-high molecular weight polyethylene (**UHMWPE**), were pyrolyzed with HZSM-5 catalyst in an ex-situ catalyst contact mode, it was found that the yield of ethylene and propylene were greatly increased from when pyrolyzed in an ex-situ contact mode, as the in-situ contact mode favored more the production of light alkanes such as ethane and propane (Xue et al., 2017).

In addition to light olefins, it is also possible to produce monoaromatic hydrocarbons, such as benzene, toluene, ethylbenzene, and xylene (BTEX) in high yields during the catalytic pyrolysis of plastic waste. Benzene is a main precursor for the production of other chemicals, toluene is used often in paint thinners and as an octane booster for gasoline, ethylbenzene is used for the production of styrene in order to make polystyrene plastic, and xylene is the main procurer to terephthalic acid and dimethyl terephthalate which acts as monomers in the production of polyethylene terephthalate (PET) plastic (Pellegrino, 2000). Table 2 shows the wt% yield of BTEX compounds that were reported for the catalytic pyrolysis of various polymers. It can be seen that the catalytic pyrolysis of both **PS** and **PET** lead to high yields of benzene, as both polymers have aromatic rings present in their structure (Du et al., 2016; Xue et al., 2017). In addition to zeolite catalysts, metal oxides catalysts such as CaO and ZnO have demonstrated their ability to decarboxylate carboxylic acids, which are formed during the non-catalytic pyrolysis of **PET**. The GC/MS area% of benzene observed during PET pyrolysis in a micro-reactor increased from 8.0% to 45.9–73.3% when CaO and ZnO were used as a catalyst (Kumagai et al., 2018, 2020). In-situ catalytic pyrolysis of polyolefins with MgO was shown to form higher amount of coke (5-23 wt%) than for zeolite and FCC catalysts (0-3wt%) (Saeaung et al., 2021), while other researchers reported low coke yield for MgO (0-2 wt%) during the microwave assisted ex-situ catalytic pyrolysis of **LDPE** (Fan et al., 2017). There are known interaction effects to occur during the catalytic co-pyrolysis of **PET** with **PE** with HZSM-5 which increase the yield of aromatic hydrocarbons (Xue et al., 2017). HY catalyst with higher amounts of Lewis acid sites have shown to be more effective than HZSM-5 for increasing the selectivity of alkylated benzenes, such as toluene and xylene, during the in-situ catalytic co-pyrolysis of **LDPE** and **PET** (Okonsky et al., 2022). HZSM-5 catalyst has been shown to be better at producing aromatics than HY catalysts during catalytic pyrolysis of polyolefins due to its shape selective nature and high acid strength (Hasan et al., 2022; Kim et al., 2016). When ZSM-5 catalyst was structured

Table 2 Reported weight% yields of benzene (B) toluene (T) ethylbenzene (E) and xylene (X) observed during in-situ (I) and ex-situ (E) catalytic pyrolysis.

Polymer (feedstock weight)	Catalyst	C/F ratio	Reactor used[a]	T (R1) °C	T (R2) °C	BTEX wt% yield	Reference
PET (0.5 mg)	HZSM-5	20:1	Tandem μ-reactor (I)	600	600	25.59% (B) 1.95% (T) 0.50% (E) 0.57% (X)	Xue et al. (2017)
PET (carpet, 0.4 g)	CaO	10:1	Fixed Bed Reactor (E)	750	–	19.58% (B)	Du et al. (2016)
PS (0.5 mg)	HZSM-5	20:1	Tandem μ-reactor (I)	600	600	39.60% (B) 5.45% (T) 2.47% (E) 0.83% (X)	Xue et al. (2017)
PP (0.5 mg)	HZSM-5	20:1	Tandem μ-reactor (I)	600	600	4.46% (B) 9.54% (T) 0.61% (E) 6.33% (X)	Xue et al. (2017)
PE + PET 1:1 (0.5 mg)	HZSM-5	20:1	Tandem μ-reactor (I)	600	600	20.36% (B) 7.05% (T) 2.37% (E) 3.31% (X)	Xue et al. (2017)
LLDPE (1 mg)	HZSM-5	1:1	Single-shot μ-reactor (I)	600	–	2.7% (B) 8.9% (T) 1.5% (E) 10.0% (X)	Kim et al. (2016)
LLDPE (1 mg)	HY	1:1	Single-shot μ-reactor (I)	600	–	0.9% (B) 1.0% (T) 0.4% (E) 5.4% (X)	Kim et al. (2016)
LDPE (6 g)	2 wt% Zn/ZSM-5	1:20	Batch reactor (4 bar)	380	–	2.71% (B) 13.23% (T) 2.84% (E) 16.25% (X)	Wang et al. (2022)

[a](E) refers to pyrolysis reactors operated in ex-situ configuration, and (I) corresponds to reactors operated in in-situ configuration.

with SiC foam, it was found to increase catalyst stability maintaining high selectivity to C_6–C_{12} aromatics during the catalytic pyrolysis of LDPE (Zhou et al., 2022). Loading ZSM-5 catalyst with 2 wt% zinc and conducting pyrolysis in batch orientation at high pressure led to high yields of xylene and toluene being obtained from LDPE pyrolysis (Wang et al., 2022).

Another common target product for plastic waste pyrolysis is fuel range hydrocarbons. It has been shown that the pyrolysis of **LDPE** with activated carbon catalysts is able to produce high yields of liquid (61.6–73.1 wt%) which have a high percentage (86.5–100 GC/MS area%) of jet fuel range hydrocarbons (C_8–C_{16}) and diesel range hydrocarbons (C_9–C_{23}) (Zhang et al., 2019). The liquid organic products obtained from **LDPE** pyrolysis with HZSM-5 can be hydrotreated with RANEY® nickel catalyst to form oil product with specifications similar to JP-5 navy fuel (Zhang and Lei, 2016). For the catalytic pyrolysis of **HDPE** it was found that Y-zeolite decreased pyrolysis temperature to a greater extent than for $MgCO_3$ catalyst, and resulted in higher gasoline fraction where $MgCO_3$ resulted in higher diesel fraction, both of which were higher than for non-catalytic pyrolysis (Kunwar et al., 2016). Incorporation of 5 wt% iron into ZSM-5 catalyst during mixed polyolefin pyrolysis resulted in a higher fraction (66.39%) of fuel range hydrocarbons (C_6–C_{12}) than for parent ZSM-5 catalyst (47.33%) and for non-catalytic pyrolysis (27.39%) (Dwivedi et al., 2021).

2.4.3 Kinetic analysis

Thermogravimetric analysis (TGA) is a popular technique to determine global pyrolysis kinetics, as it measures for a given heating rate at what temperatures the volatilization of a polymer will occur. The Friedman method, the Flynn-Wall-Ozawa (FWO) method, and the Kissinger method are all popular model-free isoconversional techniques which can determine the apparent activation energy for pyrolysis at various levels of conversion, if the pyrolysis has been conducted at multiple heating rates. It has been shown with isoconversional approach that the addition of catalysts help to decrease the activation energy during pyrolysis, as the polymer degradation occurs at lower temperatures than for non-catalytic pyrolysis (Berčič et al., 2019; Durmuş et al., 2005). Other researchers have used nth-order least-squares model fitting approach to determine the kinetic triplet of activation energy, reaction order, and pre-exponential factor for a given polymer and catalyst, which is able to model the catalytic pyrolysis for multiple heating rates and for multiple catalyst to feedstock (CF) ratios (Marcilla et al., 2007).

A difference between using the isoconversional approach and the nth-order model fitting approach proposed by Marcilla et al. is that activation energy is dependent on the CF ratio used for isoconversional approach (Eimontas et al., 2021), where for the nth-order model fitting approach, the activation energies derived for various CF ratios are the same (Marcilla et al., 2007). The nth-order model fitting approach has been adapted to determine the kinetics for the catalytic *co*-pyrolysis of **LDPE** and **PET** and it was found this approach is able to model interaction effects observed between the two polymers during their co-pyrolysis with multiple zeolite catalysts (Okonsky et al., 2022). It has been mentioned the importance of deriving "catalyst-free" TGA curves for catalytic pyrolysis by running and subtracting out TGA runs with CF ratio of 1:0, as this eliminates the ability for catalyst mass loss to be interpreted as polymer mass loss, which leads to an incorrect estimate of the activation energy for catalytic pyrolysis (Berčič et al., 2017; Okonsky et al., 2022). In addition to determining global kinetics via TGA, researchers have also used lumped kinetics to predict the product distribution during catalytic pyrolysis. Feedstocks and products present during the pyrolysis process, such as plastic waste feedstock, cracked polymer intermediate, heavy liquid, light liquid, gas, and coke can be assigned as "lumps." The amount of these lumps can vary with reaction time and temperature depending on the reaction pathways assigned and their associated kinetic parameters. Ten reactions have been used to determine lumped kinetic parameters for the in-situ catalytic pyrolysis of mixed polyolefins, subsequently upgraded via catalyst in a 2nd reactor for ZSM-5 catalyst with multiple metal loadings (Till et al., 2018).

2.5 Gasification

2.5.1 Syngas production

Gasification is a commercially available thermochemical process operated at higher temperatures than of pyrolysis (600–1000 °C) and at atmospheric pressure (Fig. 2) (Huang et al., 2022). Gasification yields in a narrower product portfolio compared to pyrolysis (Lopez et al., 2018). Carbonaceous materials are partially oxidized under air, steam, O_2 or CO_2 atmosphere. The products are syngas comprising mainly of CO and H_2 (as well as CH_4, CO_2, trace amounts of H_2S, NOx, HCN and NH_3), light olefinic/aromatic hydrocarbons, tar (gaseous) and char (solid residue) (Christopher et al., 2022; Huang et al., 2022; Lopez et al., 2018; Mark et al., 2020; Yang et al., 2022). Syngas can be utilized as a precursor of chemicals, fuels and new plastic raw materials, or as fuel gas in industry due to its high heating value (Fig. 1). Hydrogen production is improved in the existence of steam by the steam reforming reaction ($C_nH_m + nH_2O \rightarrow (n + m/2)\ H_2 + nCO$, $\Delta H > 0$). Using **Ni** based reforming catalysts in gasification enhances hydrogen production by favoring the water gas shift (WGS) reaction ($CO + H_2O \leftrightarrow CO_2 + H_2$, $\Delta H = -41\ kJ/mol$) (Lopez et al., 2018; Yang et al., 2022).

Gasification, as in pyrolysis can handle a wide range of feedstock including plastics, biomass, and coal. The suitability of mixed and contaminated plastic waste in gasification eliminates the need of sorting and pretreatment (Dogu et al., 2021; Soni et al., 2021; Yang et al., 2022). Severe operating temperatures in plastic waste gasification, coupled with fluidized bed reactor configurations enable accelerated degradation of the solid material into gaseous products. In gasification, polyolefins such as HDPE, LDPE, and PP mainly undergo end-chain scission, producing light olefins, while PET and the polyolefin PS transform into aromatics by random chain scission and subsequent secondary reactions which are triggered at high temperature and residence times (Christopher et al., 2022; Lopez et al., 2017, 2018). At 700–850 °C operating temperature, steam gasification may improve the H_2 yield compared to air gasification of **PET** or **PE**. However the $H_2/(CO + CO_2)$ ratio of the syngas drops substantially in the co-gasification of PE with PET. Besides, the formation of the polyaromatic compound (PAH) biphenyl by the combination of two benzene radicals was promoted at high temperature, steam ratio and residence time conditions (Li et al., 2021; Wilk and Hofbauer, 2013).

In mixed plastic waste recycling, raw gas derived from gasification is associated with lower amounts of toxic emissions such as polychlorinated compounds, compared to the gas emitted by direct combustion, as the former process is operated at oxygen-lean conditions (Maric et al., 2020; Weiland et al., 2021). Gasification has challenges including high capital and energy costs (Huang et al., 2022; Lopez et al., 2017). Besides, gasification products must be cleaned and upgraded to enable high value chemicals and fuels (Mark et al., 2020). Solid plastic waste gasification is associated with tar formation, especially in the case of **PET**, **PVC** and **PS** (Dogu et al., 2021). Tar is mixture of mono and polyaromatics and oxygenated hydrocarbons. Different PAH formation mechanisms are proposed, including (i) direct combination aromatic rings to form biphenyl, (ii) H_2-abstraction-C_2H_2-addition (HACA) to the aromatic rings, (iii) CO abstraction from the phenol (C_6H_6O) precursor, (iv) cyclization through intermolecular/intramolecular Diels-Alder reactions (Dogu et al., 2021; Font, 2013). Natural olivine, **$(Mg,Fe_2)SiO_4$**, has been widely employed in research studies as a fluidized bed material to enhance heat transfer and perform tar cracking in the gasification of plastic waste (Li et al., 2021; Mastellone and Arena, 2008; Wilk and Hofbauer, 2013). **Ni**, **Co** and **Fe** based catalysts with supports such as metal oxides have been used after the gasification step for the catalytic reforming of tar (Huang et al., 2022; Liu et al., 2019). Catalytic upgrading of gasification products is discussed in Section 2.6.

Syngas can be converted into alcohols (methanol, ethanol, C_3+ alcohols) and aldehydes via different catalytic routes. Biological routes provide important commodities such as acetic acid and butanol from syngas fermentation (Grim et al., 2019). Industrial scale methanol production from syngas occurs with catalytic hydrogenation over **Cu** based catalysts. Methanol can be further converted into olefins or gasoline with the catalytic Methanol to olefins (MTO) and Methanol to gasoline (MTG) processes. Alternatively, syngas can undergo Fischer Tropsch synthesis to produce liquid crudes on **Fe** or **Co** based catalysts which are then upgraded into fuels and chemicals with refinery processes such as fluid catalytic cracking and hydrocracking (Dogu et al., 2021; Huang et al., 2022; Petri, 2009; Tomasek et al., 2020). Reaction of CO and H_2 on metal sites initiates by dissociation of H_2 to atomic H and the adsorption of CO (dissociative or non-dissociative) (Cheng et al., 2017). H_2/CO stoichiometric ratio, as well as the CO_2 content, influences on the further processing of syngas. For instance, H_2/CO of 2 is common in catalytic methanol production. $H_2/CO > 2$ is required for FT process with Co-based catalyst, whereas with Fe-based catalysts $H_2/CO < 2$ is common at low temperature FT process (170–270 °C), and $H_2/(2CO + 3CO_2) = 1.05$ is applied at high temperature FT process (320–360 °C) (de Klerk, 2020; Weiland et al., 2021).

2.5.2 Syngas to methanol, methanol to olefins, methanol to gasoline routes

Conventional methanol is obtained from fossil-based syngas. Methanol is a raw material in chemical industry to produce formaldehyde, acetic acid, methyl methacrylate and dimethyl carbonate. In oil gas industry, methanol is utilized to produce methyl tertiary butyl ether (MTBE) as a gasoline additive, or dimethyl ether (DME) as a diesel alternative. Methanol can also be directly blended with gasoline or used as a feedstock for biodiesel production (Ali et al., 2015; Sarp et al., 2021).

Methanol synthesis follows with partial CO hydrogenation on **Cu** based catalysts at 200–300 °C temperature and 3.5–10 MPa pressure conditions (Fig. 2). The H_2:CO molar ratio is preferred as 2:1. The exothermicity of the equilibrium reaction ($CO+2H_2 \leftrightarrow CH_3OH$, $\Delta H=-91$ kL/mol) necessitates adequate cooling of the reactor and recycle of the products to reach high conversions (de Klerk, 2020). The typical methanol synthesis catalyst is **$Cu/ZnO/Al_2O_3$**, a heterogeneous catalyst employed in industry for more than 50 years. CO_2 in the syngas can also be converted into methanol, as the catalyst is active in the reverse water gas shift (RWGS) reaction ($CO_2+H_2 \leftrightarrow CO+H_2O$, $\Delta H=41$ kJ/mol) (de Klerk, 2020). The nature of the active sites has recently been proposed as the Cu–Zn interface sites with positively charged Zn species in vicinity of Cu (Cu^0–$Zn^{\delta+}$), using an surface-sensitive operando method under industrially relevant conditions of 200°C and 6MPa (Laudenschleger et al., 2020).

Methanol to Olefins (MTO) process achieves high C_2–C_3 olefin yields (~75wt%) over small pore size (0.4nm) zeolite **SAPO-34**, and high C_3–C_4–C_5 olefin yields (~80wt%) over medium pore zeolite **HZSM-5** (0.5–0.6nm), at 400–500°C and 0.1–0.3MPa (Fig. 2). UOP/Norsk Hydro's SAPO-34 based MTO process converts coal or natural gas-based methanol into ethylene and propylene. Methanol to gasoline (MTG) operation, such that of ExxonMobil, employs ZSM-5 catalyst at 350–400°C and atmospheric pressure to convert methanol into gasoline range liquid hydrocarbons with low sulfur and benzene content. The reaction sequence initiates with the adsorption of methanol on the Brønsted acid site, formation of the methoxy species (Zeolite$-OCH_3$), and dehydration of methanol to dimethyl ether ($2CH_3OH \leftrightarrow CH_3OCH_3+H_2O$) (Stöcker, 2010). Various reaction mechanisms have been postulated for C–C bond formation by the coupling of methanol or dimethyl ether molecules on zeolite active sites. Generally accepted mechanism consists of two stages: (i) formation of the hydrocarbon pool (HP) precursors $(CH_2)_n$ by direct C–C coupling of adsorbed methanol/dimethyl ether with incoming methanol/dimethyl ether ($CH_3OCH_3 \rightarrow (CH_2)_n+H_2O$) and, (ii) conversion of $(CH_2)_n$ hydrocarbon pool precursors in olefinic and aromatic cycles with alkylation, dealkylation, cyclization and cracking reactions on the so-called dual-cycle mechanism (($(CH_2)_n \rightarrow C_2H_4$, C_4H_8, C_3H_6, C_6+ olefins, paraffins, aromatics and coke) (Stöcker, 2010; Yarulina et al., 2018). Functioning of the olefinic and aromatic cycles depends on the number of Brønsted acid sites on zeolite surface. As shown in Fig. 4, the decrease in the acid site density in ZSM-5 with an increase of Si/Al ratio (Fig. 4C and D) limits the aromatics and paraffins formation (Fig. 4B) by suppressing the cyclization and hydride transfer reactions, while enhancing propene selectivity (Fig. 4A) (Yarulina et al., 2018). Product selectivity also depends

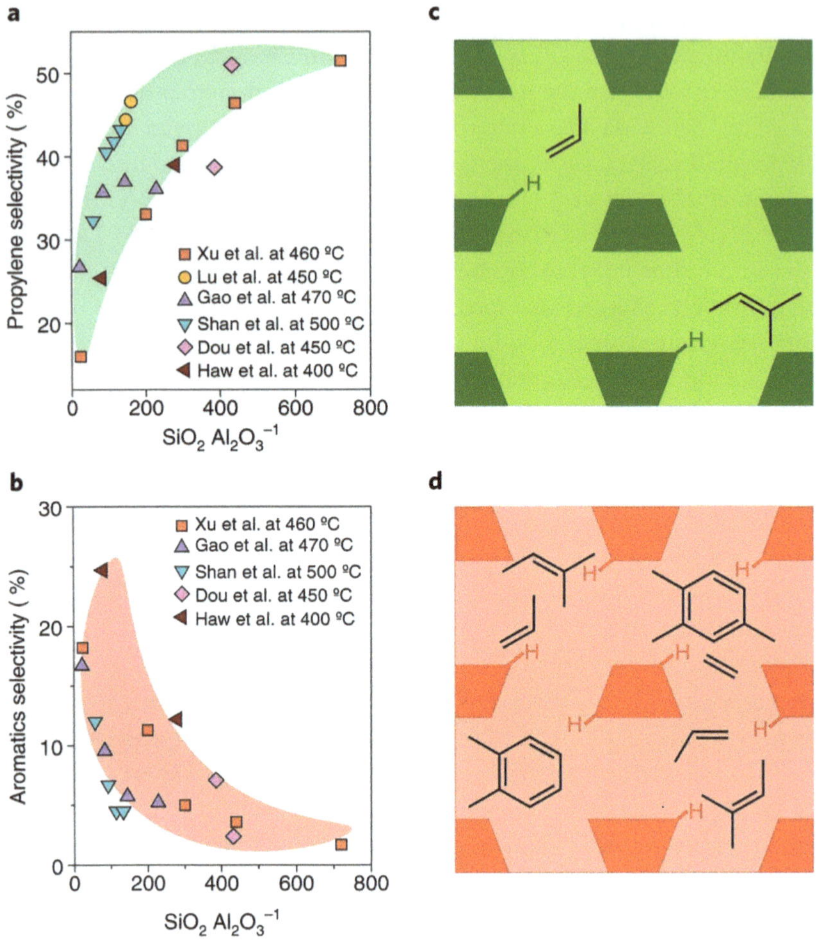

Fig. 4 Effect of the Si/Al ratio in ZSM-5 on methanol to hydrocarbons product selectivity. *Retrieved p. 406 from Yarulina I, Chowdhury AD, Meirer F, Weckhuysen BM, Gascon J: Recent trends and fundamental insights in the methanol-to-hydrocarbons process,* Nat Catal *1:398–411,2018. https://doi.org/10.1038/s41929-018-0078-5.*

on the confinement effect of zeolite cavities. Three types of cavities, *lev* (30), *cha* (36 T atoms) and *lta* (48 T atoms) (wherein T is Si or Al), was studied with density functional theory (DFT). Cavity models and the HP species were optimized at the ONIOM (B3LYP/6-31g(d,p):AM1) level. The lowest intrinsic energy barriers for methylation and olefin production were obtained for the *cha* cavity (which represents the SAPO-34 zeolite structure) with the calculated H···O distance between protonated HP species and the zeolite cavities as 2.30–2.51 Å, enabling the translation, rotation and confinement of the HP species in the void (Li et al., 2014).

2.5.3 Fischer Tropsch synthesis

Fischer and Tropsch were the first to obtain long chain hydrocarbons with CO hydrogenation on Fe/ZnO and Co/Cr_2O_3 catalysts, almost a hundred years before (Cheng et al., 2017). Being an exothermic CO polymerization and hydrogenation process (160 kJ heat released per mole of CO converted), FT produces a broad carbon number of hydrocarbons (paraffins and olefins), as well as oxygenates (alcohols, carboxylic acid, aldehydes) with the production of water (Fig. 5). The product distribution follows Anderson–Schulz–Flory (ASF) distribution, wherein the molar concentration (X) of the products with n + 1 carbon divided by the ones with n carbon is α, the chain growth probability ($\alpha = X_{n+1}/X_n$). In ASF model, the chain grows with the coupling of CH_x monomers, producing C_nH_m products. Commercial interest in FT process emanates from the production of gasoline and diesel range hydrocarbons (Fig. 1). Besides, FT crude has lower concentrations of sulfur and nitrogen compounds compared to crude oil, enabling mild hydrotreatment conditions in upgrading (Cheng et al., 2017; de Klerk, 2020).

An ideal FT catalyst has moderate hydrogenation and CO adsorption tendency to enable C–C coupling reactions. **Fe**, **Co**, and **Ru** metals are the most suitable FT catalysts. The low cost and availability of Fe and Cu metals paved the way for industrial implementation (Cheng et al., 2017). Industrial FT processes are thus classified into three groups: (i) Fe-catalyzed low temperature process (Fe-LTFT) at 200–270 °C, (ii) Co-catalyzed low temperature process (Co-LTFT) at 170–230 °C,

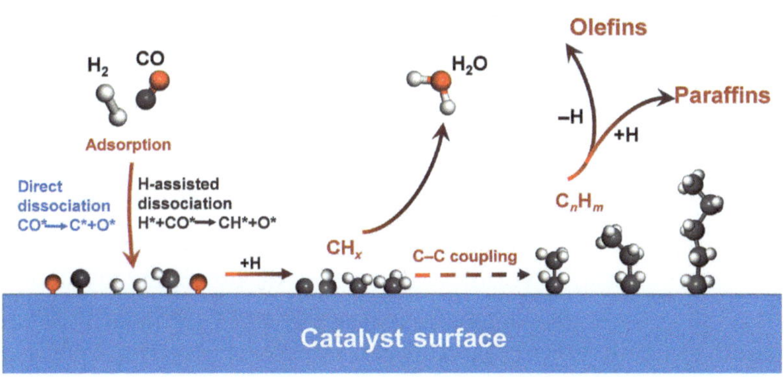

Fig. 5 Schematic drawing of Fischer Tropsch synthesis reactions. *Retrieved p. 136 from Cheng K, Kang J, King DL, et al: Chapter Three—Advances in catalysis for syngas conversion to hydrocarbons. In Song C, editor:* Advances in catalysis, *vol 60, 2017, Academic Press, pp 125–208. https://doi.org/10.1016/bs.acat.2017.09.003.*

and (iii) Fe-catalyzed high temperature process (Fe-HTFT) 320–360 °C, all processes occurring in the pressure range of 2.0–2.5 MPa (Fig. 2) (de Klerk, 2020).

Fe based catalysts operate under a wider range of temperature and at lower H_2/CO conditions. As they can catalyze the water gas shift reaction, CO can be obtained from CO_2, or H_2 can be obtained from H_2O (Cheng et al., 2017; de Klerk, 2020). Fe-HTFT operation has high selectivity to light hydrocarbons (~50 wt% of C1–C4 products), while Fe-LTFT produces large amount of paraffin wax (~50 wt%). On the other hand, **Co** based catalysts have higher selectivity to gasoline plus diesel range hydrocarbons (~40 wt%) and are more tolerant to deactivation. FT products are usually subjected to further upgrading to enable fuel grade end products. The upgrading process is hydroprocessing wherein hydrogen is used to remove the impurities (S, N, O, metals), crack and hydrogenate the hydrocarbons toward high value-added kerosene and diesel fuels (Cheng et al., 2017). **Fe** and **Co** active sites, based on experimental and theoretical proof, have been proposed to catalyze hydrogen assisted CO dissociation. DFT calculations on Fe (110) and Co (0001) surfaces were performed using DACAPO software with the generalized gradient approximation (GGA-PW91). The activation barriers for both catalysts were lower in the H assisted CO dissociation ($CO^* + H^* \leftrightarrow HCO^* + *$, $HCO^* + H^* \rightarrow HCOH^* + *$, $HCOH^* + * \rightarrow CH^* + OH^*$). The O atom in hydroxymethylene intermediate ($HCOH^*$) is rejected by formation of water ($HCOH^* + H^* \rightarrow CH^* + * + H_2O$), while hydrogenation of CH^* produces $CH_2{}^*$ and $CH_3{}^*$ species. (Ojeda et al., 2010). As depicted in Fig. 5, the CH_x–CH_y recombination reactions promote chain growth, CH_2 species are inserted into the chain during chain propagation, chain growth is terminated either through H addition or abstraction, forming paraffin and olefin products (van Santen et al., 2011).

2.6 Catalytic upgrading of thermochemical conversion products

Pyrolysis and gasification volatiles can be catalytically upgraded using steam or carbon dioxide to produce hydrogen; using air or nitrogen to favor the formation of filamentous coke (carbon nanotubes), or with hydrogen to obtain saturated hydrocarbons for fuel purposes.

2.6.1 Catalytic upgrading toward hydrogen production

A route of producing hydrogen from alternative feedstocks including plastic waste is presented as a two-step process consisting of pyrolysis and the following catalytic gasification of the pyrolysis volatiles for the maximization of hydrogen productivity. The catalytic gasification employs **Ni** based catalysts and transition metal catalysts, in which the support varies as metal oxides, zeolites and activated carbon.

2.6.1.1 Steam reforming of pyrolysis volatiles

Steam reforming, also called steam gasification, of **PP** pyrolysis vapors at 800 °C in a two-stage pyrolysis-gasification setup using different 10 wt% Ni loaded catalysts pointed out >90 wt% gas yields (H_2, CO, CO_2, CH_4, C_2–C_4) on **Ni/CeO$_2$/Al$_2$O$_3$** and **Ni/HZSM-5** catalysts with the majority of the gas product being hydrogen (63–64 vol%) (Wu and Williams, 2009). Comparison of three Ni impregnated zeolite supports of similar Si/Al ratios (25–30), **Ni/HZSM-5**, **Ni/HBeta** and **Ni/HY** in the steam gasification of **HDPE** (recycled waste plastic) pyrolysis vapors, revealed the highest gas, syngas and H_2 yields on Ni/ZSM-5 (66 mmol H_2/g plastic) with the least coke formation. The superior catalytic activity of Ni/ZSM-5 was speculated as the higher number of non-framework acid sites (Yao et al., 2018). Steam reforming of conical spouted bed reactor originated **HDPE** pyrolysis vapors on commercial **Ni** based catalyst (NiO on Al_2O_3 and doped with Ca) at 700 °C yielded 190 mmol H_2/g plastic (Barbarias et al., 2016). Support effect was studied in terms of coke formation in the steam reforming of **LDPE** pyrolysis vapors on Ni impregnated Al_2O_3, MgO, CeO_2, Y_2O_3, TiO_2, SiO_2 and ZSM-5 supports. XRD and TPR results indicated strong metal-support interactions in 10 wt% **Ni/Al$_2$O$_3$** (Ni–O–Al bond), maintaining Ni nanoparticle size at 9 nm after the reaction (resistance to sintering) thus resulting in high syngas yield (55 mmol H_2/g LDPE) and low coke formation (10 wt%) compared to other supports (Huang et al., 2021).

Dual support **Ni-CaO-C** catalyst has been investigated in the steam gasification of biomass co-pyrolysis vapors of **LDPE** (Chai et al., 2020b) and **HDPE**, **PP**, **PS** (Chai et al., 2020a) at 600–700 °C. Novelty of the catalyst comprised of Ni sites performing tar cracking and favoring the steam reforming reaction ($C + H_2O \rightarrow H_2 + CO$), CaO support adsorbing the generated carbon dioxide thus shifting the WGS reaction toward higher hydrogen production ($CO + H_2O \rightarrow CO_2 + H_2$) and activated carbon acting both as reforming catalyst and large internal surface area support. H_2 yield of 115 mmol H_2/g feedstock was achieved on 10 wt% Ni-CaO-C (CaO:C ratio 5:5) at 600 °C using **LDPE** in co-pyrolysis (Chai et al., 2020b).

Support effect towards reduced coke formation has been exemplified with a DFT study on steam methane reforming reaction over **Ni/CeO$_2$**

catalyst. The (3 × 3) CeO_2 (111) surface was modeled with an adsorbed Ni_{13} cluster (representing low Ni loading), total energies were calculated using the generalized gradient approximation (GGA) with Perdew-Burke-Ernzerhof (PBE) exchange-correlation functional. Dehydrogenation of methane ($CH_4^* \rightarrow CH_3^* \rightarrow CH_2^* \rightarrow CH^* \rightarrow C^*$) followed lower activation barriers on the supported Ni_{13} cluster, compared to the pure Ni (111) surface. The proposed pathway of CO formation from adsorbed C was the dissociation of H_2O at the Ni-CeO_2 interface ($H_2O(g) \rightarrow OH^*_{metal} + H^*_{support}$), followed by the reaction of C with the OH ($C^* + OH^*_{metal} + H^*_{support} \rightarrow COH^* + H^*_{support} \rightarrow CO + H^* + H^*_{support}$). The formation of H_2 occurs with the migration of the $H_{support}$ to the metal ($H^* + H^*_{support} \rightarrow 2H^* \rightarrow H_2(g)$), energetically facilitated by another water molecule adsorbing on the terrace sites of the Ni cluster (Salcedo et al., 2021).

2.6.1.2 Carbon dioxide (dry) reforming of pyrolysis volatiles

Dry reforming utilizes carbon dioxide instead of steam for the catalytic reforming of hydrocarbons ($C_xH_y + xCO_2 \leftrightarrow 2xCO + y/2H_2$, $\Delta H > 0$). Pyrolysis vapors of plastic waste from the first stage proceed to the second stage wherein dry reforming reactions occur over **Ni** and **Co** based catalysts. Simulated mixed plastic waste (42 wt% LDPE, 20 wt% HDPE, 16 wt% PS, 12 wt% PET and 10 wt% PP) was studied in the two stage pyrolysis-gasification system operating at 500 and 800 °C, respectively (Saad and Williams, 2016a,b, 2017b) over **Ni/Al_2O_3, Ni-Co/Al_2O_3, Ni-Co-Al** catalysts prepared by different techniques which were rising pH (co-precipitation) and impregnation. The syngas yields were comparable for all catalysts prepared by the rising pH method: 140 mmol H_2 + CO/g plastic for Ni/Al_2O_3, 149 mmol H_2 + CO/g plastic for Ni-Co/Al_2O_3, 149 mmol H_2 + CO/g plastic for Ni-Co-Al (Saad and Williams, 2016a,b), whereas the Ni/Al_2O_3 catalyst prepared by the impregnation method yielded in lower syngas productivity (127 mmol H_2 + CO/g plastic) which was linked to the larger catalyst particle size in the latter one, determined by SEM (Saad and Williams, 2017b). Adding steam to the dry reforming process (with the optimum CO_2: steam ratio of 4:1) over **Ni-Co/Al_2O_3** catalyst (prepared by rising pH technique) improved the H_2:CO molar ratio from 0.74 to 0.94 at 800 °C, thus approaching to the H_2/CO value for the industrial utilization (Saad and Williams, 2017a).

H_2/CO ratio of ~1.0 was achieved by employing bimetallic **Pt-Ni** catalyst on mixed oxide Mg-Al support, 1 wt% Pt–12 wt% Ni/Mg(Al)O, with controlled nanoparticle size (<1 nm) in methane dry reforming reaction operated at the kinetic region (elimination of internal and external diffusion limitations). DFT analysis of key elementary reactions steps on Ni (111), 1/9 Pt-Ni (111) and 3/9 Pt-Ni (111) models were performed using GGA with BEEF-vdW exchange-correlation functional. Support effect was not included in the model. The adsorption energy of H^* on the Pt-Ni bimetallic catalyst, as well as the activation barrier for H_2^* formation ($H^* + H^* = H_2^* + ^*$) were lower than the Ni catalyst. The experimental and computational results pointed out to the suppression of the reverse water gas shift reaction on the Pt-Ni bimetallic catalyst toward enhanced H_2 selectivity (Niu et al., 2021).

2.6.1.3 Chemical vapor deposition of pyrolysis and gasification vapors

Carbon nanotubes (CNTs) are high tensile strength, highly elastic and conductive materials comprising of sp^2 hybridized carbon. They are used as reinforcement nanomaterials in polymer composites for advanced industrial applications including electronics and aerospace (Eatemadi et al., 2014; Lowery and Green, 2022; Wu et al., 2016). Catalytic chemical vapor deposition (CVD) is a method to synthesize CNTs on metallic nanoparticles such as **Ni**, **Co** and **Fe** by introducing gaseous hydrocarbons such as ethylene, acetylene and methane at around 700 °C (Eatemadi et al., 2014). CVD method has recently been extended to plastic recycling research for the valorization of pyrolysis and gasification vapors.

CVD of **LDPE** pyrolysis vapors in a two-stage setup by using nitrogen as the carrier gas enabled the formation of filamentous carbon deposits in the form of CNTs on **Ni/Al_2O_3**, **Fe/Al_2O_3** and **Cu/Al_2O_3** catalysts. The metals were introduced to the γ-alumina support by the impregnation method. Both CNT and hydrogen production maximized on the Fe/Al_2O_3 catalyst (179 mg CNT/g plastics, 56 wt% gas yield with 51 vol% H_2). CNT growth was influenced by the strength of the metal-support interactions, as the metal nanoparticle size upon sintering in reaction conditions depended on the metal mobility on the support (Acomb et al., 2016). Catalyst preparation method has been shown to alter the structure and performance of Ni and Fe based catalysts in catalytic gasification of **PP** pyrolysis vapors. **Fe-Ni** bimetallic catalyst on Al_2O_3 prepared with the sol-gel method had the highest metal dispersion of 42 wt%. The same catalyst was associated with maximized hydrogen and CNT formation (25 mmol H_2/g plastic and 360 mg CNT/g plastic at 800 °C). HRTEM and XRD analysis indicated Fe_3C (iron carbide) structure on the Fe-Ni alloy

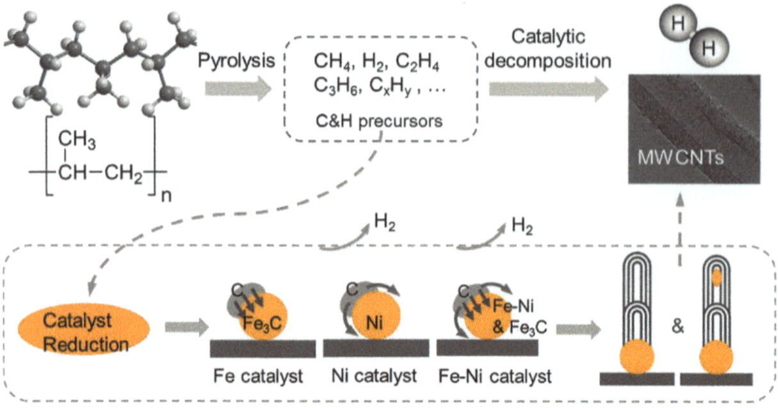

Fig. 6 Hydrogen and carbon nanotubes production from catalytic gasification of plastic pyrolysis vapors. *Retrieved p. 114819 from Yao D, Wang C-H: Pyrolysis and in-line catalytic decomposition of polypropylene to carbon nanomaterials and hydrogen over Fe- and Ni-based catalysts,* Appl Energ *265:114819, 2020. https://doi.org/10.1016/j.apenergy.2020.114819.*

as the active phase in CNT growth (Yao and Wang, 2020). The reaction pathway from polyolefin feedstock to hydrogen and carbon nanotube products via thermal pyrolysis and catalytic gasification steps is depicted in Fig. 6. According to Fig. 6, hydrocarbon species in pyrolysis vapors (C_xH_y) adsorb on the metallic active sites on the catalyst, followed by the decomposition of C_xH_y to elemental C, coupled with H_2 formation. Elemental C diffuses into the metal and nucleation starts once carbon reaches over-saturation. Growth of graphene layers is promoted from both sides of the metal nanoparticle, enabling the growth of nanotubes with diameters similar to the metal nanoparticle size (20–30 nm).

Alternative to pyrolysis vapors, hydrogen and CNTs can be produced from plastic gasification vapors. Air gasification of **PE-PP** mixture in a fluidized bed reactor at 600°C, followed by upgrading with different Ni/Al_2O_3 catalysts in a fixed bed reactor system at the same temperature. 10wt% **Ni/Al_2O_3** catalysts prepared with impregnation method were calcined at different atmospheres following as air, N_2 and H_2. The catalyst calcined at hydrogen atmosphere resulted in the highest H_2 and CNT yields (32 mmol H_2/g plastic and 22 wt% CNT yield wherein CNT is the carbon product on the used catalyst). H_2 and CNT yields were inverse proportional to the Ni particle size calculated from XRD data, and the fraction of oxidized Ni species (Ni^{+3} and Ni^{+2}) determined by XPS, indicating smaller particles with more metallic nickel (Ni^0) produce more syngas, but less

CH_4 and C_2–C_4 hydrocarbons (Yang et al., 2015). It was possible to improve the CNT yield by employing catalysts both at gasification and upgrading steps. For instance, catalytic gasification of **PP-LDPE** mixture at 600 °C on **Ni/Al-SBA-15** acidic catalyst, followed by the second catalytic gasification at 800 °C on dual **Ni-Cu/CaO-SiO$_2$** reforming catalyst enabled 48 wt% CNT yield, however hydrogen yield remained at low levels compared to the literature (0.18 m^3 H_2/kg plastic waste, or 8 mmol H_2/g plastic) (Yang et al., 2020).

2.6.1.4 Plasma steam gasification of pyrolysis volatiles

Plasma reactors have been investigated on the upgrading of pyrolysis vapors. Plasma generates high energy electrons; collision of these electrons with the gas phase molecules produces ions and radicals enabling bond cleavage and formation of new bonds. Upgrading of waste **HDPE** pyrolysis vapors in a dielectric barrier discharge (DBD) plasma reactor at 250 °C on 10 wt% **Ni/Al$_2$O$_3$** catalyst demonstrated improved H_2 yield in plasma catalysis compared to catalysis without plasma, and plasma without catalyst conditions. H_2 yield increased further by introducing steam to the plasma catalysis up to 4.5 mmol H_2/g plastic. Hydrogen yields of the tested Fe, Cu and Co-based catalysts were lower compared to **Ni/Al$_2$O$_3$** (Aminu et al., 2020). Comparing different Ni impregnated supports Al_2O_3, zeolite Y, MCM-41 mesoporous silica and $BaTiO_3$, the highest H_2 yield of 18 mmol H_2/g plastic was enabled with the 10 wt% **Ni/MCM-41** catalyst. The success with the MCM-41 support was linked to a combination of its high surface area, low dielectric constant and high porosity, improving plasma generation within the pores (Aminu et al., 2022).

2.6.2 Hydroreforming toward fuel production

Hydroreforming is the catalytic upgrading of pyrolysis products at lower temperatures compared to pyrolysis (<400 °C) under high pressure hydrogen atmosphere (2–7 MPa) (Fig. 2), mostly toward fuel range saturated hydrocarbons (Fig. 1). Hydroreforming catalyst is equivalent to the hydrocracking catalyst used in petroleum refining, an heterogeneous bifunctional catalyst with metal particles supported on a solid acid, catalyzing the hydrogenation/dehydrogenation, hydrocracking, hydroisomerization and aromatization reactions (Escola et al., 2015; Huang et al., 2022; Weitkamp, 2012). In the hydroreforming studies of plastic waste, metals such as **Ni, Pd, Pt, Fe, Ni-Mo**, and solid acids including **ZrO$_2$, Al$_2$O$_3$, SiO$_2$-Al$_2$O$_3$, activated carbon, zeolites** have been investigated

(Munir et al., 2018). Product quality has been assessed by boiling point range with distillation, isoparaffinic content with PIONA, aromatics content with HPLC and olefin content with Bromine index methods (Escola et al., 2015; Munir et al., 2018).

Large surface area acid catalysts with microporosity, including hierarchical zeolites (**h-HBeta, h-HZSM-5**) and mesoporous silica (**SBA-15, MCM-41**) were used as supports for **Ni** nanoparticles in the hydroreforming C_5–C_{40} range **LDPE** pyrolysis vapors at 310 °C and 2 MPa. NiO particles in the calcined zeolite were reduce under H_2 flow to obtain the metallic nickel nanoparticles as the active sites. In terms of product selectivity, zeolite supports were associated with high gasoline range hydrocarbons (C_5–C_{12}) (~54 wt%) including aromatics and isoparaffins produced via aromatization and hydroisomerization reactions, leading to higher research octane number (RON of 88–89, compared to 82 of the gasoline fraction of thermal pyrolysis product) (Escola et al., 2011). Larger Ni nanoparticle size (>10 nm), altered by the impregnation method, was shown to produce isoparaffins and aromatics with higher selectivity, reaching to 18 wt% total aromatics and 35% iso-paraffins in gasoline using the nickel chloride as the Ni precursor (Escola et al., 2015). Coimpregnation with nickel and ruthenium chlorides on hierarchical HBeta formed the **Ni5%-Ru1%/h-HBeta** catalyst with higher hydrogenation activity compared to **Ni5%/h-HBeta,** deduced from the lower olefinicity (low Bromine index) of the former one (Serrano et al., 2015).

Hydroreforming of post-consumer **PP** pyrolysis oil under 350 °C and 7 MPa hydrogen pressure, employing **Ni5%-Au0.5%/MOR** bimetallic catalyst of 10–20 nm nanoparticle size on dealuminated zeolite support yielded in 68 wt% paraffins (~40 wt% in the C_{10}–C_{20} diesel range, 16 wt % of them were isoparaffins) and 32 wt% aromatics (21 wt% benzene +11 wt% naphthalene). The PP pyrolysis oil was obtained with in-situ catalytic pyrolysis on **HZSM-5**. Hydroreformed PP pyrolysis oil had lower isoparaffinic and higher aromatic content than the reference commercial diesel. Testing the 10–40 vol% blends with diesel demonstrated slightly higher specific fuel consumption and engine-out emissions, up to 15% compared to reference commercial diesel (Mangesh et al., 2020).

Alternatively, hydropyrolysis operation may employ pyrolysis and hydroreforming in the same reaction environment. Model municipal waste comprising of 40 wt% **HDPE**, 10 wt% **LDPE**, 30 wt% **PP** and 20 wt% **PS** was hydropyrolyzed at 400 °C under 2 MPa bar hydrogen pressure, using the zeolite-mesoporous silica composite catalyst **HY/SBA-16**. The composite catalyst indicated higher diesel range (C_{13}–C_{18}) hydrocarbon yield compared to HY alone, because of improved mesoporosity (Munir and Usman, 2018). The catalytic pyrolysis of **PET** with **molybdenum oxides** in H_2 atmosphere at 400 °C resulted in high yields of olefins and terephthalic acid, reaching 10.1 and 36.5 wt%, respectively (Liu et al., 2022).

Hydrocracking of various polyolefins has recently been demonstrated at milder conditions of 250 °C and 3 MPa hydrogen pressure, over a physical mixture **$Pt/WO_3/ZrO_2$**, with a solid acid. Among the tested solid acids in **LDPE** (MW: 250 kDa) conversion including **HY, HBeta, HZSM-5, HMOR, Al-MCM-41** and **WO_3/ZrO_2**, HY provided the highest conversion and yield of fuel range products at short reaction time of 2 h due to its larger pore size compared to other zeolites and higher number of Brønsted acid sites compared to the sold acids tested. For instance, the yield of gasoline range (C_5–C_{12}) and diesel range (C_9–C_{22}) products were 73 wt% and 20 wt%, respectively at 94% conversion with high degree (>80%) of branching. Hydrocracking of other polyolefin feedstocks such as pure **HDPE** and **PP** (isotactic, MW: 250 kDa), as well as **LDPE** bottle, **HDPE** bottle, **HDPE** bag and **PP-PE-PS** composite tape has provided promising results in terms of obtaining gasoline, jet fuel and diesel range hydrocarbons from plastic waste (Liu et al., 2021).

2.7 Catalytic pathways on the acid and metal sites

Catalytic transformations of plastic polymers resemble hydrocarbon transformations on acid and metal catalysts realized in petroleum refining. A number of authors have proposed reaction pathways for the catalytic degradation of plastic waste based on this knowledge (Aguado et al., 2001; Okonsky et al., 2022; Xue et al., 2017; Zhang and Lei, 2016). In the fluid catalytic cracking (FCC) process in petroleum refineries, zeolite containing FCC catalysts are used for cracking high molecular weight petroleum fractions such as vacuum gas oil into gasoline (C_5–C_{12}) and light olefins (C_2–C_4) at 500–550 °C and ~0.3 MPa operating conditions (Rigutto et al., 2007). The FCC catalyst contains **zeolite Y**, and optionally zeolite **ZSM-5** if more light olefin production is targeted (Vogt and Weckhuysen, 2015). In the hydrocracking process, heavy molecular weight petroleum fractions are converted into middle distillates such as jet fuel (C_9–C_{16}) over metal containing acid catalysts at 300–450 °C and 4–20 MPa (Rigutto et al., 2007). The metal sites in a hydrocracking catalysts comprise of a Group VIII metal (**Ni, Co**) and a Group VIB metal (**Mo, W**), while the acidic support is a zeolite (**Y, ZSM-5 or Beta**) (Domokos et al., 2010). Reaction pathways for an n-paraffin R_1-$(CH_2)_5$-R_2, illustrating a linear polyethylene chain is schematized for catalytic cracking in Fig. 7A and, hydrogenolysis and hydrocracking in Fig. 7B.

2.7.1 Catalytic cracking

Acid catalysts such as zeolite Y and ZSM-5 with Brønsted and Lewis type acid sites form carbocations (carbenium and carbonium ions) during the catalytic cracking of long hydrocarbon chains into short chains. Catalytic cracking mechanism proceeds with initiation, propagation and termination steps (Corma and Orchillés, 2000), also shown in Fig. 7A. In the **initiation** step, hydride abstraction from the n-paraffin on a Lewis acid site forms a carbenium ion (carbenium ion(1) in Fig. 7A), while protonation of the n-paraffin by a Brønsted acid site forms a carbonium ion (Corma and Orchillés, 2000; Corma et al., 1985; Kotrel et al., 2000; Vogt and Weckhuysen, 2015). The **propagation** step, in which shorter hydrocarbon chains are formed, involves the β-scission of carbenium ions, protolytic cracking of carbonium ions, isomerization of carbenium ions, and hydride abstraction from a reactant paraffin to a carbenium ion.

In a carbenium, the carbon atom bonded to the positively charged carbon atom is named as the α carbon, while the carbon atom bonded to the α carbon atom is named as the β carbon (carbenium ion(2) in Fig. 7A). In β-scission on Brønsted acid sites, electrons from the C_α–C_β bond moves to the C^+–C_α bond, reducing the electron density on the C_α–C_β bond and forming a $C^+ = C_\alpha$ double bond (Fig. 7A). The products are an olefin and a shorter carbenium ion (carbenium ion(3) in Fig. 7A) (Corma and Orchillés, 2000; Mazar et al., 2013). The energy barrier for β-scission of an adsorbed tertiary carbocation is the lowest (Mazar et al., 2013). β-scissions which form secondary and tertiary carbocations are energetically favored (type A, B and C), while type D, forming primary carbocations is unfavorable (Weitkamp, 2012). Among the two types of isomerization, charge isomerization occurs through 1,2 hydrogen shift, thus do not create branching. However in branch isomerization, primary and secondary carbeniums are converted into tertiary carbeniums with a protonated cyclopropane ring (PCR) transition state (Corma and Orchillés, 2000). As tertiary carbeniums are the most stable, isomerization yields in branched chains (Vogt and Weckhuysen, 2015). A carbonium ion has excess hydrogen with a positively charged carbon ($C-CH_3^+-C$), called the pentacoordinated carbonium structure. Carbonium proceeds with the protolytic cracking of

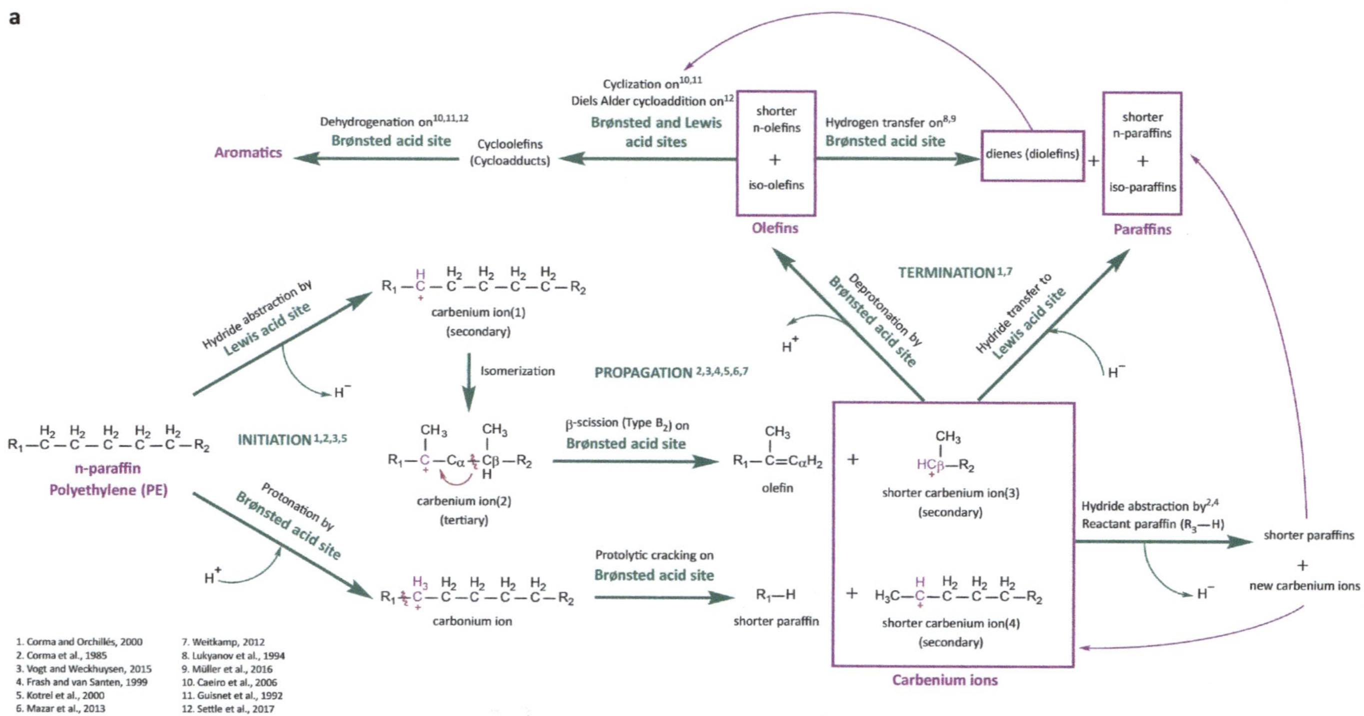

Fig. 7 Catalytic reaction pathways of an n-paraffin on acid and metal active sites: (A) Catalytic cracking (Caeiro et al., 2006; Corma et al., 1985; Corma and Orchillés, 2000; Frash and van Santen, 1999; Guisnet et al., 1992; Kotrel et al., 2000; Lukyanov et al., 1994; Mazar et al., 2013; Müller et al., 2016; Settle et al., 2017; Vogt and Weckhuysen, 2015; Weitkamp, 2012). (B) Hydrogenolysis and hydrocracking (Cheng et al., 2017; Corma et al., 1985; Ellis et al., 2021b; Goldman et al., 2006; Noh et al., 2018; Weitkamp, 2012).

(Continued)

C−C bond directly attached to the positively charged carbon, called as the Haag-Dessau cracking mechanism, forming a shorter paraffin and a carbenium ion ($H_3C{-}CH^+{-}R$) (carbenium ion(4) in Fig. 7A) (Corma and Orchillés, 2000; Kotrel et al., 2000). Protolytic cracking becomes kinetically significant at low olefin concentrations in the reaction environment, because protonation of an olefin by a Brønsted acid site is preferred over the protonation of a paraffin (Kotrel et al., 2000). Alternatively, H_2 can form from carbonium ions containing tertiary carbon, via protonation of the C−H bond of the tertiary (Corma and Orchillés, 2000). A carbenium ion formed in the propagation step may take a hydride from a reactant paraffin R_3_H, forming a paraffin and a new carbenium ion (Corma et al., 1985; Frash and van Santen, 1999), or may proceed with the termination step (Vogt and Weckhuysen, 2015). In the termination step, Brønsted and Lewis acid sites are restored, completing one catalytic cycle. The carbenium ion transfers a proton to the Brønsted acid site, or takes an hydride from the Lewis acid site, forming an olefin and a paraffin, respectively (Weitkamp, 2012).

b

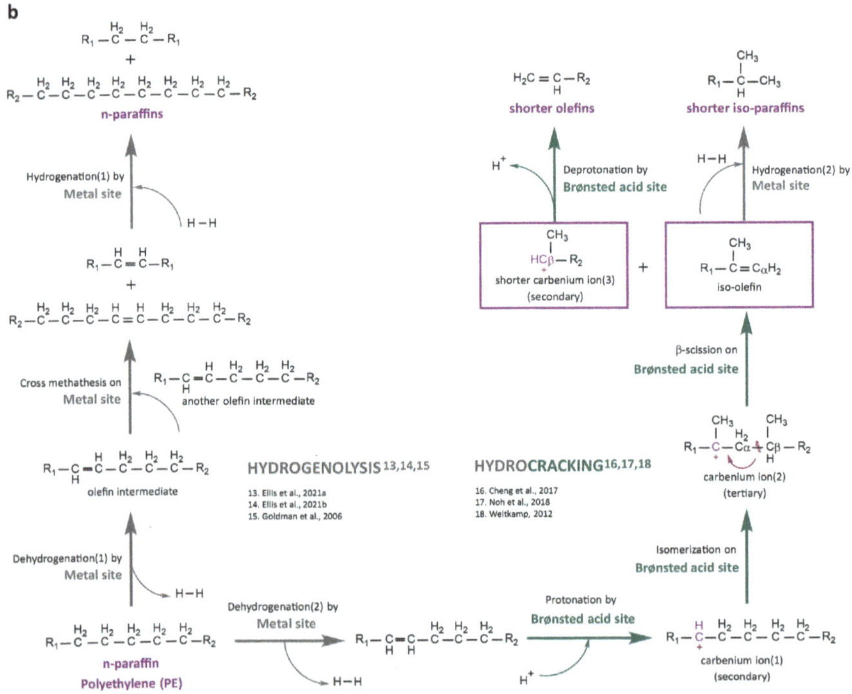

Fig. 7—Cont'd

The olefins formed in the catalytic cracking steps may react in the hydrogen transfer reaction wherein an olefin protonated on a Brønsted acid site receives a hydride from a reactant olefin to form a paraffin and a diene (olefins with two C=C bonds) (Lukyanov et al., 1994; Müller et al., 2016). Aromatics are formed from olefins via cyclization on Brønsted acid sites (Caeiro et al., 2006; Guisnet et al., 1992), or by Diels Alder cycloaddition synergistically catalyzed by Brønsted and Lewis acid sites (Settle et al., 2017). Cyclization of olefins (including dienes) occurs with a carbenium mechanism wherein a cyclic olefin with a six membered ring forms (Caeiro et al., 2006; Guisnet et al., 1992). Alternatively, in Diels Alder cycloaddition, the highest occupied molecular orbital (HOMO) of a diene (e.g., butadiene, isoprene) and the lowest unoccupied molecular orbital (LUMO) of a dienophile (e.g., ethylene, propylene butene) interacts to form two new sigma bonds, again forming a six membered cycloolefin, also called a cycloadduct. The subsequent dehydrogenation of the cycloolefin on Brønsted acid sites forms a monoaromatic molecule (Settle et al., 2017).

2.7.2 Hydrogenolysis and hydrocracking

Metal catalysts such as **Ni, Co** or **platinum** group metals catalyze the hydrogenation/dehydrogenation reactions of hydrocarbons in the hydrogenolysis and hydrocracking processes. In both processes, initially, a paraffin is converted into an olefin on the metal active sites by dehydrogenation (dehydrogenation(1) and dehydrogenation(2), respectively in Fig. 7B). The main products are paraffins in hydrogenolysis, while iso-paraffins are produced in hydrocracking (Cheng et al., 2017). Hydrogenolysis enables the production of C_9–C_{20} hydrocarbons as ideal diesel fuel components from hydrocarbon feedstocks (Goldman et al., 2006). In hydrogenolysis, C–M bonds are formed at the expense of C–H and M–H bonds in the so-called α,β-adsorbed structure on noble metals such as **Ir, Pt, Rh** and **Ru** (Flaherty et al., 2014). Hydrocarbons proceed with dehydrogenation, olefin metathesis and hydrogenation reactions on the metal sites, in the so-called tandem hydrogenation and olefin cross metathesis system (Ellis et al., 2021a,b; Goldman et al., 2006). Hydrogenation of the cross-coupled olefins (hydrogenation(1) in Fig. 7B) produces n-paraffins of various length. In hydrocracking, C–C cleavage occurs with a carbocation intermediate. Metal sites catalyze dehydrogenation of a paraffin to produce an olefin, followed by the protonation of the olefin by a Brønsted acid site to form a carbenium ion (carbenium ion(1) in Fig. 7B), which then proceeds with the propagation and termination steps shown in Fig. 7B, similar to catalytic cracking (Cheng et al., 2017). The carbenium ions are transformed into secondary and tertiary carbeniums, are mostly subjected to type A and B type fast β scissions to form iso-olefins. Hydrogenation of the iso-olefins on metal sites produces the final iso-paraffin products (hydrogenation(2) in Fig. 7B) (Noh et al., 2018; Weitkamp, 2012).

2.8 Current applications in catalytic chemical recycling of plastic waste

Catalytic chemical recycling of plastics in well-designed plants and catalytic upgrading of thermochemical recycling products in existing petrochemistry, petroleum refinery and biorefinery plants have already been introduced to the market. A list of catalytic chemical recycling plants in the commercialization route, from pilot to early commercial scale, toward monomer, chemical, and fuel production are given in Table 3. Catalytic solvolysis, hydrothermal liquefaction and pyrolysis recently have examples in pilot and early commercial applications. Pilot plants process a few tons per day of plastic waste (e.g., *Garbo* 3 tons/day, *Resynergi* 5 tons/day), while early commercial scale refers to tens of tons per day processing capacity (e.g., *Ioniqa* 10 kton/year, *Anhui Oursun* 10–500 tons/day, *Jeplan* 22 kton/year) (Closed Loop Partners, 2021).

PET depolymerization with two solvolysis techniques, glycolysis and hydrolysis, have found place in commercial applications. The companies *Garbo* and *Ioniqa* employs ionic liquids, while *Jeplan* uses sodium methylate transesterification catalyst in PET glycolysis to produce BHET. In **PET** hydrolysis to produce terephthalic acid (PTA) and ethylene glycol (EG), *Carbios* uses a *hydrolase* enzyme, while *Gr3N* benefits from microwave radiation in alkaline medium. Alternatively, *Aquafil* performs hydrolysis of the polyamide **PA6** in a 40 kton/year plant in Slovenia to obtain caprolactam monomers, in a non-catalytic process. *Licella* has developed an hydrothermal liquefaction technology, named Catalytic Hydrothermal Reactor (Cat-HTR™), to convert mixed plastic waste into an oil consisting of naphtha, kerosene, diesel, heavy gas oil and wax fractions (Humphreys and Rowlands, 2018). CAT-HTR™ is extended by *Mura Technology* with the name of Hydrothermal Plastic Recycling Solution (HydroPRS™). HydroPRS™ has received investment from *Dow* and is licensed by *Renew ELP*.

Table 3 Companies worldwide performing catalytic chemical recycling of plastic waste.

Feedstock	Process	Catalyst	Product	Current stage	Company, Country
PET	Solvolysis, Glycolysis	Ionic liquids	BHET	Pilot	Garbo, Italy
PET	Solvolysis, Glycolysis	Ionic liquids	BHET	Early Commercial	Ioniqa, Netherlands
PET, textile polyester	Solvolysis, Glycolysis	Sodium methylate	BHET	Early Commercial	Jeplan, Japan
PET, textile polyester	Solvolysis, Hydrolysis	PET-hydrolase enzyme	PTA, MEG	Pilot	Carbios, France
PET	Solvolysis, Hydrolysis	Sodium hydroxide	PTA, MEG	Pilot	Gr3n, Switzerland
Mixed plastic waste	Hydrothermal liquefaction (CAT-HTR™)	Catalyst[a]	Ethylene, naphtha, diesel, wax	Early commercial	Licella, Australia
Mixed plastic waste	Pyrolysis (Integrated Cascading Catalytic Pyrolysis)	Catalyst[a] (ex-situ)	BTEX	Pilot	BioBTX, Netherlands
PE	Pyrolysis	Dual catalyst[a]	Organic acids, BTEX	Pilot	Novoloop, USA
Mixed plastic waste	Pyrolysis (HiCOP)	Cracking catalyst	Pyrolysis oil	Pilot	Environmental Energy, Japan
Mixed plastic waste	Pyrolysis (MoReTec)	Catalytic[a]	Monomers	Pilot	Lyondellbasell, Italy
Mixed plastic waste	Pyrolysis (Carboliq)	Cracking catalyst	Pyrolysis oil	Pilot	Recenso GmbH, Germany

Continued

Table 3 Companies worldwide performing catalytic chemical recycling of plastic waste.—cont'd

Feedstock	Process	Catalyst	Product	Current stage	Company, Country
PE, PP, PS	Pyrolysis	Cracking catalyst	Gas, gasoline, diesel	Early commercial	Anhui Oursun Resource Technology, China
HDPE, LDPE, PP, PS	Pyrolysis	Nano-engineered catalyst[a]	Gasoline, diesel	Early commercial	EcoFuel Technologies, USA
Mixed plastic waste	Pyrolysis	Catalyst[a]	Propylene, BTEX, fuel	Early commercial	Encina, USA
HDPE, LDPE, PP, PS	Pyrolysis (Ceranivus)	Catalyst[a]	Diesel, grease, wax	Early commercial	GreenMantra Technologies, Canada
Mixed plastic waste	Pyrolysis	Catalyst[a]	Naphtha, Diesel	Early commercial	Klean Industries, Canada
Mixed plastic waste	Pyrolysis	Silicalumina based cracking catalyst	Fuel	Early commercial	Patpert Teknow Systems, India
PS	Pyrolysis (Catalytic Microwave Depolymerization)	Silicon carbide	Styrene monomer	Early commercial	Pyrowave, Canada
HDPE, LDPE, PP, PS	Pyrolysis (Continuous Microwave Assisted Pyrolysis)	Cracking catalyst (ex-situ)	Naphtha, diesel	Early commercial	Resynergi, USA

[a]The information on catalyst type or name is not publicly available.

As shown in Table 3, catalytic pyrolysis is an active commercialization route with SMEs and startup companies from various countries participating in the area. Olefin monomers, BTEX and/or fuel range hydrocarbons are obtained starting from mixed plastic waste, or a polyolefin mixture. Details on catalyst types are mostly obscured, notated as "cracking catalyst," or "catalyst." Electrification of pyrolysis with microwave heating is one of the novel commercial practices by *Resynergi*. The company has developed the Continuous Microwave Assisted Pyrolysis technology and currently manufactures microwave systems converting solid plastic waste into an oil comprising of naphtha/gasoline and diesel. Pyrolysis vapors can be fed into a catalytic cracking chamber to further crack the polymers into monomers, over acidic catalysts such as zeolites (Bauer et al., 2020). Oil & gas and chemical majors are actively involved in chemical plastic waste recycling research, funding of startups, and setting up partnerships with pyrolysis oil suppliers. Growing plastic pyrolysis oil producers such as *Plastic Energy* with its patented thermal pyrolysis oil named TACOIL™, has made supply agreements with *Sealed Air*, *Unilever*, *Nestle*, *SABIC*, *Total Energies* and *ExxonMobil*. *Alterra Energy*, with its thermal pyrolysis product, has set partnerships with *Neste* (Finland) and *Ravago* toward the construction of an early commercial chemical recycling plant in the Netherlands.

3. Outlook and future directions

The global plastic market is expected to grow with a compound annual growth rate (CAGR) of 3.7% for the period of 2022–2030, while the projected growth for the recycled plastics market is 5.3% from 2022 to 2027 (BCC Research, 2022; Grand View Research, 2022). Consumer brands such as *Coca Cola*, *Pepsi*, *Nestle Waters* and *Unilever* have publicly committed to use recycled materials up to 50% in their product packaging (Chemical & Engineering News, 2020).

A recent report from *McKinsey* envisages the role of chemical recycling in meeting the demand of recycled polymers as 4–8% of total polymer demand in 2030, and 6–10% in 2040, if $40 billion to $90 billion investments are located for that period. The pillars of improving the invasion of chemical recycling are mentioned as technology development, waste collection/sorting infrastructure optimization and investment by the partnerships within the value chain (Peng et al., 2022). In an up-to date market report, *Google* and *Afara* discusses the plastic waste recycling methodologies to enable irreversible socio-economic momentum toward plastics circularity. The current circularity gap of 93% (meaning only 7% of plastics are recycled) is suggested to be closed by 59% in 2040 with the implementation of five strategic actions of monomer recovery by chemical recycling, polymer recovery by mechanical recycling, taxing virgin plastic production, reducing plastic waste with inventory management and consumer education. The required investment in the period from 2020 to 2040 is estimated to be \$550 billion, much higher than the \$90 billion prediction by *McKinsey* (Google-Afara, 2022).

The pilot and early commercial scale enterprises are projecting to advance to the next step in the following years with various business strategies including (i) installation of industrial scale plants after the success of the demo plant (*Brighmark, Carbios*), (ii) licensing the patented technologies following the proof of demonstration (*Ioniqa, Pyrowave*), (iii) commercializing the recycled products (*Novoloop*), (iv) initiating supply agreements with refineries and chemical plants, called offtake agreements (*Alterra Energy, ApChemi*), (v) receiving funding from large companies or from government (*Pyrowave, ApChemi, ReNew ELP, Renew One*), (vi) partnering with engineering & construction companies (*Loop Industries*), (vii) partnering with several companies in the supply chain (*Carbios, Aquafil*), or (viii) improving the supply chain (*Cyclix*, launched by *Agilyx*) (Closed Loop Partners, 2021).

While current pyrolysis technologies mostly produce fuels from polyolefins such as PE and PS, more naphtha, the C_5–C_{12} feedstock of the petrochemical industry, will be demanded from the petrochemical plants. *Lyondellbasell* with *Karlsruhe Institute of Technology* work in improving the catalysts for MoReTec pyrolysis technology for naphtha optimization. In parallel, *Agilyx, Ineos Styrolution* and *Pyrowave* are finding solutions in breaking down PS to styrene monomers (Chemical & Engineering News, 2020). In PET solvolysis, *University of Portsmouth* and *National Renewable Energy Laboratory (NREL)* utilize computational and bioinformatics approaches to find the optimum PET degrading enzymes (Closed Loop Partners, 2021). Gasification practices are employed to treat mixed plastic waste, with PET and PVC content which are not preferred in pyrolysis (Peng et al., 2022). Companies such as *Enerkem* and *Fulcrum* are gasifying the municipal solid waste, including biomass and plastic waste, into syngas and aiming to produce jet fuel with the Fischer Tropsch route (Closed Loop Partners, 2021).

Acknowledgments

This material is based upon work supported by the U.S. Department of Energy's Office of Energy Efficiency and Renewable Energy (EERE) under the Advanced Manufacturing Office Award Number DE-EE0007897 awarded to the REMADE Institute, a division of Sustainable Manufacturing Innovation Alliance Corp. This report was prepared as an account of work sponsored by an agency of the United States Government. Neither the United States Government nor any agency thereof, nor any of their employees, makes any warranty, express or implied, or assumes any legal liability or responsibility for the accuracy, completeness, or usefulness of any information, apparatus, product, or process disclosed, or represents that its use would not infringe privately owned rights. Reference herein to any specific commercial product, process, or service by trade name, trademark, manufacturer, or otherwise does not necessarily constitute or imply its endorsement, recommendation, or favoring by the United States Government or any agency thereof. The views and opinions of authors expressed herein do not necessarily state or reflect those of the United States Government or any agency thereof.

References

Acomb JC, Wu C, Williams PT: The use of different metal catalysts for the simultaneous production of carbon nanotubes and hydrogen from pyrolysis of plastic feedstocks, *Appl Catal Environ* 180:497–510, 2016. https://doi.org/10.1016/j.apcatb.2015.06.054.

Aguado J, Serrano DP, Sotelo JL, Van Grieken R, Escola JM: Influence of the operating variables on the catalytic conversion of a polyolefin mixture over HMCM-41 and nanosized HZSM-5, *Ind Eng Chem Res* 40:5696–5704, 2001. https://doi.org/10.1021/ie010420c.

Akiya N, Savage PE: Roles of water for chemical reactions in high-temperature water, *Chem Rev* 102:2725–2750, 2002. https://doi.org/10.1021/cr000668w.

Ali KA, Abdullah AZ, Mohamed AR: Recent development in catalytic technologies for methanol synthesis from renewable sources: a critical review, *Renew Sustain Energy Rev* 44:508–518, 2015. https://doi.org/10.1016/j.rser.2015.01.010.

Al-Sabagh AM, Yehia FZ, Harding DRK, Eshaq G, ElMetwally AE: Fe3O4-boosted MWCNT as an efficient sustainable catalyst for PET glycolysis, *Green Chem* 18:3997–4003, 2016. https://doi.org/10.1039/C6GC00534A.

Aminu I, Nahil MA, Williams PT: Hydrogen from waste plastics by two-stage pyrolysis/low-temperature plasma catalytic processing, *Energy Fuel* 34:11679–11689, 2020. https://doi.org/10.1021/acs.energyfuels.0c02043.

Aminu I, Nahil MA, Williams PT: Hydrogen production by pyrolysis–nonthermal plasma/catalytic reforming of waste plastic over different catalyst support materials, *Energy Fuel* 36:3788–3801, 2022. https://doi.org/10.1021/acs.energyfuels.1c04186.

Antonopoulos I, Faraca G, Tonini D: Recycling of post-consumer plastic packaging waste in the EU: recovery rates, material flows, and barriers, *Waste Manag* 126:694–705, 2021. https://doi.org/10.1016/j.wasman.2021.04.002.

Artetxe M, Lopez G, Amutio M, Elordi G, Bilbao J, Olazar M: Cracking of high density polyethylene pyrolysis waxes on HZSM-5 catalysts of different acidity, *Ind Eng Chem Res* 52:10637–10645, 2013. https://doi.org/10.1021/ie4014869.

Artetxe M, Lopez G, Amutio M, et al.: Styrene recovery from polystyrene by flash pyrolysis in a conical spouted bed reactor, *Waste Manag* 45:126–133, 2015. https://doi.org/10.1016/j.wasman.2015.05.034.

Barbarias I, Lopez G, Alvarez J, et al.: A sequential process for hydrogen production based on continuous HDPE fast pyrolysis and in-line steam reforming, *Chem Eng J* 296:191–198, 2016. https://doi.org/10.1016/j.cej.2016.03.091.

Bauer B, TANNE J, SPOTT T, et al.: *Microwave methods for converting hydrocarbon-based waste materials into oil and gas fuels*, WO2020006512A1, 2020.

BCC Research: *Plastics recycling market size, share & growth analysis report*, 2022, BCC Publishing.

Berčič G, Djinović P, Pintar A: Procedure for generation of catalyst-free PE-TG profiles and its consequence on calculated activation energies, *J Therm Anal Calorim* 128:443–456, 2017. https://doi.org/10.1007/s10973-016-5872-1.

Berčič G, Djinović P, Pintar A: Simplified approach to modelling the catalytic degradation of low-density polyethylene (LDPE) by applying catalyst-free LDPE-TG profiles and the Friedman method, *J Therm Anal Calorim* 136:1011–1020, 2019. https://doi.org/10.1007/s10973-018-7774-x.

Boduszynski MM: Composition of heavy petroleums. 1. Molecular weight, hydrogen deficiency, and heteroatom concentration as a function of atmospheric equivalent boiling point up to 1400.degree.F (760.degree.C), *Energy Fuel* 1:2–11, 1987. https://doi.org/10.1021/ef00001a001.

Bonifazi G, Di Maio F, Potenza F, Serranti S: FT-IR analysis and hyperspectral imaging applied to postconsumer plastics packaging characterization and sorting, *IEEE Sens J* 16:3428–3434, 2016. https://doi.org/10.1109/JSEN.2015.2449867.

Caeiro G, Carvalho RH, Wang X, Lemos W, Lemos F, Guisnet M: Activation of C2–C4 alkanes over acid and bifunctional zeolite catalysts, *J Mol Catal A Chem* 255:131–158, 2006. https://doi.org/10.1016/j.molcata.2006.03.068.

Cano I, Martin C, Fernandes JA, et al.: Paramagnetic ionic liquid-coated SiO2@Fe3O4 nanoparticles—The next generation of magnetically recoverable nanocatalysts applied in the glycolysis of PET, *Appl Catal Environ* 260, 2020, 118110. https://doi.org/10.1016/j.apcatb.2019.118110.

Chai Y, Wang M, Gao N, Duan Y, Li J: Experimental study on pyrolysis/gasification of biomass and plastics for H2 production under new dual-support catalyst, *Chem Eng J* 396, 2020a, 125260. https://doi.org/10.1016/j.cej.2020.125260.

Chai Y, Gao N, Wang M, Wu C: H2 production from co-pyrolysis/gasification of waste plastics and biomass under novel catalyst Ni-CaO-C, *Chem Eng J* 382, 2020b, 122947. https://doi.org/10.1016/j.cej.2019.122947.

Chand R, Kohansal K, Toor S, Pedersen TH, Vollertsen J: Microplastics degradation through hydrothermal liquefaction of wastewater treatment sludge, *J Clean Prod* 335, 2022, 130383. https://doi.org/10.1016/j.jclepro.2022.130383.

Chemical & Engineering News: *The future of plastic*, 2020, American Chemical Society. https://cen.acs.org/sections/discovery-reports/the-future-of-plastic.html. (Accessed 9 September 2022).

Chen W-T, Jin K, Linda Wang N-H: Use of supercritical water for the liquefaction of polypropylene into oil, *ACS Sustainable Chem Eng* 7:3749–3758, 2019. https://doi.org/10.1021/acssuschemeng.8b03841.

Cheng K, Kang J, King DL, et al.: Chapter Three—Advances in catalysis for syngas conversion to hydrocarbons, In Song C, editor: *Advances in catalysis*, vol. 60, 2017, Academic Press, pp 125–208. https://doi.org/10.1016/bs.acat.2017.09.003.

Cheremisinoff NP, Rosenfeld P: The petroleum industry, In Cheremisinoff NP, Rosenfeld P, editors: *Handbook of pollution prevention and cleaner production—best practices in the petroleum industry*, Oxford, 2009, William Andrew Publishing, pp 1–97. https://doi.org/10.1016/B978-0-8155-2035-1.10001-6 (chapter 1).

Christopher FJ, Kumar PS, Vo D-VN, Christopher FC, Jayaraman L: Methods for chemical conversion of plastic wastes into fuels and chemicals. A review, *Environ Chem Lett* 20:223–242, 2022. https://doi.org/10.1007/s10311-021-01329-1.

Closed Loop Partners: *Assessing molecular recycling technologies in the United States and Canada. Closed loop partners*, 2021. https://www.closedlooppartners.com/research/transitioningto-a-circular-system-for-plastics-assessing-molecular-recycling-technologies-in-the-united-states-and-canada-2/ (accessed May 19, 2022).

Corma A, Orchillés AV: Current views on the mechanism of catalytic cracking, *Micropor Mesopor Mater* 35–36:21–30, 2000. https://doi.org/10.1016/S1387-1811(99)00205-X.

Corma A, Planelles J, Sánchez-Marín J, Tomás F: The role of different types of acid site in the cracking of alkanes on zeolite catalysts, *J Catal* 93:30–37, 1985. https://doi.org/10.1016/0021-9517(85)90148-4.

de Klerk A: 10—Transport fuel: biomass-, coal-, gas- and waste-to-liquids processes, In Letcher TM, editor: *Future energy*, 3rd Edition, 2020, Elsevier, pp 199–226. https://doi.org/10.1016/B978-0-08-102886-5.00010-4.

Dimitriadis A, Bezergianni S: Hydrothermal liquefaction of various biomass and waste feedstocks for biocrude production: a state of the art review, *Renew Sustain Energy Rev* 68:113–125, 2017. https://doi.org/10.1016/j.rser.2016.09.120.

Dogu O, Pelucchi M, Van de Vijver R, et al.: The chemistry of chemical recycling of solid plastic waste via pyrolysis and gasification: state-of-the-art, challenges, and future directions, *Prog Energy Combust Sci* 84:100901, 2021. https://doi.org/10.1016/j.pecs.2020.100901.

Domokos L, Jongkind H, Rigutto MS, Stork WHJ, EHCVD V: *Hydrocracking catalyst composition*, US7648939B2, 2010.

Du S, Valla JA, Parnas RS, Bollas GM: Conversion of polyethylene terephthalate based waste carpet to benzene-rich oils through thermal, catalytic, and catalytic steam pyrolysis, *ACS Sustainable Chem Eng* 4:2852–2860, 2016. https://doi.org/10.1021/acssuschemeng.6b00450.

Duan P, Savage PE: Hydrothermal liquefaction of a microalga with heterogeneous catalysts, *Ind Eng Chem Res* 50:52–61, 2011. https://doi.org/10.1021/ie100758s.

Durmuş A, Naci Koç S, Selda Pozan G, Kaşgöz A: Thermal-catalytic degradation kinetics of polypropylene over BEA, ZSM-5 and MOR zeolites, *Appl Catal Environ* 61:316–322, 2005. https://doi.org/10.1016/j.apcatb.2005.06.009.

Dwivedi U, Naik SN, Pant KK: High quality liquid fuel production from waste plastics via two-step cracking route in a bottom-up approach using bi-functional Fe/HZSM-5 catalyst, *Waste Manag* 132:151–161, 2021. https://doi.org/10.1016/j.wasman.2021.07.024.

Eatemadi A, Daraee H, Karimkhanloo H, et al.: Carbon nanotubes: properties, synthesis, purification, and medical applications, *Nanoscale Res Lett* 9:393, 2014. https://doi.org/10.1186/1556-276X-9-393.

EC CPA: *Guidance on waste definitions*, 2021.

Eimontas J, Striūgas N, Abdelnaby MA, Yousef S: Catalytic pyrolysis kinetic behavior and TG-FTIR-GC–MS analysis of metallized food packaging plastics with different concentrations of ZSM-5 zeolite catalyst, *Polymers* 13:702, 2021. https://doi.org/10.3390/polym13050702.

Ellis LD, Orski SV, Kenlaw GA, et al.: Tandem heterogeneous catalysis for polyethylene depolymerization via an olefin-intermediate process, *ACS Sustainable Chem Eng* 9:623–628, 2021a. https://doi.org/10.1021/acssuschemeng.0c07612.

Ellis LD, Rorrer NA, Sullivan KP, et al.: Chemical and biological catalysis for plastics recycling and upcycling, *Nat Catal* 4:539–556, 2021b. https://doi.org/10.1038/s41929-021-00648-4.

Eschenbacher A, Goodarzi F, Varghese RJ, et al.: Boron-modified mesoporous ZSM-5 for the conversion of pyrolysis vapors from LDPE and mixed polyolefins: maximizing the C2–C4 olefin yield with minimal carbon footprint, *ACS Sustainable Chem Eng* 9:14618–14630, 2021. https://doi.org/10.1021/acssuschemeng.1c06098.

Eschenbacher A, Varghese RJ, Abbas-Abadi MS, Van Geem KM: Maximizing light olefins and aromatics as high value base chemicals via single step catalytic conversion of plastic waste, *Chem Eng J* 428, 2022a, 132087. https://doi.org/10.1016/j.cej.2021.132087.

Eschenbacher A, Varghese RJ, Delikonstantis E, et al.: Highly selective conversion of mixed polyolefins to valuable base chemicals using phosphorus-modified and steam-treated mesoporous HZSM-5 zeolite with minimal carbon footprint, *Appl Catal Environ* 309, 2022b, 121251. https://doi.org/10.1016/j.apcatb.2022.121251.

Escola JM, Aguado J, Serrano DP, et al.: Catalytic hydroreforming of the polyethylene thermal cracking oil over Ni supported hierarchical zeolites and mesostructured aluminosilicates, *Appl Catal Environ* 106:405–415, 2011. https://doi.org/10.1016/j.apcatb.2011.05.048.

Escola JM, Serrano DP, Aguado J, Briones L: Hydroreforming of the LDPE thermal cracking oil over hierarchical Ni/beta catalysts with different Ni particle size distributions, *Ind Eng Chem Res* 54:6660–6668, 2015. https://doi.org/10.1021/acs.iecr.5b01160.

Fan L, Zhang Y, Liu S, et al.: Ex-situ catalytic upgrading of vapors from microwave-assisted pyrolysis of low-density polyethylene with MgO, *Energ Conver Manage* 149:432–441, 2017. https://doi.org/10.1016/j.enconman.2017.07.039.

Flaherty DW, Hibbitts DD, Iglesia E: Metal-catalyzed C–C bond cleavage in alkanes: effects of methyl substitution on transition-state structures and stability, *J Am Chem Soc* 136:9664–9676, 2014. https://doi.org/10.1021/ja5037429.

Font PC: Modelling of tar formation and evolution for biomass gasification: a review, *Appl Energy* 111:129–141, 2013. https://doi.org/10.1016/j.apenergy.2013.04.082.

Frash MV, van Santen RA: Quantum-chemical modeling of the hydrocarbon transformations in acid zeolite catalysts, *Top Catal* 9:191–205, 1999. https://doi.org/10.1023/A:1019183110705.

Fu J, Feng X, Liu Y, Shan H, Yang C: Mechanistic insights into the pore confinement effect on bimolecular and monomolecular cracking mechanisms of N-octane over HY and HZSM-5 zeolites: a DFT Study, *J Phys Chem C* 122:12222–12230, 2018. https://doi.org/10.1021/acs.jpcc.8b00995.

Goldman AS, Roy AH, Huang Z, Ajuha R, Schinski W, Brookhart M: Catalytic alkane metathesis by tandem alkane dehydrogenation-olefin metathesis, *Science* 312:257–261, 2006. https://doi.org/10.1126/science.1123787.

Google-Afara: *Closing the plastics circularity gap*, 2022. https://www.gstatic.com/gumdrop/sustainability/closing-plastics-gap-full-report.pdf. (Accessed 9 September 2022).

Grand View Research: *Plastic market size, share & trends report*, 2022, pp 2022–2030.

Grim RG, To AT, Farberow CA, Hensley JE, Ruddy DA, Schaidle JA: Growing the bioeconomy through catalysis: a review of recent advancements in the production of fuels and chemicals from syngas-derived oxygenates, *ACS Catal* 9:4145–4172, 2019. https://doi.org/10.1021/acscatal.8b03945.

Grubbs RH: Olefin metathesis, *Tetrahedron* 60:7117–7140, 2004. https://doi.org/10.1016/j.tet.2004.05.124.

Guisnet M, Gnep NS, Alario F: Aromatization of short chain alkanes on zeolite catalysts, *Appl Catal A-Gen* 89:1–30, 1992. https://doi.org/10.1016/0926-860X(92)80075-N.

Hahladakis JN, Velis CA, Weber R, Iacovidou E, Purnell P: An overview of chemical additives present in plastics: migration, release, fate and environmental impact during their use, disposal and recycling, *J Hazard Mater* 344:179–199, 2018. https://doi.org/10.1016/j.jhazmat.2017.10.014.

Hasan MM, Batalha N, Fraga G, et al.: Zeolite shape selectivity impact on LDPE and PP catalytic pyrolysis products and coke nature, *Sustain Energy Fuels* 6:1587–1602, 2022. https://doi.org/10.1039/D2SE00146B.

Helmer Pedersen T, Conti F: Improving the circular economy via hydrothermal processing of high-density waste plastics, *Waste Manag* 68:24–31, 2017. https://doi.org/10.1016/j.wasman.2017.06.002.

Hongthong S, Leese HS, Chuck CJ: Valorizing plastic-contaminated waste streams through the catalytic hydrothermal processing of polypropylene with lignocellulose, *ACS Omega* 5:20586–20598, 2020. https://doi.org/10.1021/acsomega.0c02854.

Horn HW, Jones GO, Wei DS, et al.: Mechanisms of organocatalytic amidation and trans-esterification of aromatic esters as a model for the depolymerization of poly(ethylene) terephthalate, *J Phys Chem A* 116:12389–12398, 2012. https://doi.org/10.1021/jp304212y.

Huang J, Veksha A, Chan WP, Lisak G: Support effects on thermocatalytic pyrolysis-reforming of polyethylene over impregnated Ni catalysts, *Appl Catal Gen* 622, 2021, 118222. https://doi.org/10.1016/j.apcata.2021.118222.

Huang J, Veksha A, Chan WP, Giannis A, Lisak G: Chemical recycling of plastic waste for sustainable material management: a prospective review on catalysts and processes, *Renew Sustain Energy Rev* 154, 2022, 111866. https://doi.org/10.1016/j.rser.2021.111866.

Humphreys LJ, Rowlands WN: *Method and apparatus for producing biofuel*, WO2018107204A9, 2018.

Jaydev SD, Martín AJ, Pérez-Ramírez J: Direct conversion of polypropylene into liquid hydrocarbons on carbon-supported platinum catalysts, *ChemSusChem* 14:5179–5185, 2021. https://doi.org/10.1002/cssc.202101999.

Jehanno C, Pérez-Madrigal MM, Demarteau J, Sardon H, Dove AP: Organocatalysis for depolymerisation, *Polym Chem* 10:172–186, 2018. https://doi.org/10.1039/C8PY01284A.

Joo S, Cho IJ, Seo H, et al.: Structural insight into molecular mechanism of poly(ethylene terephthalate) degradation, *Nat Commun* 9:382, 2018. https://doi.org/10.1038/s41467-018-02881-1.

Kim B-S, Kim Y-M, Lee HW, et al.: Catalytic copyrolysis of cellulose and thermoplastics over HZSM-5 and HY, *ACS Sustainable Chem Eng* 4:1354–1363, 2016. https://doi.org/10.1021/acssuschemeng.5b01381.

Kotrel S, Knözinger H, Gates BC: The Haag–Dessau mechanism of protolytic cracking of alkanes, *Micropor Mesopor Mater* 35-36:11–29, 2000. https://doi.org/10.1016/S1387-1811(99)00204-8.

Kots PA, Liu S, Vance BC, Wang C, Sheehan JD, Vlachos DG: Polypropylene plastic waste conversion to lubricants over Ru/TiO2 catalysts, *ACS Catal* 11:8104–8115, 2021. https://doi.org/10.1021/acscatal.1c00874.

Kratish Y, Li J, Liu S, Gao Y, Marks TJ: Polyethylene terephthalate deconstruction catalyzed by a carbon-supported single-site molybdenum-dioxo complex, *Angew Chem Int Ed* 59:19857–19861, 2020. https://doi.org/10.1002/anie.202007423.

Kumagai S, Yamasaki R, Kameda T, et al.: Aromatic hydrocarbon selectivity as a function of CaO basicity and aging during CaO-catalyzed PET pyrolysis using tandem μ-reactor-GC/MS, *Chem Eng J* 332:169–173, 2018. https://doi.org/10.1016/j.cej.2017.09.077.

Kumagai S, Yamasaki R, Kameda T, et al.: Catalytic pyrolysis of poly(ethylene terephthalate) in the presence of metal oxides for aromatic hydrocarbon recovery using tandem μ-reactor-GC/MS, *Energy Fuel* 34:2492–2500, 2020. https://doi.org/10.1021/acs.energyfuels.9b02915.

Kunwar B, Moser BR, Chandrasekaran SR, Rajagopalan N, Sharma BK: Catalytic and thermal depolymerization of low value post-consumer high density polyethylene plastic, *Energy* 111:884–892, 2016. https://doi.org/10.1016/j.energy.2016.06.024.

Laudenschleger D, Ruland H, Muhler M: Identifying the nature of the active sites in methanol synthesis over Cu/ZnO/Al2O3 catalysts, *Nat Commun* 11:3898, 2020. https://doi.org/10.1038/s41467-020-17631-5.

Lee N, Joo J, Lin K-YA, Lee J: Waste-to-fuels: pyrolysis of low-density polyethylene waste in the presence of H-ZSM-11, *Polymers* 13:1198, 2021. https://doi.org/10.3390/polym13081198.

Li X, Sun Q, Li Y, Wang N, Lu J, Yu J: Confinement effect of zeolite cavities on methanol-to-olefin conversion: a density functional theory study, *J Phys Chem C* 118:24935–24940, 2014. https://doi.org/10.1021/jp505696m.

Li S, Cañete Vela I, Järvinen M, Seemann M: Polyethylene terephthalate (PET) recycling via steam gasification—The effect of operating conditions on gas and tar composition, *Waste Manag* 130:117–126, 2021. https://doi.org/10.1016/j.wasman.2021.05.023.

Liu C, Li G, Hensen EJM, Pidko EA: Relationship between acidity and catalytic reactivity of faujasite zeolite: a periodic DFT study, *J Catal* 344:570–577, 2016. https://doi.org/10.1016/j.jcat.2016.10.027.

Liu L, Zhang Z, Das S, Kawi S: Reforming of tar from biomass gasification in a hybrid catalysis-plasma system: a review, *Appl Catal Environ* 250:250–272, 2019. https://doi.org/10.1016/j.apcatb.2019.03.039.

Liu S, Kots PA, Vance BC, Danielson A, Vlachos DG: Plastic waste to fuels by hydrocracking at mild conditions, *Sci Adv* 7, 2021, eabf8283. https://doi.org/10.1126/sciadv.abf8283.

Liu J, Li Y, Jia X, et al.: Catalytic pyrolysis of poly(ethylene terephthalate) with molybdenum oxides for the production of olefins and terephthalic acid, *Ind Eng Chem Res* 61:5054–5065, 2022. https://doi.org/10.1021/acs.iecr.1c04743.

Lopez G, Artetxe M, Amutio M, Bilbao J, Olazar M: Thermochemical routes for the valorization of waste polyolefinic plastics to produce fuels and chemicals. A review, *Renew Sustain Energy Rev* 73:346–368, 2017. https://doi.org/10.1016/j.rser.2017.01.142.

Lopez G, Artetxe M, Amutio M, Alvarez J, Bilbao J, Olazar M: Recent advances in the gasification of waste plastics. A critical overview, *Renew Sustain Energy Rev* 82:576–596, 2018. https://doi.org/10.1016/j.rser.2017.09.032.

Lowery JL, Green MJ: Safer carbon nanotube processing expands industrial and consumer applications. Science, *Advances* 8, 2022, eabq4853. https://doi.org/10.1126/sciadv.abq4853.

Lukyanov DB, Gnep NS, Guisnet MR: Kinetic modeling of ethene and propene aromatization over HZSM-5 and GaHZSM-5, *Ind Eng Chem Res* 33:223–234, 1994.

Mangesh VL, Tamizhdurai P, Santhana Krishnan P, et al.: Green energy: hydroprocessing waste polypropylene to produce transport fuel, *J Clean Prod* 276, 2020, 124200. https://doi.org/10.1016/j.jclepro.2020.124200.

Marafi M, Stanislaus A, Furimsky E: Hydroprocessing of petroleum, In Marafi M, Stanislaus A, Furimsky E, editors: *Handbook of spent hydroprocessing catalysts*, Amsterdam, 2010, Elsevier, pp 17–49. https://doi.org/10.1016/B978-0-444-53556-6.00003-3 (chapter 3).

Marcilla A, Gómez-Siurana A, Valdés F: Catalytic cracking of low-density polyethylene over H-Beta and HZSM-5 zeolites: influence of the external surface. Kinetic model, *Polym Degrad Stab* 92:197–204, 2007. https://doi.org/10.1016/j.polymdegradstab.2006.11.007.

Maric J, Berdugo Vilches T, Pissot S, Cañete Vela I, Gyllenhammar M, Seemann M: Emissions of dioxins and furans during steam gasification of Automotive Shredder residue; experiences from the Chalmers 2–4-MW indirect gasifier, *Waste Manag* 102:114–121, 2020. https://doi.org/10.1016/j.wasman.2019.10.037.

Mark LO, Cendejas MC, Hermans I: The use of heterogeneous catalysis in the chemical valorization of plastic waste, *ChemSusChem* 13:5808–5836, 2020. https://doi.org/10.1002/cssc.202001905.

Martín AJ, Mondelli C, Jaydev SD, Pérez-Ramírez J: Catalytic processing of plastic waste on the rise, *Chem* 7:1487–1533, 2021. https://doi.org/10.1016/j.chempr.2020.12.006.

Mastellone ML, Arena U: Olivine as a tar removal catalyst during fluidized bed gasification of plastic waste, *AIChE J* 54:1656–1667, 2008. https://doi.org/10.1002/aic.11497.

Mazar MN, Al-Hashimi S, Cococcioni M, Bhan A: β-Scission of olefins on acidic zeolites: a periodic PBE-D study in H-ZSM-5, *J Phys Chem C* 117:23609–23620, 2013. https://doi.org/10.1021/jp403504n.

Müller S, Liu Y, Kirchberger FM, Tonigold M, Sanchez-Sanchez M, Lercher JA: Hydrogen transfer pathways during zeolite catalyzed methanol conversion to hydrocarbons, *J Am Chem Soc* 138:15994–16003, 2016. https://doi.org/10.1021/jacs.6b09605.

Munir D, Usman MR: Catalytic hydropyrolysis of a model municipal waste plastic mixture over composite USY/SBA-16 catalysts, *J Anal Appl Pyrolysis* 135:44–53, 2018. https://doi.org/10.1016/j.jaap.2018.09.023.

Munir D, Irfan MF, Usman MR: Hydrocracking of virgin and waste plastics: a detailed review, *Renew Sustain Energy Rev* 90:490–515, 2018. https://doi.org/10.1016/j.rser.2018.03.034.

Niu J, Wang Y, Liland SE, et al.: Unraveling enhanced activity, selectivity, and coke resistance of Pt–Ni bimetallic clusters in dry reforming, *ACS Catal* 11:2398–2411, 2021. https://doi.org/10.1021/acscatal.0c04429.

Noh G, Shi Z, Zones SI, Iglesia E: Isomerization and β-scission reactions of alkanes on bifunctional metal-acid catalysts: consequences of confinement and diffusional constraints on reactivity and selectivity, *J Catal* 368:389–410, 2018. https://doi.org/10.1016/j.jcat.2018.03.033.

Ojeda M, Nabar R, Nilekar AU, Ishikawa A, Mavrikakis M, Iglesia E: CO activation pathways and the mechanism of Fischer–Tropsch synthesis, *J Catal* 272:287–297, 2010. https://doi.org/10.1016/j.jcat.2010.04.012.

Okonsky ST, Jonnalagedda VJK, Toraman HE: Catalytic co-pyrolysis of LDPE and PET with HZSM-5, H-Beta, and HY: experiments and kinetic modelling, *React Chem Eng* 7(10):2175–2191, 2022. https://doi.org/10.1039/D2RE00144F.

Onwudili JA, Williams PT: Catalytic supercritical water gasification of plastics with supported RuO2: a potential solution to hydrocarbons–water pollution problem, *Process Safety and Environmental Protection* 102:140–149, 2016. https://doi.org/10.1016/j.psep.2016.02.009.

Payne J, Jones MD: The chemical recycling of polyesters for a circular plastics economy: challenges and emerging opportunities, *ChemSusChem* 14:4041–4070, 2021. https://doi.org/10.1002/cssc.202100400.

Pellegrino JL: *ITP chemicals: energy and environmental profile of the U.S. Chemical Industry*, 2000, U.S. Department of Energy.

Peng Z, Jan Simons T, Wallach J, Youngman A: *Advanced recycling: opportunities for growth*, 2022, McKinsey.

Petri JA: *Fluid catalytic cracking and hydrotreating processes for fabricating diesel fuel from waxes*, US20090065393A1, 2009.

Ragaert K, Delva L, Van Geem K: Mechanical and chemical recycling of solid plastic waste, *Waste Manag* 69:24–58, 2017. https://doi.org/10.1016/j.wasman.2017.07.044.

Rorrer JE, Troyano-Valls C, Beckham GT, Román-Leshkov Y: Hydrogenolysis of polypropylene and mixed polyolefin plastic waste over Ru/C to produce liquid alkanes, *ACS Sustainable Chem Eng* 9:11661–11666, 2021a. https://doi.org/10.1021/acssuschemeng.1c03786.

Rigutto MS, van Veen R, Huve L: Chapter 24 – Zeolites in hydrocarbon processing, *Introduction to Zeolite Science and Practice*, 3rd, 168, 2007, Elsevier, pp 855–913.

Rorrer JE, Beckham GT, Román-Leshkov Y: Conversion of polyolefin waste to liquid alkanes with Ru-based catalysts under mild conditions, *JACS Au* 1:8–12, 2021b. https://doi.org/10.1021/jacsau.0c00041.

Ruppert AM, Weinberg K, Palkovits R: Hydrogenolysis goes bio: from carbohydrates and sugar alcohols to platform chemicals, *Angew Chem Int Ed* 51:2564–2601, 2012. https://doi.org/10.1002/anie.201105125.

Saad JM, Williams PT: Pyrolysis-catalytic-dry reforming of waste plastics and mixed waste plastics for syngas production, *Energy Fuel* 30:3198–3204, 2016a. https://doi.org/10.1021/acs.energyfuels.5b02508.

Saad JM, Williams PT: Catalytic dry reforming of waste plastics from different waste treatment plants for production of synthesis gases, *Waste Manag* 58:214–220, 2016b. https://doi.org/10.1016/j.wasman.2016.09.011.

Saad JM, Williams PT: Manipulating the H2/CO ratio from dry reforming of simulated mixed waste plastics by the addition of steam, *Fuel Process Technol* 156:331–338, 2017a. https://doi.org/10.1016/j.fuproc.2016.09.016.

Saad JM, Williams PT: Pyrolysis-catalytic dry (CO2) reforming of waste plastics for syngas production: influence of process parameters, *Fuel* 193:7–14, 2017b. https://doi.org/10.1016/j.fuel.2016.12.014.

Saeaung K, Phusunti N, Phetwarotai W, Assabumrungrat S, Cheirsilp B: Catalytic pyrolysis of petroleum-based and biodegradable plastic waste to obtain high-value chemicals, *Waste Manag* 127:101–111, 2021. https://doi.org/10.1016/j.wasman.2021.04.024.

Salcedo A, Lustemberg PG, Rui N, et al.: Reaction pathway for coke-free methane steam reforming on a Ni/CeO2 catalyst: active sites and the role of metal–support interactions, *ACS Catal* 11:8327–8337, 2021. https://doi.org/10.1021/acscatal.1c01604.

Sánchez-Rivera KL, Huber GW: Catalytic hydrogenolysis of polyolefins into alkanes, *ACS Cent Sci* 7:17–19, 2021. https://doi.org/10.1021/acscentsci.0c01637.

Sarp S, Gonzalez Hernandez S, Chen C, Sheehan SW: Alcohol production from carbon dioxide: methanol as a fuel and chemical feedstock, *Joule* 5:59–76, 2021. https://doi.org/10.1016/j.joule.2020.11.005.

Schweighuber A, Felgel-Farnholz A, Bögl T, Fischer J, Buchberger W: Investigations on the influence of multiple extrusion on the degradation of polyolefins, *Polym Degrad Stab* 192, 2021, 109689. https://doi.org/10.1016/j.polymdegradstab.2021.109689.

Serrano DP, Escola JM, Briones L, Medina S, Martínez A: Hydroreforming of the oils from LDPE thermal cracking over Ni–Ru and Ru supported over hierarchical Beta zeolite, *Fuel* 144:287–294, 2015. https://doi.org/10.1016/j.fuel.2014.12.040.

Seshasayee MS, Savage PE: Synergistic interactions during hydrothermal liquefaction of plastics and biomolecules, *Chem Eng J* 417, 2021, 129268. https://doi.org/10.1016/j.cej.2021.129268.

Settle AE, Berstis L, Rorrer NA, et al.: Heterogeneous Diels–Alder catalysis for biomass-derived aromatic compounds, *Green Chem* 19:3468–3492, 2017. https://doi.org/10.1039/C7GC00992E.

Silva BJB, Sousa LV, Sarmento LRA, Alencar SL, Quintela PHL, Silva AOS: Kinetic study of thermocatalytic degradation of UHMWPE over microporous and hierarchical ZSM-23 zeolite, *Appl Catal Environ* 267, 2020, 118699. https://doi.org/10.1016/j.apcatb.2020.118699.

Sinfelt JH: Catalytic hydrogenolysis on metals, *Catal Lett* 9:159–171, 1991. https://doi.org/10.1007/BF00773174.

Soni VK, Singh G, Vijayan BK, Chopra A, Kapur GS, Ramakumar SSV: Thermochemical recycling of waste plastics by pyrolysis: a review, *Energy Fuel* 35:12763–12808, 2021. https://doi.org/10.1021/acs.energyfuels.1c01292.

Sonnendecker C, Oeser J, Richter PK, et al.: Low carbon footprint recycling of post-consumer pet plastic with a metagenomic polyester hydrolase, *ChemSusChem* 15, 2022, e202101062. https://doi.org/10.1002/cssc.202101062.

Speight JG, editor: Refining processes, In *The refinery of the future*, Boston, 2011, William Andrew Publishing, pp 39–80. (Chapter 2).

Stöcker M: Methanol to olefins (MTO) and methanol to gasoline (MTG), In *Zeolites and catalysis*, 2010, John Wiley & Sons, Ltd, pp 687–711. https://doi.org/10.1002/9783527630295.ch22.

Tennakoon A, Wu X, Paterson AL, et al.: Catalytic upcycling of high-density polyethylene via a processive mechanism, *Nat Catal* 3:893–901, 2020. https://doi.org/10.1038/s41929-020-00519-4.

Thybaut JW, Marin GB: Multiscale aspects in hydrocracking: from reaction mechanism over catalysts to kinetics and industrial application, In Song C, editor: *Advances in catalysis*, vol. 59, 2016, Academic Press, pp 109–238. https://doi.org/10.1016/bs.acat.2016.10.001. (chapter Two).

Tian X, Zeng Z, Liu Z, et al.: Conversion of low-density polyethylene into monocyclic aromatic hydrocarbons by catalytic pyrolysis: comparison of HZSM-5, Hβ, HY and MCM-41, *J Clean Prod* 358, 2022, 131989. https://doi.org/10.1016/j.jclepro.2022.131989.

Till Z, Varga T, Sója J, Miskolczi N, Chován T: Kinetic identification of plastic waste pyrolysis on zeolite-based catalysts, *Energ Conver Manage* 173:320–330, 2018. https://doi.org/10.1016/j.enconman.2018.07.088.

Tomasek S, Lonyi F, Valyon J, Wollmann A, Hancsók J: Hydrocracking of Fischer–Tropsch paraffin mixtures over strong acid bifunctional catalysts to engine fuels, *ACS Omega* 5:26413–26420, 2020. https://doi.org/10.1021/acsomega.0c02711.

Tournier V, Topham CM, Gilles A, et al.: An engineered PET depolymerase to break down and recycle plastic bottles, *Nature* 580:216–219, 2020. https://doi.org/10.1038/s41586-020-2149-4.

US EPA O: *Plastics: material-specific data*, 2018 https://www.epa.gov/facts-and-figures-about-materials-waste-and-recycling/plastics-material-specific-data. (Accessed 17 May 2022).

van Santen RA, Ciobîcă IM, van Steen E, Ghouri MM: Mechanistic issues in Fischer–Tropsch catalysis, In Gates BC, Knözinger H, editors: *Advances in catalysis*, vol. 54, 2011, Academic Press, pp 127–187. https://doi.org/10.1016/B978-0-12-387772-7.00003-4. (chapter 3).

Vogt ETC, Weckhuysen BM: Fluid catalytic cracking: recent developments on the grand old lady of zeolite catalysis, *Chem Soc Rev* 44:7342–7370, 2015. https://doi.org/10.1039/C5CS00376H.

Vollmer I, Jenks MJF, Roelands MCP, et al.: Beyond mechanical recycling: giving new life to plastic waste, *Angew Chem Int Ed* 59:15402–15423, 2020. https://doi.org/10.1002/anie.201915651.

Wang W, Meng L, Yu J, Xie F, Huang Y: Enhanced hydrothermal conversion of caprolactam from waste monomer casting polyamide over H-Beta zeolite and its mechanism, *J Anal Appl Pyrolysis* 125:218–226, 2017. https://doi.org/10.1016/j.jaap.2017.03.020.

Wang J, Jiang J, Sun Y, et al.: Recycling benzene and ethylbenzene from in-situ catalytic fast pyrolysis of plastic wastes, *Energ Conver Manage* 200, 2019, 112088. https://doi.org/10.1016/j.enconman.2019.112088.

Wang Y, Cheng L, Gu J, et al.: Catalytic pyrolysis of polyethylene for the selective production of monocyclic aromatics over the zinc-loaded ZSM-5 catalyst, *ACS Omega* 7:2752–2765, 2022. https://doi.org/10.1021/acsomega.1c05401.

Weiland F, Lundin L, Celebi M, van der Vlist K, Moradian F: Aspects of chemical recycling of complex plastic waste via the gasification route, *Waste Manag* 126:65–77, 2021. https://doi.org/10.1016/j.wasman.2021.02.054.

Weitkamp J: Catalytic hydrocracking—mechanisms and versatility of the process, *ChemCatChem* 4:292–306, 2012. https://doi.org/10.1002/cctc.201100315.

Westhues S, Idel J, Klankermayer J: Molecular catalyst systems as key enablers for tailored polyesters and polycarbonate recycling concepts, *Sci Adv* 4, 2018, eaat9669. https://doi.org/10.1126/sciadv.aat9669.

Wilk V, Hofbauer H: Conversion of mixed plastic wastes in a dual fluidized bed steam gasifier, *Fuel* 107:787–799, 2013. https://doi.org/10.1016/j.fuel.2013.01.068.

Wu C, Nahil MA, Miskolczi N, Huang J, Williams PT: Production and application of carbon nanotubes, as a co-product of hydrogen from the pyrolysis-catalytic reforming of waste plastic, *Process Saf Environ Prot* 103:107–114, 2016. https://doi.org/10.1016/j.psep.2016.07.001.

Wu C, Williams PT: Hydrogen production by steam gasification of polypropylene with various nickel catalysts, *Appl Catal Environ* 87:152–161, 2009. https://doi.org/10.1016/j.apcatb.2008.09.003.

Xue Y, Johnston P, Bai X: Effect of catalyst contact mode and gas atmosphere during catalytic pyrolysis of waste plastics, *Energ Conver Manage* 142:441–451, 2017. https://doi.org/10.1016/j.enconman.2017.03.071.

Yang R-X, Chuang K-H, Wey M-Y: Effects of nickel species on Ni/Al2O3 catalysts in carbon nanotube and hydrogen production by waste plastic gasification: bench- and pilot-scale tests, *Energy Fuel* 29:8178–8187, 2015. https://doi.org/10.1021/acs.energyfuels.5b01866.

Yang R-X, Wu S-L, Chuang K-H, Wey M-Y: Co-production of carbon nanotubes and hydrogen from waste plastic gasification in a two-stage fluidized catalytic bed, *Renew Energy* 159:10–22, 2020. https://doi.org/10.1016/j.renene.2020.05.141.

Yang R-X, Jan K, Chen C-T, Chen W-T, Wu K-C: Thermochemical conversion of plastic waste into fuels, chemicals, and value-added materials: a critical review and outlooks, *ChemSusChem* 15:1–39, 2022. https://doi.org/10.1002/cssc.202200171.

Yang R-X, Bieh Y-T, Chen CH, et al.: Heterogeneous metal azolate framework-6 (MAF-6) catalysts with high zinc density for enhanced polyethylene terephthalate (PET) conversion, *ACS Sustainable Chem Eng* 9:6541–6550, 2021. https://doi.org/10.1021/acssuschemeng.0c08012.

Yao D, Wang C-H: Pyrolysis and in-line catalytic decomposition of polypropylene to carbon nanomaterials and hydrogen over Fe- and Ni-based catalysts, *Appl Energy* 265, 2020, 114819. https://doi.org/10.1016/j.apenergy.2020.114819.

Yao D, Yang H, Chen H, Williams PT: Investigation of nickel-impregnated zeolite catalysts for hydrogen/syngas production from the catalytic reforming of waste polyethylene, *Appl Catal Environ* 227:477–487, 2018. https://doi.org/10.1016/j.apcatb.2018.01.050.

Yao L, King J, Wu D, Chuang SSC, Peng Z: Non-thermal plasma-assisted hydrogenolysis of polyethylene to light hydrocarbons, *Catal Commun* 150, 2021, 106274. https://doi.org/10.1016/j.catcom.2020.106274.

Yarulina I, Chowdhury AD, Meirer F, Weckhuysen BM, Gascon J: Recent trends and fundamental insights in the methanol-to-hydrocarbons process, *Nat Catal* 1:398–411, 2018. https://doi.org/10.1038/s41929-018-0078-5.

Yu H, Li F, He W, et al.: Synthesis of micro–mesoporous ZSM-5 zeolite with microcrystalline cellulose as co-template and catalytic cracking of polyolefin plastics, *RSC Adv* 10:22126–22136, 2020. https://doi.org/10.1039/D0RA03082A.

Zhang X, Lei H: Synthesis of high-density jet fuel from plastics via catalytically integral processes, *RSC Adv* 6:6154–6163, 2016. https://doi.org/10.1039/C5RA25327F.

Zhang Y, Duan D, Lei H, Villota E, Ruan R: Jet fuel production from waste plastics via catalytic pyrolysis with activated carbons, *Appl Energy* 251, 2019, 113337. https://doi.org/10.1016/j.apenergy.2019.113337.

Zhang F, Zeng M, Yappert RD, et al.: Polyethylene upcycling to long-chain alkylaromatics by tandem hydrogenolysis/aromatization, *Science* 370:437–441, 2020. https://doi.org/10.1126/science.abc5441.

Zhou N, Dai L, Lyu Y, et al.: A structured catalyst of ZSM-5/SiC foam for chemical recycling of waste plastics via catalytic pyrolysis, *Chem Eng J* 440, 2022, 135836. https://doi.org/10.1016/j.cej.2022.135836.

CHAPTER FIVE

Characterization of SPW pyrolysis oils: Products spectra and opportunities

Barbara Alejandra Perez[a], Varaha Jayarama Krishna Jonnalagedda[b], and Hilal Ezgi Toraman[a,b,c,*]

[a]Department of Chemical Engineering, The Pennsylvania State University, University Park, PA, United States
[b]Department of Energy and Mineral Engineering, The Pennsylvania State University, University Park, PA, United States
[c]Institutes of Energy and the Environment, The Pennsylvania State University, University Park, PA, United States
*Corresponding author: e-mail address: hzt5148@psu.edu

Contents

Abstract

Pyrolysis is a promising technique for conversion of solid plastic waste (SPW) to fine chemicals and fuels both via thermal and catalytic cracking routes. The quality of the oil generated from pyrolysis of SPW largely depends upon on process operating parameters and feedstock characteristics. The contamination of pyrolysis oil with hetero-atom impurities like oxygen, nitrogen, sulfur, etc., is a growing concern because of its inability to be used for various applications. In order to identify the presence of possible contaminants and upgrade the pyrolysis oil produced from SPW, a detailed knowledge of their chemical composition is extremely vital. This knowledge can thus facilitate the process design required for upgrading pyrolysis oil obtained from SPW. The current chapter highlights the important techniques adopted for characterizing the plastic pyrolysis

Advances in Chemical Engineering, Volume 60
ISSN 0065-2377
https://doi.org/10.1016/bs.ache.2022.09.007

oil (PPO) from SPW. The various physicochemical properties discussed in this chapter include heating value, viscosity, density, water content, flash point, fire point, and pour point. Besides, the chemical composition of the pyrolysis oil obtained from thermal and catalytic processes analyzed using Fourier transform infrared spectroscopy (FTIR), nuclear magnetic resonance (NMR), one dimensional gas chromatography (1D-GC) and two-dimensional gas chromatography (GC × GC) coupled to different detectors is also presented.

1. Introduction

The world-wide production of plastics has reached 380 million metric tons (MMT) in 2015 and is expected to increase fourfold by 2050 (Ritchie and Roser, 2018). It is important to mention that ~75% of plastics produced world-wide ends up as waste. The top five countries that produce highest amount of solid plastic waste (SPW) in 2016 include USA (34.02 MMT), India (26.33 MMT), China (21.6 MMT), Brazil (10.68 MMT), and Indonesia (9.13 MMT) (Law et al., 2020). The major sources of SPW include municipal solid waste (MSW), construction and demolition waste (CDW), waste from electric and electronic equipment (WEEE), and end-of-life vehicles (ELV) (Lahtela et al., 2019). The most common plastics found from domestic and household items of MSW include polyethylene terephthalate (PET), high density polyethylene (HDPE), polyvinyl chloride (PVC), linear low density polyethylene (LLDPE), low density polyethylene (LDPE), polylactic acid (PLA), polypropylene (PP), and polystyrene (PS) (US Environmental Protection Agency, 2020).

Plastics being non-biodegradable cannot easily enter the natural carbon cycle. As a result, waste disposal facilities become the end for the life cycle of plastic materials. There is growing demand universally to recycle SPW to promote sustainable living and reduce carbon footprint. It is imperative to improve ways to efficiently recycle the post-consumer materials. Currently, less than 10% of SPW generated world-wide is being recycled (Geyer et al. (2017)). The key to become more resource efficient is to use SPW as a valuable resource to produce wide range of platform chemicals or fuel molecules. Plastic recycling involves producing useful products from different sources of SPW. Recycling is a viable solution to address the issue of minimizing the SPW and helps to create a sustainable circular economy for plastics.

Plastics, toward the end of their service life, can still be converted to new products or raw materials to recover energy as a substitute for virgin fossil fuels. Fig. 1 shows the different routes of polymer waste management. Of all the available routes, pyrolysis, a promising thermochemical recycling technique proves to be a potential solution to convert SPW to value-added chemicals and fuels. It involves thermal decomposition of the organic matter in inert ambience at elevated temperatures of 300–800 °C (Anuar Sharuddin et al., 2016). This technique is well suited for handling virgin polymers as well as mixed, unwashed, and unsorted polymers. Pyrolysis of polymers involves converting long-chain molecules to shorter and simple molecules by applying heat and pressure in an inert atmosphere with or without catalyst. The process provides great flexibility in terms of varying the process parameters and obtaining the required composition of the products. The liquid oil generated through pyrolysis of SPW can be utilized for multiple industrial applications, viz., turbines, boilers, and furnaces. The upgradation of SPW pyrolysis oils can also serve as a drop-in alternative to commercially available transportation fuels like gasoline and diesel. The chemicals that can be produced from pyrolysis of SPW include benzene,

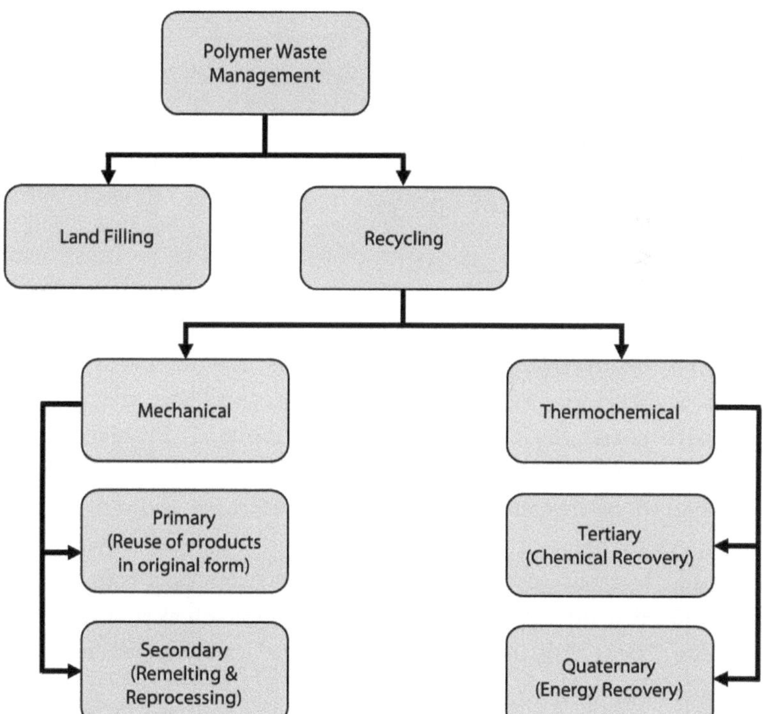

Fig. 1 Different routes of polymer waste management.

toluene, ethylbenzene, xylene (BTEX), a range of alkanes, alkenes, alkadienes, and monomers like ethene, propene, styrene, methyl methacrylate, etc. (Anuar Sharuddin et al., 2016; Jayarama Krishna et al., 2019; Kaminsky, 2021). A great deal of research has been carried out to study the potential of pyrolysis using different types of plastic feedstocks for liquid production. It was observed that the composition and the yield of the liquid produced from SPW depend on various parameters like reactor type, temperature, pressure, residence time, heating rate, catalyst and carrier gas (Anuar Sharuddin et al., 2016; Czajczyńska et al., 2017; Lam and Chase, 2012).

A thorough characterization of pyrolysis oils generated from SPW is required to understand their chemical composition and properties and hence for further upgrading strategies. This chapter reviews various analytical techniques used to obtain physicochemical properties such as heating value, viscosity, density, water content, etc., and chemical composition knowledge of plastic pyrolysis oil (PPO) from SPW. The latter is discussed in detail based on the applied chemical characterization techniques such as Fourier transform infrared spectroscopy (FTIR), nuclear magnetic resonance spectroscopy (NMR), one dimensional gas chromatography (1D-GC) and two-dimensional gas chromatography (GC × GC) coupled to different detectors such as mass spectrometry (MS), flame ionization detector (FID), and element selective detectors.

2. Physicochemical analysis techniques

The physicochemical properties of PPOs need to be investigated to compare their properties with commercial oil and fuels and test their suitability to be used in specific applications such as diesel engines and automobile fuel. The characterization of fuel and oils uses standards such as the American Society for Testing and Materials (ASTM), the Institute of Petroleum (IP), and the United States Environmental Protection Agency (EPA). Table 1 shows the methods used to conduct the physicochemical characterization of fuel and oils. The study of these properties of PPOs allows the comparison of the physicochemical properties between PPOs, fuel, and oils. In addition, the physicochemical properties help to comprehend how the feedstock and pyrolysis conditions, such as pyrolysis temperature, heating rate, residence time, reactor, and catalyst, affect the quality of the PPO. Typical analyses include but are not limited to elemental analysis, calorific value, distillation, and determination of density, viscosity, water content, and flash, fire, and pour point.

Table 1 Methods used for the analysis of fuel, oils, and PPO.

Physicochemical analysis	Standard
Elemental analysis	ASTM D5251—ASTM D5708—ASTM D2622—ASTM D8291—ASTM D5291
Heating value	ASTM 24
Density	IP160/87—ASTM D1298-17—EN ISO 12185:2001
Viscosity	ASTM D446—ASTM D446—IP 71/87—EN ISO 3104:2020
Distillation	ASTM D86
Simulated distillation by GC	ASTM D5307
Pour point	ASTM D97—IP 15/67
Flash point	ASTM D1711—ASTM D09—ASTM D1695—IP 34/85
Cloud point	ASTM D5773
Hydrocarbon analysis	ASTM D5445
Chlorine	EPA method 5050—ASTM D2361

2.1 Elemental analysis

The elemental analysis is a technique that involves the determination of the weight percentage of carbon, hydrogen, nitrogen, sulfur, and oxygen (Speight, 2015). The carbon, hydrogen, nitrogen, and sulfur are determined from the plastic sample. However, the oxygen content in the sample can be determined using the difference, i.e., $\%O = 100 - \%C - \%H - \%N - \%S - \%Ash$ or it can also be determined absolutely using elemental analyzer (De et al., 2018). In the elemental analysis of PPOs, CHNS/O elemental analyzers have been used (Banar et al., 2012; Hassan et al., 2020). The elemental analyzers combust the sample, whether liquid or solid, at around 1000 °C under an oxygen-rich atmosphere. The amount of sample used in the elemental analyzer is between 0.01 and 100 mg (Mettler-Toledo International Inc, 2022). The carbon in the sample is converted to carbon dioxide (CO_2), hydrogen to water (H_2O), nitrogen to nitrogen gas (N_2) or oxides of nitrogen (NO_x), and sulfur to sulfur dioxide (SO_2).

The combustion gases flow out of the combustion chamber with the help of a carrier gas (e.g., helium) to a copper bed. The copper bed is used to remove any oxygen not consumed and convert the NO_x to N_2. Then, the gases can be sent to GC coupled with a thermal conductivity detector (TCD) or frontal chromatography with TCD or a series of separate infra-red and thermal conductivity cells for the detection of the individual compounds (Thompson, 2008).

The elemental analysis has been investigated for the physicochemical analysis of PPOs for a wide variety of virgin plastics, plastic mixtures, and plastic waste. Table 2 shows a summary of the elemental analysis conducted for PPOs. The elemental analysis carried out for PPOs obtained from polyolefins shows that they are mainly composed of carbon (C) and hydrogen (H) (Lee et al., 2015). Nonetheless, some traces of nitrogen (N), sulfur (S), and oxygen (O) can be found in the elemental analysis due to the addition of additives to the polymers feedstock (Adrados et al., 2012; Constantinescu et al., 2019). On the other hand, the elemental analysis of PPOs from mixed plastic waste reveals C, H, N, S, and O in the PPOs composition which originates not only from the polymer feedstocks that contain O and N, such as PET and polyamide (PA), but also from the additives added to the polymer and trace amounts of biomass present in the plastic waste mixture. Baena-González et al. (2020) conducted the elemental analysis of plastic waste obtained from the rejected fraction of the packaging waste separation which contains 82.23 wt% of plastics and 17.7 wt% of other materials. The PPO contained C, H, and O besides N, S, and Cl in trace quantities. Similar results were observed by other authors as well (Anuar Sharuddin et al., 2017; Kalargaris et al., 2017; López et al., 2010; Shah et al., 1999; Wongkhorsub and Chindaprasert, 2013). Quesada et al. (2019) used elemental analysis to evaluate the PPO from plastic waste obtained at different pyrolysis conditions varying in terms of temperature, residence time, and heating rate. The authors found that the elemental analysis is similar for all the conditions tested. The average values of C, H, N, and O in PPO were observed to be 81.24%, 14.69%, 0.13%, and 3.95%, respectively.

2.2 Heating value

The heating value represents the quantity of energy released by a unit mass of a sample during the chemical reaction for complete combustion producing CO_2 and H_2O (Wauquier, 1995). The heating value is an important characteristic of petroleum products because it is an indicator of the amount of energy released per amount of fuel (Saptoadi and Pratama, 2015).

Table 2 Ultimate analysis for PPO and diesel.

PPO feedstock	C (%)	H (%)	O[a] (%)	N (%)	S (%)	Cl (%)	Others (%)	Reference
HDPE	85.37 ± 1.13	14.14 ± 0.67	n.a.	<0.01	<0.01	n.a.	n.a.	Dobó et al. (2019)
	84.4	12.7	1.99	0.79	0.12	n.a.	n.a.	Kumar et al. (2013)
	85.1	14.5	0	0.4	n.a.	n.a.	n.a.	Kunwar et al. (2016)
	85.77	14.23	0	0	n.a.	n.a.	n.a.	Chen et al. (2016)
LDPE	85.47 ± 0.37	14.25 ± 0.32	n.a.	<0.01	<0.01	n.a.	n.a.	Dobó et al. (2019)
PP	89.63 ± 2.47	9.60 ± 0.43	n.a.	<0.01	<0.01	n.a.	n.a.	Dobó et al. (2019)
PUR	74.50 ± 0.20	8.04 ± 0.14	n.a.	9.11 ± 0.21	<0.01	n.a.	n.a.	Dobó et al. (2019)
Plastic mixture (HDPE, LDPE, and PP)	88.82	7.18	3.83	0.17	0.01	n.a.	n.a.	Anuar Sharuddin et al. (2017)
Plastic mixture (HDPE, LDPE, PP and PS)	90.85	9.12	0	0.3	0.09	n.a.	n.a.	Anuar Sharuddin et al. (2017)
Plastic waste (82.23 wt% of plastic (e.g., PP, PS, LDPE, HDPE, and PET) and 17.77 wt% of other materials (e.g., wood, food, paperboard, textile, glass, paper, ferric, and aluminum))	86.67	10.62	2.3	0.21	0.03	0.04	n.a.	Baena-González et al. (2020)
Plastic Waste (PE, PP, PS, PET, and PVC)	77.3	9.5	n.a.	0.3	n.a.	0.2	12.7	Adrados et al. (2012)
Plastic waste (plastic film (PE))	81.24	14.69	3.95	0.13	0	n.a.	n.a.	Quesada et al. (2019)
Plastic waste (PE and PP)	83.79	11.36	2	n.a.	–	0.03	n.a.	Wongkhorsub and Chindaprasert (2013)

[a]Calculated by difference.

The measure of the higher heating value (HHV) can be done using the ASTM 24 procedure in which the oil sample is burned in an oxygen bomb calorimeter under controlled conditions. The heat of combustion is computed from the temperature observations before, during, and after the combustion (ASTM, 2019). To calculate the net heating value (NHV), the HHV and hydrogen content of the sample must be known.

Table 3 shows the heating values for PPOs from single polymer and plastic waste feedstocks. Several authors used calorimetric bombs to get the HHV from PPO samples. The heating value of PPO sample changes according to the pyrolysis operational conditions such as pyrolysis temperature and time of pyrolysis. Cahyono and Fenti (2017) pyrolyzed LDPE from waste plastic bags at 250 °C, 350 °C, and 450 °C. The heating value for PPO from LDPE fluctuated from 26.18 MJ kg^{-1} to 46.20 MJ kg^{-1} when the pyrolysis temperature increased from 250 °C to 450 °C. The authors claim that higher heating values are related to lighter fractions of hydrocarbons in the oil that may be produced at higher temperatures due to the secondary cracking reactions. The PPO from HDPE has heating values around 41–46 MJ kg^{-1}. Similar values were found for PPOs from LDPE and PP. Furthermore, HHV of the PPO from polyolefin pyrolysis is similar to the HHV of conventional liquid fuels, which makes them an alternative to fossil fuels.

Another approach used to calculate the HHV of the PPO has been conducted using the percentage of C, H, and O obtained from the elemental analysis (Anuar Sharuddin et al., 2017; Kunwar et al., 2016; Sharma et al., 2014). The estimation of the HHV can be computed using Dulong's equation, as follows:

$$HHV\left(MJ\ kg^{-1}\right) = \frac{338.2C + 1442.8(H - O/8)}{1000} \tag{1}$$

where C, H, and O correspond to the weight percentage of carbon, hydrogen, and oxygen obtained in the elemental analysis, respectively. Anuar Sharuddin et al. (2017) found that the PPO from a mixture of HDPE, LDPE, and PP had an HHV of 39.7 MJ kg^{-1} which was improved by the addition of PS to the pyrolysis mixture. The PPO obtained from a mixture of HDPE, LDPE, PP, and PS had a maximum value of 43.94 MJ kg^{-1}. The HHV from feedstocks containing mixture of these four plastics can be compared with the HHV of diesel which corresponds to 44.94 MJ kg^{-1}. The HHV of PPO obtained by thermal pyrolysis of PE was estimated to be 49.3 MJ Kg^{-1}. Y-zeolite and $MgCO_3$ were used to pyrolyze HDPE from storage boxes to produce PPO in the gasoline and diesel range, respectively.

Table 3 Heating values for PPO and fuel samples.

PPO feedstock	Instrument	HV[a] ($MJ\,kg^{-1}$)	HHV ($MJ\,kg^{-1}$)	LHV ($MJ\,kg^{-1}$)	Reference
HDPE	Calorimetric bomb	43.65	n.a.	n.a.	Budsaereechai et al. (2019)
	Calorimetric bomb	n.a.	44.50	n.a.	Constantinescu et al. (2019)
	n.a.	n.a.	45.8	n.a.	Dobó et al. (2019)
	Calorimetric bomb	n.a.	n.a.	41.50	Joshi and Seay (2020)
	n.a.	n.a.	40.17	n.a.	Kumar et al. (2013)
	Calorimetric bomb	43.65	n.a.	n.a.	Budsaereechai et al. (2019)
LDPE	Calorimetric bomb	43.39	n.a.	n.a.	Budsaereechai et al. (2019)
	n.a.	At 250 °C 26.18 At 350 °C 45.84 At 450 °C 46.20	n.a.	n.a.	Cahyono and Fenti (2017)
	Calorimetric bomb	n.a.	44.23	n.a.	Constantinescu et al. (2019)
	n.a.	n.a.	45.3	n.a.	Dobó et al. (2019)
	Calorimetric bomb	n.a.	n.a.	44.94	Joshi and Seay (2020)
PE	Calorimetric bomb	n.a.	37.69	n.a.	Ghenai et al. (2019)

Continued

Table 3 Heating values for PPO and fuel samples.—cont'd

PPO feedstock	Instrument	HV[a] ($MJ\,kg^{-1}$)	HHV ($MJ\,kg^{-1}$)	LHV ($MJ\,kg^{-1}$)	Reference
PP	Calorimetric bomb	43.70	n.a.	n.a.	Budsaereechai et al. (2019)
	Calorimetric bomb	n.a.	44.63	n.a.	Constantinescu et al. (2019)
	n.a.	n.a.	47.1	n.a.	Dobó et al. (2019)
	Calorimetric bomb	n.a.	35.05	n.a.	Ghenai et al. (2019)
	Calorimetric bomb	n.a.	n.a.	40.03	Joshi and Seay (2020)
	Calorimetric bomb	44.95	n.a.	n.a.	Martynis et al. (2019)
PS	Calorimetric bomb	44.55	n.a.	n.a.	Budsaereechai et al. (2019)
	Calorimetric bomb	n.a.	40.18	n.a.	Constantinescu et al. (2019)
	n.a.	n.a.	41.5	n.a.	Dobó et al. (2019)
	Calorimetric bomb	n.a.	39.09	n.a.	Ghenai et al. (2019)
	Calorimetric bomb	n.a.	n.a.	39.14	Joshi and Seay (2020)
PU	n.a.	n.a.	27.9	n.a.	Dobó et al. (2019)
Plastic waste	n.a.	40.25	n.a.	n.a.	Khandal et al. (2021)
Plastic waste	Calorimetric bomb	n.a.	n.a.	38.3	Kalargaris et al. (2017)
Plastic waste (HDPE, PP, expanded PS, PET, PVC, PE film, PP film, paper, glass)	Calorimetric bomb	n.a.	41.2	n.a.	López et al. (2010)
Diesel	n.a.	n.a.	44.94	n.a.	Anuar Sharuddin et al. (2017)
Diesel	n.a.	n.a.	43	n.a.	Ahmad et al. (2015)
Gasoline	n.a.	n.a.	42.5	n.a.	Ahmad et al. (2015)

[a]Heating value.

The estimation of HHV for the HDPE oil using Y-zeolite and $MgCO_3$ was approximately 50 MJ kg^{-1} for both catalysts used. The HHV value of the catalytic pyrolysis oil was similar to the thermal pyrolysis oil (49.3 MJ kg^{-1}), in which diesel and vacuum gas oil fractions were obtained (Kunwar et al., 2016).

2.3 Distillation

Distillation is a technique used to separate the light and heavy fractions of hydrocarbons from PPOs. The distillation fractionates the oil by heating, evaporating, and condensing its fractions at different temperatures (Fahim et al., 2009). The distillation of oils can be conducted using atmospheric distillation, ASTM distillation, and simulated distillation by GC (Gala et al., 2020; Heydariaraghi et al., 2016; Miskolczi et al., 2009). Fig. 2 shows the principal distillate petroleum products with the carbon number, boiling points, and fractions found for the distillation of PPOs. The PPOs obtained from different polymer feedstocks present similar fractions. The PPO distillate fractions can be compared with naphtha, light gas oil, heavy gas oil, gasoline, and diesel fractions of crude and petroleum products, as shown in Fig. 2 (Heydariaraghi et al., 2016; Miskolczi et al., 2009).

Heydariaraghi et al. (2016) found that the proportions of light, medium, and heavy fractions in the PPO from polyolefins and plastic waste did not meet the standard limits established for gasoline and diesel. Light naphtha (<93 °C) and heavy naphtha (93–204 °C) are the main constituents of gasoline and light gas-oil (204–343 °C) is the main constituent of diesel while the main fraction for the PPOs was heavy naphtha (93–204 °C). Miskolczi et al. (2009) studied the thermal and catalytic pyrolysis of HDPE and PP in a continuous reactor with a separation unit at 520 °C. The distillation unit separates the heavy and light oils, gasoline, and gases from the pyrolysis of HDPE and PP. The thermal pyrolysis of HDPE yielded gases (5.1%), gasoline (18.2%), light oil (17.9%), and heavy oil (58.8%). The use of ZSM-5 in the pyrolysis of HDPE yielded 12.2% of gases, 34.5% of light gases, 24.1% of gasoline, and 29.2% of heavy oil. The thermal pyrolysis of PP yielded 4.5% gases, 21% light oil, 31.4% gasoline, and 43.1% heavy oil. The use of ZSM-5 reduced the yield of heavy oil to 4%, increasing the yield of gasoline, light oil, and gases to 51.2%, 27.2%, and 14.2%, respectively. An increase in the yield of gasoline, light oil, and gas products for both polymers produced a decrease in the heavy oil fraction when the catalyst was used, which was explained by the cracking reactions that occur in the microporous and macroporous surfaces of the catalyst. Singh et al. (2020) studied

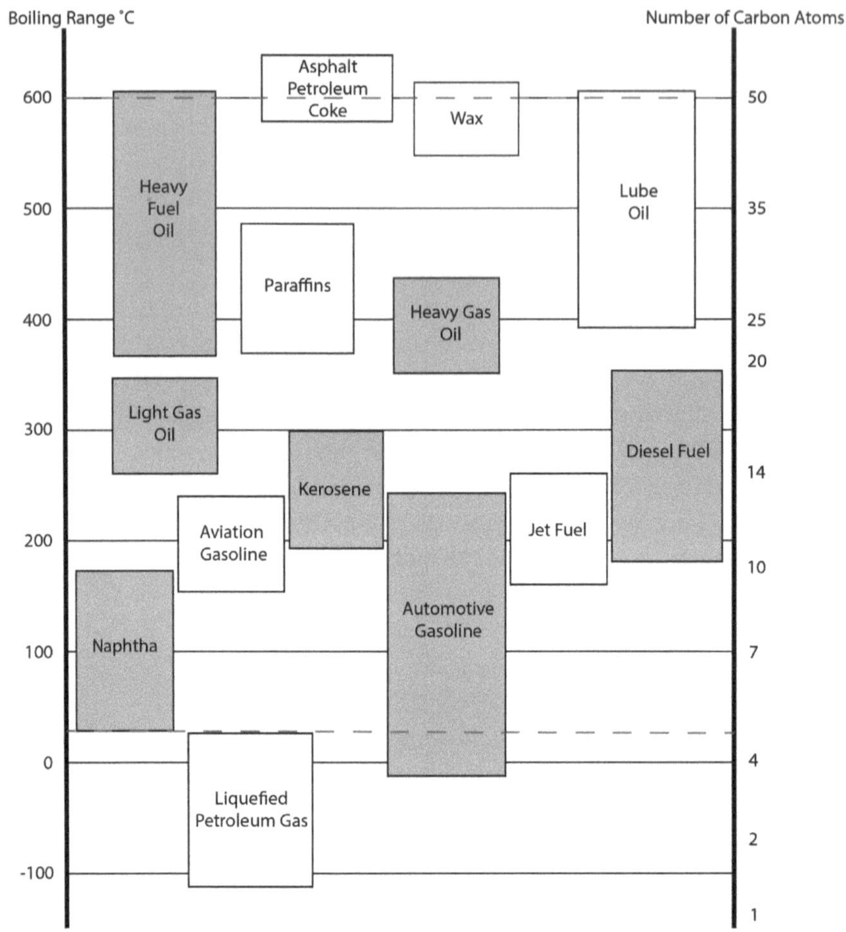

Fig. 2 Petroleum products with their carbon number and boiling point (Fahim et al., 2009; Wauquier, 1995).

the carbon distribution of PPO from HDPE, PP, PS, and their mixture and the effect of the pyrolysis temperature on the carbon fraction distribution. The authors found that the increase in reaction temperature from 450 to 550 °C during pyrolysis of HDPE led to a decrease in the amount of lower hydrocarbon fraction oil (C_5–C_{12}) from 45.06% to 28.75% and an increase in the high hydrocarbon fraction (>C_{19}) from 5.09% to 30.02%. The lower recovery of light hydrocarbons at higher pyrolysis temperatures was explained by the conversion of the light hydrocarbons to non-condensable gases. PPO from PP resulted in high fraction of C_5–C_{12}, (65.41–68.80%) for all the investigated pyrolysis temperatures which are 450, 500, and 550 °C.

The major carbon fraction for PPO from PS was C_5–C_{12} at 450 °C. However, the carbon fraction (C_5–C_{12}) decreased from 57.93% to 47.41% when the temperature was increased from 450 to 550 °C. The pyrolysis oil obtained from the real plastic waste mixture had a gasoline fraction (C_6–C_{12}) of 71.68% at 450 °C followed by a 70.32% at 500 °C and 68.75% at 550 °C. The lower hydrocarbons (C_5–C_{12}) in PPO showed the trend plastic mixture > PP > PS > HDPE, while for the heavier hydrocarbons (> C_{19}) it is HDPE > PP > PS > plastic mixture.

The PPOs from polyolefins contain more volatile components than conventional diesel oil, which means that the initial boiling point (IBP) of the PPOs is lower than diesel, as shown in Fig. 3. The final boiling point (FBP) for the PPOs from LDPE, PS, and PP is 368 °C, 359 °C, and 370 °C, respectively. The FBP of LDPE, PS, and PP oils are in the range of diesel fuels (320–370 °C). However, HDPE oil had a final boiling point of 398 °C, which is higher than the range for diesel fuels. A high FBP may produce

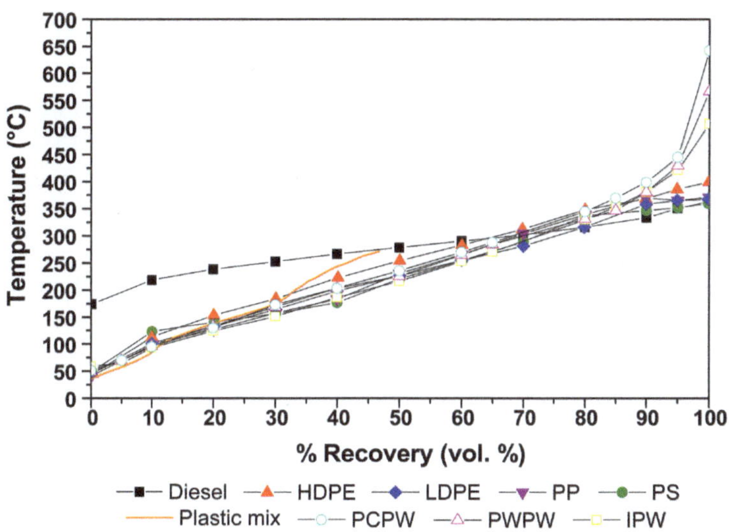

Fig. 3 Distillation curves for PPOs from HDPE, LDPE, PP, PS, plastic mix, PCPW, PWPW, and IPW. *Data from Kizza R, Banadda N, Seay J: Qualitative and energy recovery potential analysis: plastic-derived fuel oil versus conventional diesel oil,* Clean Techn Environ Policy *24(3):789–800, 2022. doi: 10.1007/s10098-021-02028-9; Gala A, Guerrero M, Guirao B, Domine ME, Serra JM: Characterization and distillation of pyrolysis liquids coming from polyolefins segregated of MSW for their use as automotive diesel fuel,* Energy Fuel *34(5):5969–5982, 2020. doi: 10.1021/acs.energyfuels.0c00403; Paradela F, Pinto F, Gulyurtlu I, Cabrita I, Lapa N: Study of the co-pyrolysis of biomass and plastic wastes,* Clean Techn Environ Policy *11(1):115–122, 2009. doi: 10.1007/s10098-008-0176-1.*

dilution of the lubricating oil produced by the incomplete volatilization of the less volatile fuel components. The higher FBP can be explained by the larger hydrocarbons present in the PPO from HDPE. PPOs obtained from LDPE, PS, and PP had the same hydrocarbon range as the diesel fuels (Kizza et al., 2022). The distillation curve for a mixture of plastics (PE, PP, and PS) showed that the IBP is lower than 50 °C, while recovery percentage of 50 vol% was obtained at around 275 °C (Paradela et al., 2009). Industrial plastic waste (IPW), post-consumer colored plastic waste (PCPW), and post-consumer white plastic waste (PWPW) had higher FBP in comparison with the PPO obtained from polyolefins, as shown in Fig. 3. The light range of the distillation curve (up to 65 vol%) was recovered below 250 °C for IPW, PCPW, and PWPW oils, which fulfills the requirement to use the PPO as diesel fuel. However, FBP for IPW, PCPW, and PWPW were higher than the FBP for diesel oil. IPW presented an FBP of 365 °C for 91.5 vol% of the liquid recovered while PCPW and PWPW had FBP over 420 °C for the 95 vol% of the liquid recovered in the distillation (Gala et al., 2020). Kumar et al. (2013) pyrolyzed HDPE using kaolin as catalyst with a heating rate of 20 °C min^{-1} until 450 °C. The authors found that the PPO had an IBP and FBP of 58 °C and 376 °C, respectively. It can be inferred by the boiling point range that the PPO was a mixture of different oil components, such as gasoline, kerosene, and diesel (Kumar et al., 2013).

2.4 Other techniques

The other techniques for physical characterization of PPO include density, viscosity, water content, flash point, and pour point, among others, as shown in Table 4. Density and viscosity are important properties to assess the use of PPO in diesel engines. It is important to mention that density and viscosity of PPOs are function of temperature. A decrease in the temperature will increase the viscosity and density, which affects the application of the PPO. For example, PPO with high viscosity can block the atomization system of an engine. Thus, their study is important to characterize and evaluate their use as an alternative to fuels (Miteva et al., 2016). The fuel density affects the fuel performance as well as engine properties such as cetane number, heating value, and viscosity (Eze et al., 2020). The density of PPO from polyolefins and their mixture range between 0.7 and 0.9 g cm^{-3} (Ahmad et al., 2015; Al-Salem et al., 2021; Anuar Sharuddin et al., 2017; Heydariaraghi et al., 2016; Kizza et al., 2022). The PPO density depends on the pyrolysis parameters such as temperature, catalyst, and type of reactor used as well as the method of evaluation. Al-Salem et al. (2021) studied the PPO properties of LDPE and LLDPE at different pyrolysis temperatures ranging between 500 °C and 800 °C using a heating rate of 5 °C min^{-1}. The density of the PPO from LDPE was relatively constant with values from 0.79 to 0.80 g cm^{-3} when the pyrolysis temperature increased from 500 °C to 800 °C. The density of PPO from LLDPE changed from 0.87 to 0.92 g cm^{-3} in the pyrolysis temperature range of 500–800 °C. The higher density of the LLDPE pyrolysis oil may be attributed to the nature of the polymer which contains less branches than LDPE. Kizza et al. (2022) determined the density of PPO from HDPE, LDPE, PP, and PS waste using a glass hydrometer according to the ASTM D1298 standard. The density of PPO produced from PP, LDPE, and HDPE was 0.778, 0.772, and 0.781 g cm^{-3}, respectively. The density of the PS oil was 0.783 g cm^{-3}. Park et al. (2020) pyrolyzed PS using two-stage pyrolysis which consists of an auger and fluidized bed reactors to produce BTEX. The temperature of the auger reactor was set between 15.2 °C and 300 °C, while the temperature of the fluidized bed reactor ranged between 505.1 °C and 782.2 °C. Lower temperatures in the auger reactor (15.2 °C) maximized the yield of styrene to 73.6 wt%. On the other hand, a temperature of 782.2 °C in the fluidized bed reactor increased the yield of BTEX in the PPO from PS with a maximum yield of 23.3 wt%. The density of the PPO from PS at 697.6 °C and 782.2 °C was 0.95 and 0.98 g cm^{-3}, respectively. Miandad et al. (2016) studied the influence of temperature and reaction time for pyrolysis of PS. Temperatures and reaction times between 400–500 °C and 60–120 min, respectively were investigated. The authors found that the temperature and reaction time of 450 °C and 75 min, respectively maximized the yield of oil to 80.8%. The density of the PPO at different temperatures and reaction time of 75 min was obtained. The density for the PPO from PS was 0.94, 0.92, and 0.93 g cm^{-3} at 400 °C, 450 °C, and 500 °C, respectively. The density of PPO from plastic waste varies from 0.79 to 0.98 g cm^{-1}, depending on factors such as the proportion of plastics and other components that are not plastics (e.g., biomass, and metals) (Baena-González et al., 2020; Eze et al., 2020; Heydariaraghi et al., 2016; Kalargaris et al., 2017). As can be observed from Table 4, the density of the PPOs from different feedstocks and mixed plastic waste had a range between 0.75 and 0.92 g cm^{-1}, which meet the density found for commercial gasoline (0.78 g cm^{-1}) and diesel (0.807 g cm^{-1}).

The determination of oil viscosity is fundamental because higher values of viscosity affect the fuel injection system and lead to poor engine performance (Eze et al., 2020). In addition, viscosity is affected by the temperature, and hence lower ambient temperatures and high viscosities produce viscous oil which affects the injection system (Miandad et al., 2016). Dynamic viscosity and kinematic viscosity have been reported for PPOs. Dynamic viscosity is defined as the ratio of shear resistance to the shear velocity, and it is expressed in centipoise (cP). On the other hand, kinematic viscosity is the ratio between dynamic viscosity and density. It is generally expressed in centistoke (cSt) (Wauquier, 1995). The PPO viscosity values depend on the type of feedstock, pyrolysis conditions, and the method of evaluation. The viscosity of PPO from polyolefins (e.g., HDPE, LDPE, and PP) ranged between 1 and 3 cSt at 40 °C (Khan et al., 2016; Kizza et al., 2022; Miandad et al., 2016; Olalo, 2021; Park et al., 2020). However, Ahmad et al. (2015) found that the viscosity of PPOs from PP and HDPE was 4.09 and 5.09 cSt at 40 °C. Cahyono and Fenti (2017) studied the PPO from LDPE at different pyrolysis temperatures. The viscosity decreased from 12.1 to 2.8 cSt when the pyrolysis temperature increased from 250 °C to 450 °C. The viscosity of PPO from PS ranged from 0.97 to 1.93 cSt (Kizza et al., 2022; Park et al., 2020). The viscosity of PPO from plastic waste was between 1.9 and 2.6 cSt according to the composition of the waste and pyrolysis conditions, as shown in Table 4. Singh et al. (2020) evaluated the effect of the pyrolysis temperature (450–550 °C) on mixed plastic waste. The composition of the mixed plastic was 58.6% of PE (HDPE and LDPE), 26.7% PP, 8.7% PS, 5.6% PET, and 0.4% of others (thermoset). The viscosity of mixed plastic waste was 1.91, 2.8, and 3.38 cSt at 450 °C, 500 °C, and 550 °C, respectively.

Determining the water content in oils has relevant importance since water is a critical factor for estimating the higher heating value (HHV) of the oil (Mohammed et al., 2015). The water content of pyrolysis oils has been determined using Karl-Fischer (KF) titration (Alvarez et al., 2017; Chen et al., 2016). KF titration has been reported as a fast, precise, and accurate analytical technique for the water determination (Smets et al., 2011). The technique is based on the Bunsen reaction between iodine and sulfur dioxide in an aqueous or organic media (Mohammed et al., 2015). KF titration has several advantages, such as covering a wide range of water content, determining the free and bond water in a sample, and allowing the use of different solvents and analysis temperatures for relatively unstable samples prone to degradation. However, certain chemicals, such as aldehydes and ketones, may interfere with the KF reagents due to the reaction with the alcohol used as a solvent and the iodine-reducing agents (Popescu et al., 2020). Anuar Sharuddin et al. (2017) reported the PPO water content using Karl Fischer. The water content of PPO from a mixture of HDPE, LDPE, and PP was 0.92 wt%. The addition of PS to the pyrolysis feedstock reduced the water content of the PPO to 0.31 wt%. Baena-González et al. (2020) found that the water content of the PPO from the rejected fraction of packaging waste was 1.53 wt%. The higher water content was explained by the presence of biomass residues in the pyrolysis feedstock. Kalargaris et al. (2017) used the method ASTM D6304 to measure the water content in plastic waste oil. The authors found that the PPO had a water content of $1190\,mg\,kg^{-1}$. On the other hand, Syamsiro et al. (2014) used the method ASTM D95 to determine the water content of the oil. The ASTM D95 covers the determination of the water content in a range from 0 to 25 vol % using a distillation system (ASTM, 2013). The water content for PE and HDPE waste oil was 0.1 and 0.5 vol%, respectively.

Flash, fire, and pour points are other important physicochemical properties that need to be determined for the PPOs because of their importance for safety considerations, storage, and transportation of the products (Riazi, 2005). The flashpoint measures the tendency of a petroleum material to form a flammable mixture when it is brought in contact with air (Wauquier, 1995). The fire point is defined as the temperature at which a fuel continues burning for at least 5 s after the source of ignition is removed (Eze et al., 2020). The pour point is the lowest temperature at which the oil can still flow while being cooled without stirring under standardized conditions (Riazi, 2005). In addition to flash, fire, and pour points, aniline point is another important physicochemical property. The aniline point is defined as the temperature that characterizes the level of aromatics in a non-formulated oil. Low aromatic content is represented by higher aniline values (Wauquier, 1995). Singh et al. (2020) studied the flashpoint, fire point, and pour point for PPO from HDPE, PP, PS, and plastic mixture containing 58.6% PE, 26.7% PP, 8.7% PS, 5.6% PET, and 0.4% others (thermoset) at different pyrolysis temperatures (450–550 °C). The PPO from PP had the lowest flashpoint and fire point when the pyrolysis temperature increased from 450 to 500 °C, in a range of 26–32 °C and 31–37 °C, respectively. The PS pyrolysis oil presented the highest flashpoint (55 °C), fire point (69 °C), and pour point (5 °C) at a pyrolysis temperature of 550 °C. The flashpoint and fire point for PPO from HDPE and plastic mixture were similar and were observed to be in the range of 42–50 °C and 47–56 °C.

Table 4 Physical characterization of PPOs, diesel, and gasoline.

PPO feedstock	Density (g cm^{-3})	Viscosity (cSt)	Water content (vol%)	Flashpoint (°C)	Fire point (°C)	Pour point (°C)	Aniline point (°C)	Cetane number	Reference
HDPE	0.89	5.08	n.a.	48	n.a.	−5	45	46	Ahmad et al. (2015)
	0.916	2.5[a]	n.a.	50	n.a.	24	n.a.	n.a.	Budsaereechai et al. (2019)
	0.781	1.93	<0.01	25	n.a.	−23	n.a.	n.a.	Kizza et al. (2022)
	0.7991	2.319	0.5	<10	n.a.	27	n.a.	n.a.	Syamsiro et al. (2014)
HDPE waste	0.761–0.765	1.72–2.41	n.a.	42–50	47–56	−18 to −7	n.a.	n.a.	Singh et al. (2020)
LDPE	0.911	2.5[a]	n.a.	45	n.a.	24	n.a.	n.a.	Budsaereechai et al. (2019)
	0.772	1.71	<0.01	25	n.a.	−21	n.a.	n.a.	Kizza et al. (2022)
	n.a.	1.7	n.a.	10	n.a.	n.a.	n.a.	n.a.	Yuliansyah et al. (2015)
PE	0.8544	1.739	0.1	<10	n.a.	24	n.a.	n.a.	Syamsiro et al. (2014)
	0.824	1.838	Trace	<10	n.a.	24	n.a.	n.a.	Syamsiro et al. (2014)
PP	0.86	4.09	n.a.	30	n.a.	−9	40	49.3	Ahmad et al. (2015)
	0.905	2.3[a]	n.a.	40	n.a.	15	n.a.	n.a.	Budsaereechai et al. (2019)
	0.778	1.93	<0.01	25	n.a.	−23	n.a.	n.a.	Kizza et al. (2022)
	0.768	1.64	n.a.	n.a.	n.a.	2.7	n.a.	n.a.	Uebe et al. (2022)
PP waste	0.753	1.66	n.a	n.a.	n.a.	10.7	n.a.	n.a.	Uebe et al. (2022)
	0.730–0.752	1.86–2.45	n.a.	26–32	31–37	−19 to −11	n.a.	n.a.	Singh et al. (2020)

PS	0.855	2[a]	n.a.	48	n.a.	19	n.a.	n.a.	Budsaereechai et al. (2019)
	0.783	1.93	<0.01	25	n.a.	−30	n.a.	n.a.	Kizza et al. (2022)
PS waste	0.864–0.884	2.98–4.43	n.a.	48–55	53–69	−1 to 5	n.a.	n.a.	Singh et al. (2020)
Plastic waste	0.9813	1.918	1190[b]	13	n.a.	n.a.	n.a.	n.a.	Kalargaris et al. (2017)
Plastic waste	0.8355	2.52	n.a.	40	43	n.a.	n.a.	46	Khandal et al. (2021)
Plastic waste	0.924	2.69[a]	n.a.	100	n.a.	n.a.	n.a.	n.a.	Wongkhorsub and Chindaprasert (2013)
Plastic waste (58.6% PE, 26.8% PP, 8.8% PS, 5.6% PET, 0.1% other plastics)	0.712–0.759	1.91–3.38	n.a.	42–47	47–50	−16 to −12	n.a.	n.a.	Singh et al. (2020)
Diesel	0.807	1.9–4.1	n.a.	52	n.a.	6	77.5	<40	Ahmad et al. (2015)
Gasoline	0.78	1.17	n.a.	42	n.a.	n.a.	71	n.a.	Ahmad et al. (2015)

[a]Dynamic viscosity (cP).
[b]$mg\ kg^{-1}$.

3. Composition analysis techniques

The product composition of SPW derived pyrolysis oil is quite complex in nature. Thus, it is imperative to study the different compounds present in PPO which may include aliphatics, aromatics, oxygenates, and some impurities. The impurities that can be found include traces of sulfur, oxygen, and nitrogen containing compounds. These impurities contaminate PPO and thus considered to be undesirable as they can cause problems in the feedstock processing thereby affecting the quality of the final products (Fahim et al., 2009). The techniques such as 1D-GC and GC × GC coupled with different detectors have been used to determine the composition of PPO. Detectors such as FID, MS, electron capture detector (ECD), nitrogen chemiluminescence detector (NCD), and atomic emission detector (AED) have been employed to analyze the hydrocarbon, nitrogen, oxygen, and halogen content in the PPOs. Besides, 1D-GC and GC × GC, other techniques widely used to characterize PPO include gel permeation chromatography (GPC), FTIR, and NMR spectroscopy.

3.1 1D-GC

1D-GC is an analytical technique used to separate and analyze compounds present in PPO generated from SPW. The separation of compounds in 1D-GC is achieved due to their relative vapor pressures and affinity toward the stationary phase (Bartle and Myers, 2002). For the composition analysis of PPO, nonpolar (Williams and Williams, 1999) and mid-polar (Baena-González et al., 2020) columns have been employed. Several detectors like MS, FID, ECD, and AED have been used for the qualitative and quantitative analysis of PPO as shown in Fig. 4 (Ahmad et al., 2015; Al-Salem et al., 2021; Fekhar et al., 2019; Jangid et al., 2022; Kumar et al., 2013).

The composition of PPOs has been studied using GC coupled to different detectors (Ahmad et al., 2015; Budsaereechai et al., 2019; Constantinescu et al., 2019; Das and Tiwari, 2018a,b; Dobó et al., 2019; Kumar et al., 2013; Kumar and Singh, 2011). Table 5 shows the operational conditions used to analyze PPO from several plastic feedstocks. Ahmad et al. (2015) used GC/MS to analyze the liquid product from the pyrolysis of HDPE and PP. A GC/MSQP2010 Shimadzu system equipped with an auto-injector and a DB5MS column (25 m × 0.5 mm i.d. 0.25 μm) was utilized to analyze the oil samples. Kumar et al. (2013) used GC/FID/MS to identify the chemical compounds of PPO from HDPE. The system employed for the analysis was an MS-QP 2010 Shimadzu coupled with FID and MS detectors, and a

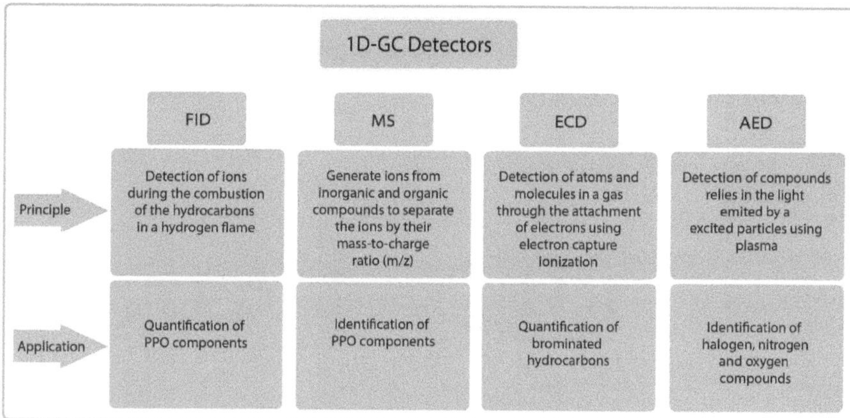

Fig. 4 Principle and application of GC detectors.

capillary DB-5 column (30 m × 0.25 mm i.d. 0.25 μm) was used. An Agilent 8860 GC coupled to Agilent 5977B MSD MS with HP5MS-UI column (30 m × 0.25 μm) was used to evaluate the chemical composition of PPO from LDPE and LLDPE (Al-Salem et al., 2021). In all three works, the authors used the NIST library to identify the oil products (Ahmad et al., 2015; Al-Salem et al., 2021; Kumar et al., 2013).

Tables 6 and 7 show the product distribution by carbon number and group type for PPOs, respectively. The GC/MS analysis of the PPO from HDPE showed a wide product distribution with carbon numbers from C_6 to C_{21} (Budsaereechai et al., 2019a; Chen et al., 2016; Ahmad et al., 2015). PPO obtained from thermal pyrolysis of HDPE was enriched in low boiling naphtha range hydrocarbons (C_6–C_{16}) with 63% of hydrocarbons in the gasoline and diesel range. Moreover, hydrocarbons from C_{17} to C_{30} constituted about 34% of the oil (Ahmad et al., 2015). Paraffins, olefins (Kumar et al., 2013; Kumar and Singh, 2011), naphthenes (Ahmad et al., 2015), and aromatics (Das and Tiwari, 2018a,b; Budsaereechai et al., 2019) have been identified in the HDPE oil. Kumar et al. (2013) obtained PPO from the catalytic pyrolysis of HDPE using modified kaolin as a catalyst. The authors found that the components present in waste HDPE liquid fuel were mostly aliphatic hydrocarbons (alkanes and alkenes) with carbon numbers between C_{10}–C_{25}. Nearly 25 compounds were observed for catalytic pyrolysis of waste HDPE at 450 °C. The major compounds with area% greater than 5% include tetradecane (5.79%), pentadecane (6.52%), n-hexadecane (7.44%), n-heptadecane (8.07%), n-octadecane (7.87%), nonadecane (6.7%), and eicosane (5.97%). The catalytic pyrolysis of waste plastics in presence of bentonite clay was studied by Budsaereechai and co-authors.

Table 5 Operational condition for GC analysis.

System	Feedstock	Column	Carrier gas	Carrier gas flow rate (ml min^{-1})	Split ratio	Injector temp. (°C)	Sample injection (μL)	Oven program	Reference
GC/MS	HDPE PP	DB–5 MS column (25 m × 0.25 mm i.d., 0.25 μm film thickness)	He	1.3	50	300	1	Initial temp.: 35 °C Ramp 1: 5 °C min^{-1} Temp. 1: 100 °C Hold time 1: 5 min Ramp 2: 10 °C min^{-1} Temp. 2: 150 °C Hold time: 10 min Ramp 3: 2.5 °C min^{-1} Temp. 3: 290 °C Hold time: 10 min	Ahmad et al. (2015)
GC/FID/MS	HDPE	DB–5 capillary column (30 m × 0.25 mm i.d., 0.25 μm film thickness)	He	1.51	1:10	200	n.a.	Initial temp.: 70 °C Hold time: 2 min Ramp 1: 10 °C min^{-1} Temp. 1: 300 °C Hold time 1: 7 min	Kumar et al. (2013)
GC/MS	LDPE LLDPE	HP5MS–UI column (30 m × 0.25 μm film thickness)	He	1.2	50:1	280	n.a.	n.a.	Al-Salem et al. (2021)

GC/MS	HDPE LDPE	HP–5MS (30 m × 0.25 mm i.d.)	He	n.a.	n.a.	n.a.	5	Initial temp.: 60 °C Ramp 1: 10 °C min^{-1} Temp. 1: 200 °C Ramp 2: 5 °C min^{-1} Temp. 2: 280 °C Hold time: 10 min Ramp 3: 2.5 °C min^{-1} Temp. 3: 290 °C Hold time 3: 10 min	Jangid et al. (2022)
GC/MS	Plastic waste (HDPE, LDPE, PP, and PS)	HP–5 (50 m × 0.32 mm i.d., 0.25 μm film thickness)	He	1.0	n.a.	n.a.	1	Initial temp.: 50 °C Hold time: 1 min Ramp 1: 7 °C min^{-1} Temp. 1: 270 °C Hold time 1: 10 min	Anuar Sharuddin et al. (2016)
GC/MS	Real waste sample	Phenomenex ZB–624 column (30 m × 0.25 mm i.d., 0.25 μm film thickness)	He	1	n.a.	225	n.a.	Initial temp.: 50 °C Hold time: 5 min Ramp 1: 20 °C min^{-1} Temp. 1: 220 °C	Baena-González et al. (2020)
GC/FID	Plastic mixture	Rtx PONA column (100 m × 0.25 mm i.d., 0.5 μm film thickness)	n.a.	n.a.	n.a.	350	n.a.	Initial temp.: 40 °C Hold time: 5 min Ramp 1: 10 °C min^{-1} Temp. 1: 350 °C Hold time 1: 20 min	Fekhar et al. (2019)

Continued

Table 5 Operational condition for GC analysis.—cont'd

System	Feedstock	Column	Carrier gas	Carrier gas flow rate (ml min^{-1})	Split ratio	Injector temp. (°C)	Sample injection (μL)	Oven program	Reference
GC/MS	Plastic mixture (PE, PP, PS, PET, PVC)	HP–5 MS	He	1	100:1	280	1	Initial temp.: 40 °C Hold time: 10 min Ramp 1: 8 °C min^{-1} Temp. 1: 280 °C Hold time 1: 10 min	López et al. (2011) Lopez-Urionabarrenechea et al. (2012)
GC/MS	HDPE LDPE PP PS	DB 35–MS capillary column (30 m × 0.25 mm i.d., 0.25 μm film thickness)	He	1	n.a.	n.a.	n.a.	Initial temp.: 70 °C Ramp 1: 6 °C min^{-1} Temp. 1: 260 °C	Mangesh et al. (2020)
GC/MS	PS, PE, PP and their mixture	HP–5 capillary column (30 m × 0.25 mm i.d., 0.25 μm film thickness)	n.a.	n.a.	n.a.	290 °C	n.a.	Initial temp.: 50 °C Hold time: 2 min Ramp 1: 5 °C min^{-1} Temp. 1: 290 °C Hold time 1: 10 min	Miandad et al. (2017) Miandad et al. (2016)
GC/MS	LDPE, HDPE, PET, PP, PS and their mixture	Cp–Sil low bleed ®—5 (30 m × 0.25 mm i.d., 0.25 μm film thickness)	He	1.0	20:1	250 °C	n.a.	Initial temp.: 50 °C Hold time: 2.0 min Ramp 1: 10 °C min^{-1} Temp. 1: 140 °C Ramp 2: 20 °C min^{-1} Temp. 2: 250 °C Hold time: 20 min	Siddiqui and Redhwi (2009)

GC/AED	WEEE	HP-1 (25 m × 0.32 mm i.d., 0.17 μm film thickness)	n.a.	n.a.	n.a.	n.a.	n.a.	n.a.	Vasile et al. (2006)
GC/ECD	Polymer mixture (PE, PP, and PS)	CP-SIL 5CB (15 m × 0.25 mm i.d., 0.25 mm film thickness)	n.a.	n.a.	n.a.	300	n.a.	Initial temp.: 100 °C Hold time: 1 min Ramp 1: 2 °C min^{-1} Temp. 1: 140 °C Ramp 2: 4 °C min^{-1} Temp. 2: 220 °C Ramp 3: 8 °C min^{-1} Temp. 2: 280 °C Hold time: 36.5 min	Bhaskar et al. (2007)

The pyrolysis of four plastics viz., PS, PP, LDPE, and HDPE was studied using a fixed bed reactor. It was observed that in the absence of catalyst, highly active pyrolysis oils were formed that underwent polymerization to form solid materials. On the other hand, in the presence of bentonite clay the plastics produced low viscous liquids which indicated that there was cracking of plastic wastes that led to production of low molecular weight liquid hydrocarbons. The major products from PS pyrolysis were styrene (20.12%), ethylbenzene (15.07%), benzene, 1,10-(1,3-propanediyl)bis-(11.17%), α-methylstyrene (10.38%), and toluene (7.93%). The pyrolysis of PP mainly produced 2,4-dimethyl-1-heptene (15.08%), 1-tricosene (14.98%), 3-octadecene, (*E*)-(7.78%), 1-pentadecene (7.76%), and 1-decene, 2,4-dimethyl-(4.33%). For LDPE and HDPE pyrolysis, the major pyrolysis products were aliphatic hydrocarbons with carbon number greater than C_{13} with area% of 62.08% (LDPE) and 62.73% (HDPE) (Budsaereechai et al., 2019).

The 1D-GC/MS analysis of PPO from LDPE and LLDPE showed that both oils have similar product distribution at different pyrolysis temperatures (500 °C, 600 °C, 700 °C, and 800 °C). The product distribution corresponded to diesel range hydrocarbons predominantly with carbon number range of C_{10}–C_{19}. The area% of C_{10}–C_{19} hydrocarbons for LDPE was in the range of 60–70% while for LLDPE, it was in the range of 65–70%. The oil yield was observed to be maximum at 600 °C with 29% and 27% for LDPE and LLDPE, respectively (Al-Salem et al., 2021). Dobó et al. (2019) analyzed the distillate obtained between 20 °C and 200 °C for LDPE oil, showing that the fraction contained mostly paraffins and olefins. The yield of transportation fuel obtained from LDPE is 695 g/kg_{oil} in temperature range of 20–200 °C.

The composition of PPO from PP has been investigated by several authors using 1D-GC (Budsaereechai et al., 2019; Constantinescu et al., 2019; Das and Tiwari, 2018a,b; Dobó et al., 2019; Mangesh et al., 2020; Miandad et al., 2017). Miandad et al. (2017) studied the pyrolysis of PP at 450 °C and pyrolysis time of 75 min. The authors reported that PP pyrolysis resulted in production of α-methylstyrene (4.9%), ethylbenzene (7.3%), benzene (7.8%), propenylbenzene (23.4%), xylene (7.8%), naphthalene (5.3%), methylnaphthalene (8.4%), phenylnapthalene (3.3%), and phenanthrene (7.6%). Budsaereechai et al., 2019 studied the pyrolysis of PP at 500 °C with pyrolysis time of 10 min. The product composition of PPO from PP included both aromatic and non-aromatic compounds. The carbon distribution of aromatic compounds include C_5–C_9 (6.26%), C_{10}–C_{13} (2.2%), and $>C_{13}$ (13.5%) while the non-aromatic compounds include C_5–C_9 (18.65%), C_{10}–C_{13} (7.76%), and $>C_{13}$ (61.63%). The product distribution of PP derived pyrolysis oil at 300 °C from Ahmad et al. (2015) showed that the hydrocarbon distribution of PP varied as follows: C_6–C_{12} (15.16%), C_{13}–C_{16} (33.04%), C_{17}–C_{20} (12.99%), C_{20}–C_{30} (24.97%), and >C_{30} (8.05%) (Ahmad et al., 2015). The products found in the PP oil were paraffins (66%), olefins (25%), naphthenes (7%) (Ahmad et al., 2015), and aromatics (~1–2%) (Das and Tiwari, 2018a), such as benzene (~1.44%) (Mangesh et al., 2020).

1D-GC was also used to study the composition of the PPO obtained through pyrolysis of plastic mixtures that are composed of different polyolefins such as LDPE, HPDE PP, PS, PET, and PVC (Adrados et al., 2012; Das and Tiwari, 2018a; Fekhar et al., 2019; López et al., 2011). The pyrolysis of mixture of LDPE (30 wt%), HDPE (30 wt%), and PP (40 wt%) were conducted at 350 °C, 375 °C, and 400 °C. The product composition varied in the range of 12.44–27.24, 60.39–69.04, 3.72–27.17 wt% for light (C_6–C_{11}), middle (C_{12}–C_{20}), and heavy (C_{21}–C_{32}) fractions, respectively in the temperature range of 350–400 °C (Das and Tiwari, 2018a).

Table 6 Distribution of PPO by carbon number.

PPO feedstock	Pyrolysis	Distribution by carbon number	Reference
HDPE	Thermal	At 350 °C C6–C12: 32.56% age C13–C16: 30.80% age C17–C20: 14.19% age C20–C30 12.35% age >C30: 10.95% age	Ahmad et al. (2015)
LDPE	Thermal	At 500 °C C4–C9: 5.5 area% C10–C19: 61.7 area% >C19: 37.1 area% At 600 °C C4–C9: 8.9 area% C10–C19: 61.7 area% >C19: 22.7 area% At 700 °C C4–C9: 6.3 area% C10–C19: 69.6 area% >C19: 28.6 area% At 800 °C C4–C9: 9.2 area% C10–C19: 67.2 area% >C19: 27.7 area%	Al-Salem et al. (2021)

Continued

Table 6 Distribution of PPO by carbon number.—cont'd

PPO feedstock	Pyrolysis	Distribution by carbon number	Reference
LLDPE	Thermal	At 500 °C C4–C9: 5.2 area% C10–C19: 65.9 area% >C19: 30.1 area% At 600 °C C4–C9: 9.2 area% C10–C19: 65.9 area% >C19: 20.7 area% At 700 °C C4–C9: 5 area% C10–C19: 63.9 area% >C19: 32.1 area% At 800 °C C4–C9: 6.8 area% C10–C19: 68.7 area% >C19: 26.2 area%	Al-Salem et al. (2021)
PP	Thermal	At 300 °C C6–C12: 15.16% age C13–C16: 33.0% age C17–C20: 12.99% age C20–C30 24.97% age >C30: 8.05% age	Ahmad et al. (2015)
PS	Thermal	At 500 °C Aromatics: C5–C9: 60.22 area% C10–C13: 1 area% >C13: 34.77 area% Non-aromatic: C5–C9: 0.44 area% C10–C13: 0 area% >C13: 3.57 area%	Budsaereechai et al. (2019)
PS	Catalytic (Bentonite clay)	At 500 °C Aromatics: C5–C9: 61.8 area% C10–C13: 1 area% >C13: 31.53 area% Non-aromatic C5–C9: 1.54 area% C10–C13: 0.43 area%% >C13: 3.5 area%%	Budsaereechai et al. (2019)

Table 6 Distribution of PPO by carbon number.—cont'd

PPO feedstock	Pyrolysis	Distribution by carbon number	Reference
Plastic waste (HDPE, PP, expanded PS, PET, PVC, PE film, PP film, paper, glass)	Thermal	C5–C9: Aromatics: 71.0 area% Non-aromatics: 8.7 area% C10–C13: Aromatics: 1.6 area% Non-aromatics: 5.9 area% >C13: Aromatics: 2.3 area% Non-aromatics: 6.4 area%	López et al. (2010)
Plastic waste (40 wt% PE, 35 wt % PP, 18 wt% PS, 4 wt% PET, and 3 wt% PVC)	Thermal	C5–C9: Aromatics: 68.0 area% Non-aromatics: 10.1 area% C10–C13: Aromatics: 1.2 area% Non-aromatics: 6.3 area% >C13: Aromatics: 2.5 area% Non-aromatics: 6.0 area%	López et al. (2011)
Plastic waste (PE, PP, Ps, PET, and PVC)	Thermal	C5–C9: Aromatics: 80.7 area% Non-aromatics: 1.2 area% C10–C13: Aromatics: 10.6 area% Non-aromatics: 3.3 area% >C13: Aromatics: 2 area% Non-aromatics: 2.1 area%	Adrados et al. (2012)

The pyrolysis of mixture of PS with PP in 1:1 ratio resulted in production of nearly the same compounds, as were produced from pyrolysis of virgin PS. The produced compounds in liquid oil were styrene (24.5%), methylstyrene (52.0%), benzene (10.6%), ethylbenzene (6.6%), and propylbenzene (6.3%) (Miandad et al., 2017). Furthermore, the pyrolysis of mixture of plastics viz., LDPE (35%), HDPE (32%), PP (24%), PVC (4%), ethylene-propylene dimer (3%), and polystyrene (2%) was studied using mixture of four catalysts viz., Ni/ZSM-5, Ni/SAPO-11, red mud, and $Ca(OH)_2$ in varying proportions. In the absence of catalyst, the concentrations of n-paraffin, n-olefin, branched hydrocarbons (saturated and unsaturated), and aromatic were 30.8%, 27.0%, 40.3%, and 1.9%, respectively. The aromatic content increased to 11.4–17.6% when Ni/ZSM-5 containing catalysts were used while it increased to 8.1–11% when Ni/SAPO-11 containing catalysts were used. These aromatics included benzene, toluene, xylenes, and short chain substituted aromatic hydrocarbons. In the absence of catalyst, the ratio of branched and non-branched aliphatic hydrocarbons was 0.70 in pyrolysis oil while it increased to 1.31–2.17 in presence of Ni/ZSM-5 containing catalysts and 0.97–1.75 in presence of Ni/SAPO-11 containing catalysts. The ratio of n-paraffin to n-olefin was observed to be 1.09–1.18 in presence of Ni/ZSM-5 containing catalysts, while it was 1.31–1.43 using Ni/SAPO-11 containing catalysts. The higher ratio of n-paraffin to n-olefin using Ni/SAPO-11 containing catalysts can be attributed to local hydrogenation reaction. The hydrogen source is essentially from cyclization and aromatization reactions of aliphatics and isomerization reactions when non-branched hydrocarbons transform to branched hydrocarbons (Fekhar et al., 2019).

The composition of the PPO generated from real sample, simulated sample, and simulated sample with red mud was studied by Adrados et al. (2012). The real sample consisted of few plastics, metals, glass, paper, and wood. The simulated sample consisted of mixture of plastics viz., PE, PP, PS, PET, and PVC. Interestingly, the aromatics were high in all three samples (>70%) even though both real and simulated sample contained >70% polyolefins. The total aromatic content in PPO was observed to be 93.4%, 73.9%, and 89.6% for real sample, simulated sample, and simulated sample with red mud, respectively. The use of red mud showed catalytic effect which can be accounted for aromatization reactions. Besides, the high aromatic content in real sample was due to the catalytic role of metals and glasses. The presence of red mud and inorganics in real plastic samples can promote cracking which causes generation of free radicals. For stabilization, the free radicals can essentially cycle or capture H^+ from other molecules, thus paving way for cyclization and dehydrogenation reactions to form aromatics.

The pyrolysis of industrial plastic waste comprising of plastics (82.23%) and other materials (17.77%) like paper, aluminum, wood, and textile was studied by Baena-González et al. (2020). The GC/MS analysis showed that aromatics (75.68%) constituted the largest chemical group in the pyrolysis oil. Styrene (33.37%) was the major aromatics found in the pyrolysis oil. The high aromatic content in the PPO was primarily observed due to two reasons. The first reason is the presence of PS (6.22 wt%) in the plastic waste which upon cracking results in direct styrene production besides other aromatics. The second reason is that aromatics can also be produced from PP (30.28 wt%) and film bags (PE or PP, 40.72 wt%) via two reaction pathways. The first pathway is that aromatics can be produced from Diels-Alder reaction followed by dehydrogenation. The second pathway is through formation of cyclic compounds via intramolecular cyclization reactions which undergo dehydrogenation to produce aromatics. Besides aromatics, the PPO from industrial plastic waste also constituted 16.97% olefins (C_7–C_{23}). The main compound of the olefinic fraction was 2,4-dimethyl-1-heptene (4.15%), which is formed by the decomposition of PP.

GC/ECD has been used to analyze polybrominated diphenyl ethers (PBDEs) from the pyrolysis oil obtained from the mixture of PE, PP, PS, and high impact polystyrene (HIPS) with decabromo diphenylethane as flame retardant. Bhaskar et al. (2007) reported that ECD detectors were quite effective in identifying the presence of PBDEs due to their excellent capability to respond to halogen, oxygen, and nitrogen containing compounds in the pyrolysis oil. The detection of PBDEs by ECD was possible as no clean-up was conducted on the oil. Besides GC/ECD, another technique used for the detection of halogen, nitrogen, and oxygen containing compounds is GC/AED. Vasile et al. (2006) analyzed the PPO obtained from the pyrolysis of thermoplastic and thermoset fractions from waste electric and electronic equipment (WEEE) using GC/AED. It was found that the PPO from printed circuit boards (PCBs) had the highest amount of oxygen compounds. The main products found in the PPO were acetone, phenol, and phenol derivatives. The oil from keyboards constituted a higher amount of light aliphatic nitriles. Furthermore, the PCB oil comprised of a higher concentration of halogen compounds while casing and keyboard oils contained relatively low amount of halogens.

Table 7 Product distribution by group type for PPOs.

PPO feedstock	Pyrolysis	Group type	Reference
HDPE	Thermal	Paraffins: 59.7% age Olefins: 31.9% age Naphthenes: 8.4% age	Ahmad et al. (2015)
PP	Thermal	Paraffins: 66.5% age Olefins: 25.87% age Naphthenes: 7.58% age	Ahmad et al. (2015)
Plastic waste (40 wt% PE, 35 wt% PP, 18 wt% PS, 4 wt% PET, and 3 wt% PVC)	Thermal	Naphthenes: 0.9 area% Lineal olefins: 14.4 area% Branched olefins: 7.0 area% Mono-aromatics: 68 area% PAH and derivatives: 3.7 area%	López et al. (2011)
Plastic waste (35% LDPE, 32% HDPE, 24% PP, 4% PVC, 3% ethylene-propylene dimer, and 2% PS)	Thermal	n-Paraffin: 30.8% n-Olefins: 26.9% Branched: 40.3% Aromatics: 2.0%	Fekhar et al. (2019)
Plastic waste	Catalytic	Aromatics: 0.80 area% Cycloalkanes: 3.6% Iso-alkanes: 27.3 area% n-alkanes: 24.9 area% Olefins: 9.0 area% Higher alkanes: 2.8 area% Others: 32.9 area%	Lee et al. (2015)

3.2 GC × GC

The composition of PPO can be analyzed in much more detail using a comprehensive GC × GC which is a powerful analytical technique that allows superior separation of the peaks compared to 1D-GC. The separation of the compounds in PPO is enhanced using GC × GC consisting of two columns connected in series by a modulator. The modulator is the heart of GC × GC and allows the flow of the analytes from the first-dimension column to the second-dimension column for further separation. The compounds in first column are usually separated by difference in boiling point while the compounds in second column are separated based on difference in polarity. This column combination is referred as normal phase column combination. Table 8 shows the different column combinations used in GC × GC for PPO analysis. GC × GC systems are coupled with detectors such as FID,

Table 8 Column combination used in GC × GC for PPO analysis.

PPO feedstock	Configuration	First dimension column	Second dimension column	Reference
Plastic waste	Normal	RTX–1 PONA (50 m × 0.25 mm × 0.25 μm)	BPX–50 (2 m × 0.18 mm × 0.18 μm)	Dao Thi et al. (2021)
	Reverse	Stabilwax (30 m × 0.25 mm × 0.25 μm)	Rxi–5 ms (2 m × 0.1 mm × 0.1 μm)	
	Reverse	IL–59 (60 m × 0.25 mm × 0.25 μm)	Rxi–5 ms (2 m × 0.1 mm × 0.1 μm)	
	Reverse	BPX–50 (60 m × 0.25 mm × 0.25 μm)	Rxi–5 ms (2 m × 0.1 mm × 0.1 μm)	
Plastic waste	Normal	MXT–1/RTX–1 PONA (60 m × 0.25 mm × 0.25 μm/ 50 m × 0.25 mm × 0.5 μm)	BPX–50 (2 m × 0.18 mm × 0.15 μm)	Toraman et al. (2014)
	Reverse	BPX–50 (60 m × 0.25 mm × 0.25 μm)	Rxi–5 ms (2 m × 0.1 mm × 0.1 μm	
	Reverse	Stabilwax (60 m × 0.25 mm × 0.25 μm)	Rxi–5 ms (2 m × 0.1 mm × 0.1 μm	
	Reverse	IL–59 (50 m × 0.25 mm × 0.25 μm)	Rxi–5 ms (2 m × 0.1 mm × 0.1 μm	

MS, sulfur chemiluminescence detector (SCD), and nitrogen chemiluminescence detector (NCD) for the identification and quantification of hydrocarbons, oxygenated compounds, sulfur containing compounds, and nitrogen containing compounds (Dao Thi et al., 2021; Kusenberg et al., 2022; Toraman et al., 2014).

The evaluation of the composition of PPO from SPW has been studied less extensively using GC × GC compared to 1D-GC. Toraman et al. (2014) provided the first comprehensive compositional analysis of PPO using GC × GC coupled to four different detectors viz., FID, SCD, NCD, and TOF-MS. The pyrolysis oil was derived from SPW comprising of PP, PE, PS, PA, etc., besides trace amounts of food residuals. It was found that the identification and quantification of the PPO compounds from SPW required the use of FID, SCD, NCD, and TOF-MS detectors due to the complexity of the oil matrix. The authors used 3-chlorothiophene for FID and SCD detector while 2-chloropyridine for NCD detector as internal standards. A normal column combination, either RTX-1 PONA or MXT-1/BPX-50, provided an adequate peak resolution for the nitrogen and sulfur containing compounds. It was found that the pyrolysis oil contained aromatics, paraffins, isoparaffins, olefins, and naphthenes. The separation of the olefins and naphthenes was improved using a more polar second-dimension column such as an IL-50. The major composition of PPO included monoaromatics (51.3 wt%), diaromatics (7.5 wt%), and naphteno (di/tri)aromatics (5.9 wt%). It was observed that the PPO also contained nitrogen containing compounds in significant amounts (6.4 wt%) while sulfur containing compounds were only 0.6 wt%. Nearly 2.5 wt% of oxygen containing compounds were identified which included phenols, ketones, ethers, nitrogenated oxygen containing compounds, aldehydes, and acids in 1, 1, 0.2, 0.1, 0.05, and 0.05 wt%, respectively.

Dao Thi et al. (2021) studied light and heavy fractions of PPO from plastic waste. The authors found that the ramp temperature has an important role in the separation of isoparaffins and isoolefins. The resolution of the peaks improved when the oven ramp was decreased from 3 °C min^{-1} to 2 °C min^{-1}. The reverse column combination (Stabilwax/Rxi-5 ms) improved the resolution of the peaks in the diesel fraction (C_7–C_{23}) and helped in identification of oxygen containing compounds while the separation of the naphtha fraction (C_5–C_{11}) was enhanced using a normal-phase column combination (RTX-1 PONA and BPX-50). The major compounds of diesel fraction were n-paraffins, α-olefins, mononaphthenes + unsaturated mononaphthenes in 28.36, 36.39, and 14.62 wt%, respectively.

Besides, the diesel fraction also consisted of isoparaffins, isoolefins, diolefins, and dinaphthenes in 4.05, 8.77, 4.21, and 2.30 wt%, respectively. On the other hand, the naphtha fraction constituted 15.22, 35.08, 19.87 wt% of n-paraffins, α-olefins, and mononaphthenes, respectively as the prominent chemical groups.

Kusenberg et al. (2022) performed a comprehensive study of the PPO from post-consumer plastic packaging waste fractions, mixed polyolefins (MPO), PE, and PP. The authors used GC × GC coupled with FID to quantify the hydrocarbon composition of the PPO. In addition, a GC × GC/quadrupole MS (qMS) was used to identify the products in the pyrolysis oil. The nitrogen and sulfur containing compounds were analyzed using NCD and SCD detectors, respectively. The GC × GC was connected to two detectors viz., FID/SCD and qMS/NCD. It was observed that more than 1500 peaks were separated and identified for MPO. The PPO from the pyrolysis of PE-rich waste produced linear paraffins (~34 wt%) and olefins (~26 wt%). The oil from the PP-rich waste produced substantial amount of branched olefins (~63 wt%) and diolefins (~20 wt%) and only a few peaks of linear paraffins and olefins. GC × GC/NCD detected traces of nitrogen containing compounds in the PPO from the plastic feedstocks. The nitrogen containing compounds can be found in the PPOs due to contamination of the feedstocks with nitrogen-containing polymers, such as PA or polyurethane (PUR). The nitrogen analysis of the PPO was conducted using an elemental analyzer. However, no nitrogen compounds were detected in the PPOs using the elemental analyzer. Furthermore, even though no sulfur was detected in the plastic feedstock, the authors detected traces of sulfur containing compounds in the PPOs using GC × GC/SCD.

3.3 FTIR

FTIR is a technique used to obtain the infrared spectrum of absorption for the identification of characteristic functional groups from a PPO sample (Sindhu et al., 2015). Table 9 displays FTIR bands that may be found in PPOs. The technique has been used to analyze the functional groups of the PPO from polyolefins, PS, plastic mixture, and real samples. Moreover, FTIR is typically examined together with GC to confirm and understand the composition of PPOs. The FTIR spectra for the PPO from polyolefins and their mixture (e.g., HDPE-PPO, LDPE-PPO, and PP-PPO) are similar. Peaks around 3000 to 2840 cm^{-1} demonstrated the presence of aliphatic C—H stretching. The peaks around 1675–1540 cm^{-1} and 875–950 cm^{-1} indicated the presence of C=C stretches (Das and Tiwari, 2018a). The C—H bending vibrations in methyl groups can be found between 1460 and 1350 cm^{-1} (Heydariaraghi et al., 2016). The presence of alkenes can be detected by C—H bending vibrations at 966 cm^{-1}, C—H stretching vibrations at 910 and 990 cm^{-1}, and the vinylidene at 890 cm^{-1} (Kumar et al., 2013; Miskolczi et al., 2009). The phenyl rings substitution bands can be identified by the C—H bending vibration at 723 cm^{-1} (Kumar et al., 2013). A comparison of the polyolefin oils and commercial diesel showed that both have several functional groups in common which indicates the presence of alkanes and alkenes (Miteva et al., 2016). The FTIR spectra of pyrolysis oils obtained from PP, LDPE, and HDPE both in absence and presence of catalyst (bentonite clay) look similar whereas for PS, it looks different. The spectrum showed the presence of =C—H stretches and aromatics between 3500 and 3000 cm^{-1}. The peaks between 3000 and 2800 cm^{-1} are evident for PP, LDPE, and HDPE which shows the presence of $—CH_3$, $—CH_2$, and C—H groups illustrating the strong aliphatic nature of the oil obtained from these plastics (Budsaereechai et al., 2019).

Table 9 FTIR bands found in FTIR of PPO (Budsaereechai et al., 2019; Kumar and Singh, 2011).

Wave number	Type of vibration	Functional group
>3000	=C—H stretching	Aromatics
2800–3000	C—H stretching	Alkane
~1640	C=C stretching	Alkene/fingerprint region for phenyl ring substitution overtone
~1600	C—C stretching in ring	Aromatics
1490–1500	C—C stretching in ring	Aromatics
1450–1460	C=C stretching	Alkene/fingerprint region for phenyl ring substitution overtone
1370–1380	C—H scissoring and bending	Alkane
~900	=C—H bending	Alkene
870–890	C—H out of plane bending	Alkene
770–780	C—H “oop”	Aromatics
720–730	C—H rock	Alkane
690–700	C—H “oop”	Aromatics

3.4 Other techniques

The molecular weight distribution of oil derived from SPW can also be determined using GPC. The mechanism of separation of components is based on hydrodynamic volume or size of analytes. The accessibility of pore volume is higher for molecules with smaller size in comparison to larger size molecules. As a result, the elution time of smaller molecules is higher than larger molecules. The drawback of this technique is the presence of secondary separation effects that may affect the elution times. A secondary separation effect is the formation of hydrogen bond between the mobile phase and the sample which can lead to lower elution times. On the other hand, the interaction between the stationary phase and the solute may increase the elution time (Hoekstra et al., 2011). The system consists of an autosampler, a pump, a column, and a detector (Harman-Ware et al., 2020). The various detectors used in GPC include viscosimeters, ultraviolet detectors, refractive index detectors, and light scattering detector (low angle, LALS; dual angle, DALS; multi angle, MALS) (Polymer Char, 2021).

Several authors have used GPC to determine the molecular weight of PPO (Arabiourrutia et al., 2012; Çit et al., 2010; Miskolczi et al., 2008; Williams and Williams, 1997). Arabiourrutia et al. (2012) used GPC to characterize the PPO from polyolefins (HDPE, LDPE, and PP) obtained using a conical spouted bed reactor. The authors used a high-performance liquid chromatograph (HPLC) with a Tosoh-Hass column filled with a mixture of PS/divinylbenzene of 5 μm particle size at 40 °C. The column length was 30 cm and 7.8 mm of internal diameter. Tetrahydrofurane (THF) was used as mobile phase and the detector employed was a diode array detector (DAD) in the ultraviolet-visible range. The calibration of the column was conducted using a kit of PS patterns with weight-average molecular weight between 500 and 1.09×10^6. The authors studied the effect of three pyrolysis temperatures (450 °C, 500 °C, and 600 °C) on the molecular weight of the PPO. An increase in the temperature produced lighter waxes. Therefore, lower pyrolysis temperatures produce PPO with wider molecular weight distribution and a higher value of polydispersity. The mass average molecular weight (M_w) of HDPE, LDPE, and PP at 450 °C was 1721, 1734, and 1609, respectively while at 600 °C, M_w of HDPE, LDPE, and PP decreased to 1310, 1460, and 1279, respectively. The M_w decreased with pyrolysis temperature, which can be explained by the production of shorter chain compounds of low molecular weight due to tertiary cracking reactions. Similarly, there was a decrease in polydispersity (PD) with temperature for the waxes. The PD values decreased from 1.58–1.55 at 450 °C to 1.46–1.44 at 600 °C. This implies that there would be formation of more uniformly sized products at higher temperatures which results from cracking of waxes to produce short chain products with narrow range molecular weight.

Miskolczi et al. (2008) used GPC to analyze the heavy oil fraction obtained through the pyrolysis of flame retarded high impact polystyrene (HIPS) in a tube reactor between 360 °C and 440 °C. The analysis was conducted in a Waters type chromatography system with THF as mobile phase. The detector used was a Waters 2414 type refraction index. The calibration of the system was carried out using PS standards. The authors observed that higher pyrolysis temperatures decreased the average molecular weight, favoring the secondary reactions. The M_w of HIPS decreased from 735 at 360 °C to 635 at 440 °C. Çit et al. (2010) performed a comparative study of the pyrolysis of PP, LDPE, and PET and characterized the PPO using GPC. The samples were dissolved in THF, which also was used as mobile phase. The temperature of the column oven was 30 °C, and a refractive index detector was used. PET-PPO presented the lowest mass average compared to the PPO from polyolefins which was explained by the formation of smaller molecules due to the degradation of PET to gasses. The M_w of PET was found to be 226 g/mol while M_w of LDPE and PP was 277 and 280 g/mol, respectively.

Another technique used to identify and analyze organic structures is proton nuclear magnetic resonance spectroscopy (^{1}H NMR). ^{1}H NMR provides information about the number of protons in the PPO. It is based on magnetic nuclear spin of ^{1}H and gives information about aromatic structures and the existence of methyl groups attached to CH_2 or CH_3 groups (Balci, 2005). used ^{1}H NMR for the quantification and determination of the research octane number (RON) of PPO obtained from polyolefins and their mixture (HDPE, LDPE, and PP). Eqs. (2–5) shows the correlations to obtain the volume% of aromatics, paraffins, olefins, and H/C. Eq. (6) shows the isoparaffin index, which is a measure of the amount of branching and is calculated by the ratio CH_3/CH_2 (Myers et al., 1975b). The research octane number (RON) is presented in Eq. (7). The RON study the correlation between the amount of branching in the paraffinic hydrocarbon and the octane number (Albahri, 2003).

$$Aromatics\,(vol.\%) = \frac{(A + C/3)\cdot 0.97 \times 10^2}{(A + C/3)\cdot 0.97 + (D - 2B + E/2 + F/3)\cdot 1.02 + 3.33B} \quad (2)$$

$$Paraffins\,(vol.\%) = \frac{(D - 2B + E/2 + F/3)\cdot 1.02 \times 10^2}{(A + C/3)\cdot 0.97 + (D - 2B + E/2 + F/3)\cdot 1.02 + 3.33B} \quad (3)$$

$$Olefins\,(vol.\%) = \frac{3.33B \times 10^2}{(A + C/3)\cdot 0.97 + (D - 2B + E/2 + F/3)\cdot 1.02 + 3.33B} \quad (4)$$

$$H/C = \frac{A + B + C + D + E + F}{(A + C/3)\cdot 1.28 + (D - 2B + E/2 + F/3)\cdot 1.02 + 3.42B} \quad (5)$$

$$Isoparaffin\ index = \frac{CH_3}{CH_2} = \frac{(F/3)}{(E/2)} \quad (6)$$

$$RON = 80.2 + 8.9\cdot Isoparaffin\ index + 0.107\cdot Aromatics\,(vol.\%) \quad (7)$$

where the values A, B, C, D, E, and F represent the integrated area of the spectral regions, parts per million (ppm) from tetramethylsilane reference peaks, as shown in Table 10. Das and Tiwari (2018a,b) suggested that the hydrocarbon composition of the PPO from polyolefins and their mixture is mainly dominated by paraffinic and olefinic groups at different pyrolysis temperatures. The aromatics, paraffins, and olefins calculated using Eqs. (2)–(4) for PPO obtained from LDPE were 2.33, 65.58, 32.09 v/v%, respectively at a pyrolysis time of 437 min. The aromatics, paraffins, and olefins for PPO from PP were 2.82, 49.49, and 47.69 v/v%, respectively at a pyrolysis time of 390 min.

Table 10 NMR spectral region (Das and Tiwari, 2018a; Myers et al., 1975a,b).

	Proton type	Chemical shift region
A	Ring aromatic	6.6–8.0
B	Olefin	4.5–6.0
C	α-methyl	2.0–3.0
D	Methane (paraffins)	1.5–2.0
E	Methylene (paraffins)	1.0–1.5
F	Methyl (paraffins)	0.6–1.0

Siddiqui and Redhwi (2009) used ^{13}C NMR and ^{1}H NMR to study the chemical content of the PPO from different plastics viz., LDPE, HDPE, PET, PP, and PS. The PPO obtained from the catalytic pyrolysis of PET-PS showed very high aromatic hydrogens (82.64%) compared with HDPE-PS (56.5%), and LDPE-PS (65.79%). The higher percentage of aromatic hydrogens implies greater extent of interaction between PS and other plastics. In addition, the ^{13}C NMR spectra showed that the aromatic carbon content decreased in PPO produced through catalytic pyrolysis of mixture of PP and PS (75.2%) compared to the one from thermal pyrolysis (85.6%). This implies that there is strong interaction between PS and PP that paves way for breakage or isomerization of aromatic ring of PS.

4. Conclusions

The utilization of PPOs has aroused interest since pyrolysis has become an attractive thermochemical technique to recycle SPW. Therefore, the study of their physicochemical properties and composition is an important step to understand how the pyrolysis conditions and feedstock properties affect the PPO quality and properties. The physicochemical characterization techniques include elemental analysis, distillation, and the determination of calorific value, water content, viscosity, and density. These characterization techniques are important to evaluate the suitability of the PPO to be used in applications related to commercial automobile fuel, diesel engines, furnaces, boiler applications, etc. One of the techniques widely used to conduct physicochemical characterization is elemental analysis. The elemental analysis can also be used to calculate the HHV through Dulong's equation. Viscosity, water content, and density are obtained by using ASTM methodology, which is also used to calculate the physicochemical properties of fuel and diesel.

The chemical analysis of the PPOs has been investigated using different techniques. FTIR, NMR, and GPC are techniques that provide information about the chemistry of the PPO. A widely used technique is 1D-GC coupled to different detectors. The FID is extensively used for the quantification of compounds present in PPOs such as paraffins, olefins, oxygenates, cyclic compounds, and aromatics. Other detectors such as ECD and AED have been used specifically to identify halogenated compounds like nitrogen, sulfur, and bromine containing compounds present in PPOs. 1D-GC is a separation technique that is commonly used for the composition analysis of PPOs. However, 1D-GC can show poor resolution of the PPO products because of the complexity of the PPO and the use of only one dimension to separate the analytes. Thus, comprehensive GC × GC has been introduced to analyze PPO in great detail. It allows for separation of the compounds by the use of two columns, thereby increasing the resolution of the peaks. One of the columns of GC × GC is utilized to separate compounds in PPO using difference in boiling point while the other column is used to separate compounds based on the difference in polarity. In addition, it has been shown that GC × GC can be coupled with several detectors like FID, SCD, NCD, and TOF-MS for complete in-depth identification of all types of hydrocarbons besides trace amounts of nitrogen, sulfur, oxygen, and halogen containing compounds that may be found in the PPOs. For these reasons, GC × GC is a technique that is gaining tremendous significance in characterization of PPOs.

Acknowledgments

This material is based upon work supported by the U.S. Department of Energy's Office of Energy Efficiency and Renewable Energy (EERE) under the Advanced Manufacturing Office Award Number DE-EE0007897 awarded to the REMADE Institute, a division of Sustainable Manufacturing Innovation Alliance Corp. This report was prepared as an account of work sponsored by an agency of the United States Government. Neither the United States Government nor any agency thereof, nor any of their employees, makes any warranty, express or implied, or assumes any legal liability or responsibility for the accuracy, completeness, or usefulness of any information, apparatus, product, or process disclosed, or represents that its use would not infringe privately owned rights. Reference herein to any specific commercial product, process, or service by trade name, trademark, manufacturer, or otherwise does not necessarily constitute or imply its endorsement, recommendation, or favoring by the United States Government or any agency thereof. The views and opinions of authors expressed herein do not necessarily state or reflect those of the United States Government or any agency thereof. We would like to gratefully acknowledge the support of Dow and ANID BIO-Fulbright for the scholarship granted to Barbara Alejandra Perez (Grant 56170014).

References

Adrados A, de Marco I, Caballero BM, López A, Laresgoiti MF, Torres A: Pyrolysis of plastic packaging waste: a comparison of plastic residuals from material recovery facilities with simulated plastic waste, *Waste Manag* 32(5):826–832, 2012. https://doi.org/10.1016/j.wasman.2011.06.016.

Ahmad I, Khan MI, Khan H, et al.: Pyrolysis study of polypropylene and polyethylene into premium oil products, *Int J Green Energy* 12(7):663–671, 2015. https://doi.org/10.1080/15435075.2014.880146.

Albahri TA: Structural group contribution method for predicting the octane number of pure hydrocarbon liquids, *Ind Eng Chem Res* 42(3):657–662, 2003. Scopus https://doi.org/10.1021/ie020306+.

Al-Salem SM, Chandrasekaran SR, Dutta A, Sharma BK: Study of the fuel properties of extracted oils obtained from low and linear low density polyethylene pyrolysis, *Fuel* 304, 2021, 121396. https://doi.org/10.1016/j.fuel.2021.121396.

Alvarez J, Lopez G, Amutio M, et al.: Evaluation of the properties of tyre pyrolysis oils obtained in a conical spouted bed reactor, *Energy* 128:463–474, 2017. https://doi.org/10.1016/j.energy.2017.03.163.

Anuar Sharuddin SD, Abnisa F, Wan Daud WMA, Aroua MK: A review on pyrolysis of plastic wastes, *Energy Convers Manag* 115:308–326, 2016. https://doi.org/10.1016/j.enconman.2016.02.037.

Anuar Sharuddin SD, Abnisa F, Wan Daud WMA, Aroua MK: Energy recovery from pyrolysis of plastic waste: study on non-recycled plastics (NRP) data as the real measure of plastic waste, *Energy Convers Manag* 148:925–934, 2017. https://doi.org/10.1016/j.enconman.2017.06.046.

Arabiourrutia M, Elordi G, Lopez G, Borsella E, Bilbao J, Olazar M: Characterization of the waxes obtained by the pyrolysis of polyolefin plastics in a conical spouted bed reactor, *J Anal Appl Pyrolysis* 94:230–237, 2012. https://doi.org/10.1016/j.jaap.2011.12.012.

ASTM: *Test method for water in petroleum products and bituminous materials by distillation*, 2013, ASTM International. https://doi.org/10.1520/D0095-13E01.

ASTM: *Standard Test Method for Heat of Combustion of Liquid Hydrocarbon Fuels by Bomb Calorimeter*, 2019 http://www.astm.org/d0240-19.html.

Baena-González J, Santamaria-Echart A, Aguirre JL, González S: Chemical recycling of plastic waste: bitumen, solvents, and polystyrene from pyrolysis oil, *Waste Manag* 118:139–149, 2020. https://doi.org/10.1016/j.wasman.2020.08.035.

Balci M: *Basic 1H- and 13C-NMR spectroscopy*, 2005, Elsevier.

Banar M, Akyıldız V, Özkan A, Çokaygil Z, Onay Ö: Characterization of pyrolytic oil obtained from pyrolysis of TDF (tire derived fuel), *Energy Convers Manag* 62:22–30, 2012. https://doi.org/10.1016/j.enconman.2012.03.019.

Bartle KD, Myers P: History of gas chromatography, *TrAC Trends Anal Chem* 21(9):547–557, 2002. https://doi.org/10.1016/S0165-9936(02)00806-3.

Bhaskar T, Hall WJ, Mitan NMM, Muto A, Williams PT, Sakata Y: Controlled pyrolysis of polyethylene/polypropylene/polystyrene mixed plastics with high impact polystyrene containing flame retardant: effect of decabromo diphenylethane (DDE), *Polym Degrad Stab* 92(2):211–221, 2007. https://doi.org/10.1016/j.polymdegradstab.2006.11.011.

Budsaereechai S, Hunt AJ, Ngernyen Y: Catalytic pyrolysis of plastic waste for the production of liquid fuels for engines, *RSC Adv* 9(10):5844–5857, 2019. https://doi.org/10.1039/C8RA10058F.

Cahyono MS, Fenti UI: Influence of heating rate and temperature on the yield and properties of pyrolysis oil obtained from waste plastic bag, *Conserve J Energy Environ Stud* 1(1):1–8, 2017. https://doi.org/10.30588/cjees.v1i1.248.

Chen W, Shi S, Zhang J, Chen M, Xiaoyan Z: Co-pyrolysis of waste newspaper with high-density polyethylene: synergistic effect and oil characterization, *Energy Convers Manag* 112:41–48, 2016. https://doi.org/10.1016/j.enconman.2016.01.005.

Çit İ, Sınağ A, Yumak T, Uçar S, Mısırlıoğlu Z, Canel M: Comparative pyrolysis of polyolefins (PP and LDPE) and PET, *Polym Bull* 64(8):817–834, 2010. https://doi.org/10.1007/s00289-009-0225-x.

Constantinescu M, Bucura F, Ionete EI, et al.: From plastic to fuel—new challenges, *Mater Plast* 56(4):721–729, 2019. https://doi.org/10.37358/MP.19.4.5259.

Czajczyńska D, Anguilano L, Ghazal H, et al.: Potential of pyrolysis processes in the waste management sector, *Therm Sci Eng Prog* 3:171–197, 2017. https://doi.org/10.1016/j.tsep.2017.06.003.

Dao Thi H, Djokic MR, Van Geem KM: Detailed group-type characterization of plastic-waste pyrolysis oils: by comprehensive two-dimensional gas chromatography including linear, branched, and di-olefins, *Separations* 8(7):103, 2021. https://doi.org/10.3390/separations8070103.

Das P, Tiwari P: Valorization of packaging plastic waste by slow pyrolysis, *Resour Conserv Recycl* 128:69–77, 2018a. https://doi.org/10.1016/j.resconrec.2017.09.025.

Das P, Tiwari P: The effect of slow pyrolysis on the conversion of packaging waste plastics (PE and PP) into fuel, *Waste Manag* 79:615–624, 2018b. https://doi.org/10.1016/j.wasman.2018.08.021.

De S, Agarwal AK, Moholkar VS, Thallada B, editors: *Coal and biomass gasification: recent advances and future challenges*, Singapore, 2018, Springer. https://doi.org/10.1007/978-981-10-7335-9.

Dobó Z, Jakab Z, Nagy G, Koós T, Szemmelveisz K, Muránszky G: Transportation fuel from plastic wastes: production, purification and SI engine tests, *Energy* 189, 2019, 116353. https://doi.org/10.1016/j.energy.2019.116353.

Eze WU, Madufor IC, Onyeagoro GN, Obasi HC: The effect of Kankara zeolite-Y-based catalyst on some physical properties of liquid fuel from mixed waste plastics (MWPs) pyrolysis, *Polym Bull* 77(3):1399–1415, 2020. https://doi.org/10.1007/s00289-019-02806-y.

Fahim MA, Al-Sahhaf TA, Elkilani A: *Fundamentals of petroleum refining*, 2009, Elsevier.

Fekhar B, Gombor L, Miskolczi N: Pyrolysis of chlorine contaminated municipal plastic waste: in-situ upgrading of pyrolysis oils by Ni/ZSM-5, Ni/SAPO-11, red mud and Ca(OH)2 containing catalysts, *J Energy Inst* 92(5):1270–1283, 2019. https://doi.org/10.1016/j.joei.2018.10.007.

Gala A, Guerrero M, Guirao B, Domine ME, Serra JM: Characterization and distillation of pyrolysis liquids coming from polyolefins segregated of MSW for their use as automotive diesel fuel, *Energy Fuel* 34(5):5969–5982, 2020. https://doi.org/10.1021/acs.energyfuels.0c00403.

Geyer R, Jambeck JR, Law KL: Production, use, and fate of all plastics ever made, *Sci Adv* 7(3), 2017, e1700782. https://doi.org/10.1126/sciadv.1700782.

Ghenai C, Alamara K, Inayat A: Solar assisted pyrolysis of plastic waste: pyrolysis oil characterization and grid-tied solar PV power system design, *Energy Procedia* 159:123–129, 2019. https://doi.org/10.1016/j.egypro.2018.12.029.

Harman-Ware AE, Orton K, Deng C, Kenrick S, Carpenter D, Ferrell JR: Molecular weight distribution of raw and catalytic fast pyrolysis oils: comparison of analytical methodologies, *RSC Adv* 10(7):3789–3795, 2020. https://doi.org/10.1039/C9RA09726K.

Hassan H, Hameed BH, Lim JK: Co-pyrolysis of sugarcane bagasse and waste high-density polyethylene: synergistic effect and product distributions, *Energy* 191, 2020, 116545. https://doi.org/10.1016/j.energy.2019.116545.

Heydariaraghi M, Ghorbanian S, Hallajisani A, Salehpour A: Fuel properties of the oils produced from the pyrolysis of commonly-used polymers: effect of fractionating column, *J Anal Appl Pyrolysis* 121:307–317, 2016. https://doi.org/10.1016/j.jaap.2016.08.010.

Hoekstra E, Kersten SRA, Tudos A, Meier D, Hogendoorn KJA: Possibilities and pitfalls in analyzing (upgraded) pyrolysis oil by size exclusion chromatography (SEC), *J Anal Appl Pyrolysis* 91(1):76–88, 2011. https://doi.org/10.1016/j.jaap.2011.01.006.

Jangid CJ, Miller KM, Seay JR: Analysis of plastic-derived fuel oil produced from high- and low-density polyethylene, *Recycling* 7(3):29, 2022. https://doi.org/10.3390/recycling7030029.

Jayarama Krishna JV, Korobeinichev OP, Vinu R: Isothermal fast pyrolysis kinetics of synthetic polymers using analytical pyroprobe, *J Anal Appl Pyrolysis* 139:48–58, 2019. https://doi.org/10.1016/j.jaap.2019.01.008.

Joshi CA, Seay JR: Total generation and combustion emissions of plastic derived fuels: a trash to tank approach, *Environ Prog Sustain Energy* 39(5), 2020.

Kalargaris I, Tian G, Gu S: Combustion, performance and emission analysis of a DI diesel engine using plastic pyrolysis oil, *Fuel Process Technol* 157:108–115, 2017. https://doi.org/10.1016/j.fuproc.2016.11.016.

Kaminsky W: Chemical recycling of plastics by fluidized bed pyrolysis, *Fuel Commun* 8, 2021, 100023. https://doi.org/10.1016/j.jfueco.2021.100023.

Khan MZH, Sultana M, Al-Mamun MR, Hasan MR: Pyrolytic waste plastic oil and its diesel blend: fuel characterization, *J Environ Public Health* 2016, 2016, e7869080. https://doi.org/10.1155/2016/7869080.

Khandal SV, Gadwal SB, Raikar VA, Yunus Khan TM, Badruddin IA: An experimental-based artificial neural network performance study of common rail direct injection engine run on plastic pyrolysis oil, *Int J Sustain Eng* 14(2):137–146, 2021.

Kizza R, Banadda N, Seay J: Qualitative and energy recovery potential analysis: plastic-derived fuel oil versus conventional diesel oil, *Clean Techn Environ Policy* 24(3):789–800, 2022. https://doi.org/10.1007/s10098-021-02028-9.

Kumar S, Singh RK: Recovery of hydrocarbon liquid from waste high density polyethylene by thermal pyrolysis, *Braz J Chem Eng* 28(4):659–667, 2011. https://doi.org/10.1590/S0104-66322011000400011.

Kumar S, Prakash R, Murugan S, Singh RK: Performance and emission analysis of blends of waste plastic oil obtained by catalytic pyrolysis of waste HDPE with diesel in a CI engine, *Energy Convers Manag* 74:323–331, 2013. https://doi.org/10.1016/j.enconman.2013.05.028.

Kunwar B, Moser BR, Chandrasekaran SR, Rajagopalan N, Sharma BK: Catalytic and thermal depolymerization of low value post-consumer high density polyethylene plastic, *Energy* 111:884–892, 2016. https://doi.org/10.1016/j.energy.2016.06.024.

Kusenberg M, Zayoud A, Roosen M, et al.: A comprehensive experimental investigation of plastic waste pyrolysis oil quality and its dependence on the plastic waste composition, *Fuel Process Technol* 227, 2022, 107090. https://doi.org/10.1016/j.fuproc.2021.107090.

Lahtela V, Hyvärinen M, Kärki T: Composition of plastic fractions in waste streams: toward more efficient recycling and utilization, *Polymers* 11(1):69, 2019. https://doi.org/10.3390/polym11010069.

Lam SS, Chase HA: A review on waste to energy processes using microwave pyrolysis, *Energies* 5(10):4209–4232, 2012. https://doi.org/10.3390/en5104209.

Law KL, Starr N, Siegler TR, Jambeck JR, Mallos NJ, Leonard GH: The United States' contribution of plastic waste to land and ocean, *Sci Adv* 6(44):eabd0288, 2020. https://doi.org/10.1126/sciadv.abd0288.

Lee S, Yoshida K, Yoshikawa K: Application of waste plastic pyrolysis oil in a direct injection diesel engine: for a small scale non-grid electrification, *Energy Environ Res* 5(1), 2015, p18. https://doi.org/10.5539/eer.v5n1p18.

López A, de Marco I, Caballero BM, Laresgoiti MF, Adrados A: Pyrolysis of municipal plastic wastes: influence of raw material composition, *Waste Manag* 30(4):620–627, 2010. https://doi.org/10.1016/j.wasman.2009.10.014.

López A, de Marco I, Caballero BM, Laresgoiti MF, Adrados A: Influence of time and temperature on pyrolysis of plastic wastes in a semi-batch reactor, *Chem Eng J* 173(1):62–71, 2011. https://doi.org/10.1016/j.cej.2011.07.037.

Lopez-Urionabarrenechea A, de Marco I, Caballero BM, Laresgoiti MF, Adrados A: Catalytic stepwise pyrolysis of packaging plastic waste, *J Anal Appl Pyrolysis* 96, 2012. https://doi.org/10.1016/j.jaap.2012.03.004.

Mangesh VL, Padmanabhan S, Tamizhdurai P, Ramesh A: Experimental investigation to identify the type of waste plastic pyrolysis oil suitable for conversion to diesel engine fuel, *J Clean Prod* 246, 2020, 119066. https://doi.org/10.1016/j.jclepro.2019.119066.

Martynis M, Mulyazmi, Winanda E, Harahap AN: Thermal pyrolysis of polypropylene plastic waste into liquid fuel: reactor performance evaluation, *IOP Conf Ser: Mater Sci Eng* 543(1), 2019. https://doi.org/10.1088/1757-899X/543/1/012047.

Mettler-Toledo International Inc: *CHNSO elemental analysis—sample preparation*, 2022. https://www.mt.com/us/en/home/applications/Laboratory_weighing/chnso_elemental_analysis.html.

Miandad R, Nizami AS, Rehan M, et al.: Influence of temperature and reaction time on the conversion of polystyrene waste to pyrolysis liquid oil, *Waste Manag* 58:250–259, 2016. https://doi.org/10.1016/j.wasman.2016.09.023.

Miandad R, Barakat MA, Aburiazaiza AS, Rehan M, Ismail IMI, Nizami AS: Effect of plastic waste types on pyrolysis liquid oil, *Int Biodeterior Biodegradation* 119:239–252, 2017. https://doi.org/10.1016/j.ibiod.2016.09.017.

Miskolczi N, Hall WJ, Angyal A, Bartha L, Williams PT: Production of oil with low organobromine content from the pyrolysis of flame retarded HIPS and ABS plastics, *J Anal Appl Pyrolysis* 83(1):115–123, 2008. https://doi.org/10.1016/j.jaap.2008.06.010.

Miskolczi N, Angyal A, Bartha L, Valkai I: Fuels by pyrolysis of waste plastics from agricultural and packaging sectors in a pilot scale reactor, *Fuel Process Technol* 90(7):1032–1040, 2009. https://doi.org/10.1016/j.fuproc.2009.04.019.

Miteva K, Aleksovski S, Bogoeva-Gaceva G, Dimitrovski D: Evaluation of density and viscosity of blends of waste plastic pyrolysis oil and commercially diesel fuel, *J Environ Prot Ecol* 17:1142–1151, 2016.

Mohammed IY, Kazi FK, Abakr YA, Yusuf S, Razzaque MA: Novel method for the determination of water content and higher heating value of pyrolysis oil, *Bioresources* 10(2):2681–2690, 2015.

Myers ME, Stollsteimer J, Wims AM: Determination of hydrocarbon-type distribution and hydrogen/carbon ratio of gasolines by nuclear magnetic resonance spectrometry, *Anal Chem* 47(12):2010–2015, 1975a. https://doi.org/10.1021/ac60362a020.

Myers ME, Stollsteimer J, Wims AM: Determination of gasoline octane numbers from chemical composition, *Anal Chem* 47(13):2301–2304, 1975b. https://doi.org/10.1021/ac60363a015.

Olalo J: Characterization of pyrolytic oil produced from waste plastic in Quezon City, Philippines using non-catalytic pyrolysis method, *Chem Eng Trans* 86:1495–1500, 2021. https://doi.org/10.3303/CET2186250.

Paradela F, Pinto F, Gulyurtlu I, Cabrita I, Lapa N: Study of the co-pyrolysis of biomass and plastic wastes, *Clean Techn Environ Policy* 11(1):115–122, 2009. https://doi.org/10.1007/s10098-008-0176-1.

Park K-B, Jeong Y-S, Guzelciftci B, Kim J-S: Two-stage pyrolysis of polystyrene: pyrolysis oil as a source of fuels or benzene, toluene, ethylbenzene, and xylenes, *Appl Energy* 259, 2020, 114240. https://doi.org/10.1016/j.apenergy.2019.114240.

Polymer Char: *How to choose the most appropriate combination of detectors for high temperature SEC/GPC analysis of polyolefins*, 2021, January 26, Polymer Char. https://polymerchar.com/blog/how-to-choose-the-most-appropriate-detectors-for-size-exclusion-chromatography-analysis-of-polyolefins.

Popescu G, Radulov I, Iordănescu OA, et al.: Karl Fischer water titration—principal component analysis approach on bread products, *Appl Sci* 10(18):6518, 2020. https://doi.org/10.3390/app10186518.

Quesada L, Calero M, Martín-Lara MA, Pérez A, Blázquez G: Characterization of fuel produced by pyrolysis of plastic film obtained of municipal solid waste, *Energy* 186, 2019, 115874. https://doi.org/10.1016/j.energy.2019.115874.

Riazi MR: *Characterization and properties of petroleum fractions*, 2005, ASTM International.

Ritchie H, Roser M: *Plastic pollution. Our world in data*, 2018. https://ourworldindata.org/plastic-pollution.

Saptoadi H, Pratama N: Utilization of plastics waste oil as partial substitute for kerosene in pressurized cookstoves, *Int J Environ Sci Dev* 6(4):363–368, 2015. https://doi.org/10.7763/IJESD.2015.V6.619.

Shah N, Rockwell J, Huffman GP: Conversion of waste plastic to oil: direct liquefaction versus pyrolysis and hydroprocessing, *Energy Fuel* 13(4):832–838, 1999. https://doi.org/10.1021/ef9802431.

Sharma BK, Moser BR, Vermillion KE, Doll KM, Rajagopalan N: Production, characterization and fuel properties of alternative diesel fuel from pyrolysis of waste plastic grocery bags, *Fuel Process Technol* 122:79–90, 2014. https://doi.org/10.1016/j.fuproc.2014.01.019.

Siddiqui MN, Redhwi HH: Pyrolysis of mixed plastics for the recovery of useful products, *Fuel Process Technol* 90(4):545–552, 2009. https://doi.org/10.1016/j.fuproc.2009.01.003.

Sindhu R, Binod P, Pandey A: Chapter 17—Microbial poly-3-hydroxybutyrate and related copolymers, In Pandey A, Höfer R, Taherzadeh M, Nampoothiri KM, Larroche C, editors: *Industrial biorefineries & white biotechnology*, 2015, Elsevier, pp 575–605. https://doi.org/10.1016/B978-0-444-63453-5.00019-7.

Singh RK, Ruj B, Sadhukhan AK, Gupta P: Thermal degradation of waste plastics under non-sweeping atmosphere: part 2: effect of process temperature on product characteristics and their future applications, *J Environ Manag* 261, 2020, 110112. https://doi.org/10.1016/j.jenvman.2020.110112.

Smets K, Adriaensens P, Vandewijngaarden J, et al.: Water content of pyrolysis oil: comparison between Karl Fischer titration, GC/MS-corrected azeotropic distillation and 1H NMR spectroscopy, *J Anal Appl Pyrolysis* 90(2):100–105, 2011. https://doi.org/10.1016/j.jaap.2010.10.010.

Speight JG: Ultimate analysis, In *Handbook of coal analysis*, 2015, John Wiley & Sons, Ltd, pp 144–169. https://doi.org/10.1002/9781119037699.ch6.

Syamsiro M, Saptoadi H, Norsujianto T, et al.: Fuel oil production from municipal plastic wastes in sequential pyrolysis and catalytic reforming reactors, *Energy Procedia* 47:180–188, 2014. https://doi.org/10.1016/j.egypro.2014.01.212.

Thompson J: *CHNS elemental analysers. AMC technical brief*, 2008, The Royal Society of Chemistry, p 2.

Toraman HE, Dijkmans T, Djokic MR, Van Geem KM, Marin GB: Detailed compositional characterization of plastic waste pyrolysis oil by comprehensive two-dimensional gas-chromatography coupled to multiple detectors, *J Chromatogr A* 1359:237–246, 2014. https://doi.org/10.1016/j.chroma.2014.07.017.

Uebe J, Kryzevicius Z, Majauskiene R, Dulevicius M, Kosychova L, Zukauskaite A: Use of polypropylene pyrolysis oil in alternative fuel production, *Waste Manag Res*, 2022. https://doi.org/10.1177/0734242X211068243.

US Environmental Protection Agency: *Advancing sustainable materials management: 2018 fact sheet*, 2020, https://www.epa.gov/sites/default/files/2020-11/documents/2018_ff_fact_sheet.pdf.

Vasile C, Brebu MA, Karayildirim T, Yanik J, Darie H: Feedstock recycling from plastic and thermoset fractions of used computers (I): pyrolysis, *J Mater Cycles Waste Manag* 8(2):99–108, 2006. https://doi.org/10.1007/s10163-006-0151-z.

Wauquier J-P, editor: *Petroleum refining*, vol. 1, 1995, Editions OPHRYS.

Williams EA, Williams PT: The pyrolysis of individual plastics and a plastic mixture in a fixed bed reactor, *J Chem Technol Biotechnol* 70(1):9–20, 1997. https://doi.org/10.1002/(SICI)1097-4660(199709)70:1<9::AID-JCTB700>3.0.CO;2-E.

Williams PT, Williams EA: Fluidised bed pyrolysis of low density polyethylene to produce petrochemical feedstock, *J Anal Appl Pyrolysis* 51(1):107–126, 1999. https://doi.org/10.1016/S0165-2370(99)00011-X.

Wongkhorsub C, Chindaprasert N: A comparison of the use of pyrolysis oils in diesel engine, *Energy Power Eng* 05(04):350, 2013. https://doi.org/10.4236/epe.2013.54B068.

Yuliansyah AT, Prasetya A, Ramadhan MAA, Laksono R: Pyrolisis of plastic waste to produce pyrolytic oil as an alternative fuel, *Int J Technol* 6(7), 2015. https://doi.org/10.14716/ijtech.v6i7.1241.

CHAPTER SIX

PET recycling: Review of the current available technologies and industrial perspectives

Lorenzo Brivio[a,b,*] and Flavio Tollini[b]

[a]While True s.r.l., Urgnano, Italy
[b]Department of Chemistry, Materials, and Chemical Engineering "Giulio Natta", Politecnico di Milano, Milan, Italy
*Corresponding author: e-mail address: lorenzo.brivio@whiletrue.eco

Contents

Abstract

Due to the increasing production of plastic soft drink bottles, accompanied by the accumulation of plastic waste in landfills, our society is encouraging the development of recycling industries around the world, with a special focus on recycling post-consumer PET (polyethylene terephthalate, the world's third most dominant plastic) for food contact applications (i.e., for single-use items). Plastics have indeed become a threat to the environment because of the lack of recycling technologies, which could instead enable the production of high-quality polymers from scrap materials, at an equal or lower cost compared to the production of the corresponding virgin polymer from crude oil.

Advances in Chemical Engineering, Volume 60
ISSN 0065-2377
https://doi.org/10.1016/bs.ache.2022.09.003

PET has a great recycling potential if compared to the other most diffused plastics, and it can be treated in many different ways. In this chapter, the very broad spectrum of all the available technologies for PET recycling is presented (from the zero-order to the fourth-order recycling) with particular attention to mechanical and chemical recycling. Indeed, the former is currently the best available technology for PET recycling at the industrial level, while the latter represents the best perspective for PET recycling in terms of circular economy, allowing to get back the monomer building blocks from complex waste materials and using the monomers to produce a new recycled polymer. Advantages and disadvantages of the current state of the art are highlighted, aiming to identify the viability of every process at the industrial scale.

1. Introduction

1.1 Plastics, society, and the disposal problem

The Greeks were the first ones who applied the concept of "*shape = function*" introducing the concept of "*brand*" and consequently the "*mark of origin*". The Romans improved this system by making amphorae having special geometries to guarantee a reduction of fraud, thus protecting the origin of the content. In the Middle Ages, the jars changed their shapes, they became smaller and they were destined to be used directly. Earthenware was replaced with metal and glass, and the concept of "*container for final use*" did spread. Historically, technology was born just to satisfy social needs and to improve our living standards. Indeed, the change from earthenware to metal allowed to satisfy the need of having durable containers, allowing safer transportation by both sea and land. However, the main disadvantage of such metal jars became their cost. This last problem fades away with the onset of plastic materials for the production of food containers. Plastics are the perfect materials to be used for drink containers, but in the last decades, their extreme cheapness and the high production rates did introduce and cement the concept of "disposable," while the historical concept of "*reusable*" was set apart, since it was born, originally, due to the high cost and time required for manufacturing items in metal, glass or even earthenware (Reike et al., 2018). This technical and social transformation has therefore led to a huge change in markets and then, to a disastrous environmental impact, encouraging the birth of new technologies and companies focused on recycling.

In the last decades, an astonishing increase in the production rates of almost every kind of item took place, and the low prices allowed the distribution of these goods to a huge slice of the population. The consequence was that the living standards during the last century did increase at rates that are not at all comparable with any other period in history, and the economic growth experienced an exponential boost (Nye, 2018).

To summarize, the point is that plastics have a significant impact on our actual life: our societies would have never been able to get to the actual living standards without plastics, and at present, they cannot manage to survive without them. Today, plastics are everywhere: from the keys on which this manuscript was typed to the printer which printed it, or to the device on which it is now being read. In particular, PET is the leading actor for food and drink containers.

However, the big problem is that the excellent properties of plastics (i.e., high chemical resistance and long-lasting in time) make them extremely problematic when it comes to disposal (Stark, 2019). Moreover, plastics (PET in particular) are often the cheapest alternative on the market for single-use products, such as food and drink containers, bags, and clothes. In other words, materials that are potentially stable in the environment for centuries are used to manufacture items designed to be used for minutes-long (or at most for days-long) applications.

The plastic end-life and their disposal are recognized as one of the biggest issues of our century, and since a world that is completely free from any plastics is not only impossible but not even imaginable, recycling has to be the solution (Recycling Magazine, 2019). The good news is that PET has great potential in terms of recycling, and it can be treated in several ways, which will be presented in detail in this chapter.

1.2 Global plastic production and municipal plastic management

Plastic production and recycling are two major industrial activities, which employ loads of workers and treat huge volumes. Indeed, as shown in Fig. 1, the overall production almost reached 359 million tons in 2018 and 367 million tons in 2020, and it is increasing year by year (PlasticsEurope, 2019).

Focusing on Europe, the total demand is 57.9 million tons of plastics distributed in different applications. As shown in Fig. 2, the largest slice is represented by packaging, which covers approximately 40% of the total demand followed by Building & Construction (20.4%), Automotive (9.6%), Electrical & Electronic (6.2%), Household, Leisure & Sports (4.1%), Agriculture (3.4%), and Other (16.7%).

On the other hand, as shown in Fig. 3, recent data shows that concerning the market share of the different polymer types the most demanded plastics

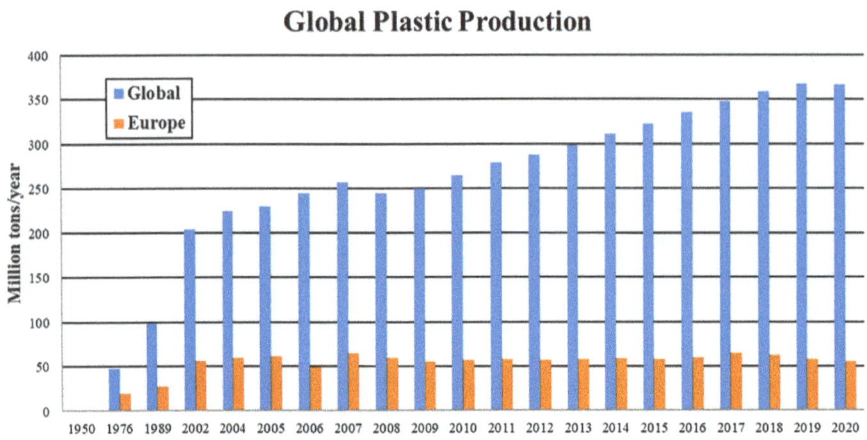

Fig. 1 Plastics production per year in Europe and in the world from 1950 to 2020.

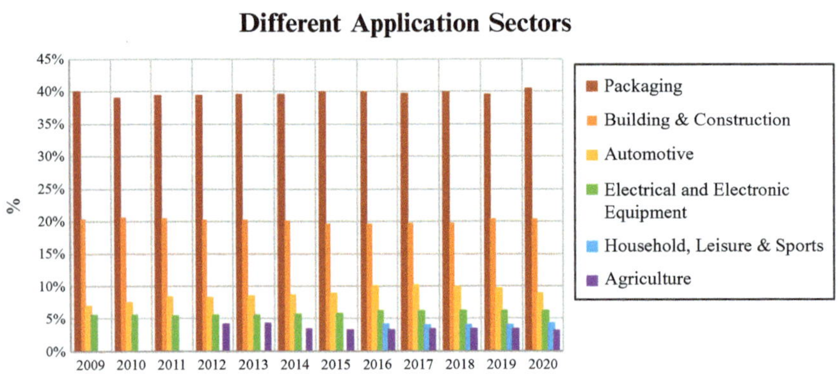

Fig. 2 Plastics share by industrial sectors from 2009 to 2020.

in Europe are polyolefins, in particular, PP (19.7%), LDPE (17.4%), and HDPE (12.9%), and they are followed by PVC (9.6%), PET (8.4%), PUR (7.8%), PS (6.6%) and others (18.1%).

To understand the life cycle of plastic products, it is important to understand that not all plastics have the same properties and not all of them have the same service life. Some plastics have a shelf life of less than 1 year, while some others have a lifespan of more than 15 years and some have a service life of 50 years or even more (Marathe et al., 2019). Thus, from production to waste, different plastics can exhibit different life cycles, and therefore the volume of collected waste cannot match, in a single year, the volume of production or consumption, but since PET is mainly employed for a very short

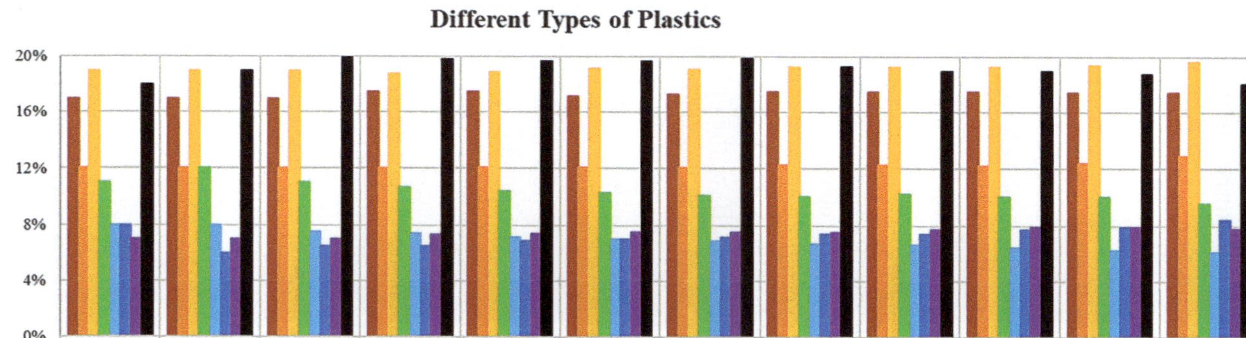

	2009	2010	2011	2012	2013	2014	2015	2016	2017	2018	2019	2020
PE-LD, PE-LLD	17.0%	17.0%	17.0%	17.5%	17.5%	17.2%	17.3%	17.5%	17.5%	17.5%	17.4%	17.4%
PE-HD, PE-MD	12.0%	12.0%	12.0%	12.0%	12.1%	12.1%	12.1%	12.3%	12.3%	12.2%	12.4%	12.9%
PP	19.0%	19.0%	19.0%	18.8%	18.9%	19.2%	19.1%	19.3%	19.3%	19.3%	19.4%	19.7%
PVC	11.0%	12.0%	11.0%	10.7%	10.4%	10.3%	10.1%	10.0%	10.2%	10.0%	10.0%	9.6%
PS, PS-E	8.0%	8.0%	7.5%	7.4%	7.1%	7.0%	6.9%	6.7%	6.6%	6.4%	6.2%	6.1%
PET	8.0%	6.0%	6.5%	6.5%	6.9%	7.0%	7.1%	7.4%	7.4%	7.7%	7.9%	8.4%
PUR	7.0%	7.0%	7.0%	7.3%	7.4%	7.5%	7.5%	7.5%	7.7%	7.9%	7.9%	7.8%
Others	18.0%	19.0%	20.0%	19.8%	19.7%	19.7%	19.9%	19.3%	19.0%	19.0%	18.8%	18.1%

Fig. 3 Plastics share by polymer type from 2009 to 2020.

shelf-life product, these volumes are quite similar. The percentage of PET bottles recycled is different from country to country, but the extent and form of recycling are mostly dictated by the prevailing regulations and economy. For example, in the United States, nearly 43% of PET bottles are recycled to fibers and only around 25% of food and beverage bottles. Differently, Germany recycles 34% of new bottles (Gopalakrishna and Reddy, 2019).

As shown in Fig. 4, from 2006 to 2020 the plastic waste collected in Europe increased by almost 20.4%, from 24.5 to 29.5 Mt. The volumes of plastic waste collected for recycling increased by 117%, and energy recovery increased by 77% and landfills decreased by 47%. In 2016, for the first time, recycling overtook landfills (PlasticsEurope, 2006; PlasticsEurope, 2021).

The overall annual demand for PET accounts for 8% by weight and 12% by volume of the global collected solid wastes (George and Kurian, 2014). In the following sections, the entire life cycle of PET bottles is going to be analyzed, comparing different valorization paths, for example, the closed-loop recycling (as the bottle to bottle recycling—B2B) or the open-loop recycling (as the bottle to fibers—B2F or as the waste to fuel) (Marathe et al., 2019).

The management of Municipal Plastic Waste (MPW) is indeed structured into the three main pathways reported in the scheme shown in Fig. 5:

The recovery route of municipal waste offers the potential to save fossil fuels while recycling the wastes and producing new materials. In particular, five different recycling approaches can be identified:

- **Zero Order Recycling**: the direct reuse of PET bottles without manipulation.
- **First Order Recycling** (or in-plant recycling): the damaged products, or what cannot satisfy the market specification, are reused as raw material.

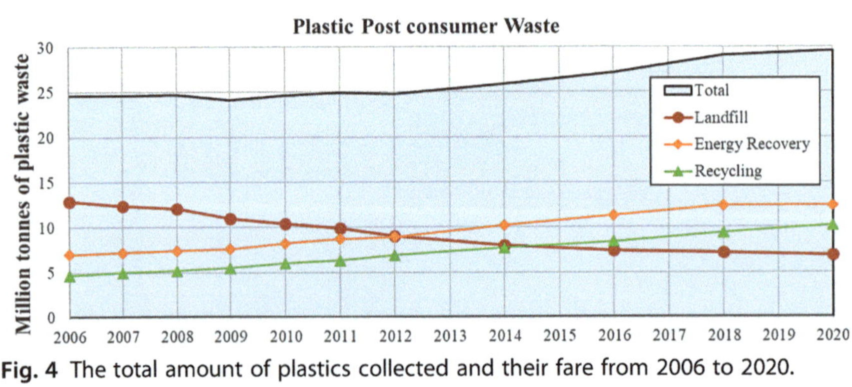

Fig. 4 The total amount of plastics collected and their fare from 2006 to 2020.

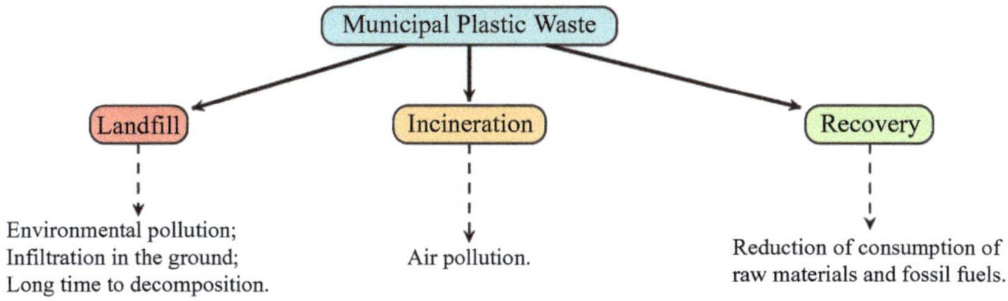

Fig. 5 Pathways for the treatment of the municipal mixed plastic wastes.

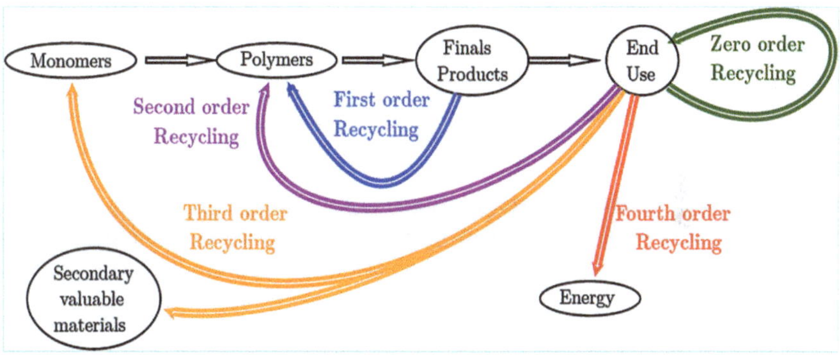

Fig. 6 Different recycling approaches, from zero to fourth order recycling.

- **Second Order Recycling** (or Mechanical Recycling): the materials at end-life are used for new applications wihout changing their chemical structure.
- **Third Order Recycling** (or Chemical Recycling): the manipulation of the chemical structure of the polymer to get new chemicals that can be used to produce again the original material.
- **Fourth Order Recycling** (or Energy Recovery): the recovery of the energy content of plastics.

In Fig. 6 the different pathways are depicted, while the advantages and disadvantages of the three 'real recycling' techniques (mechanical, chemical, and energy recovery, i.e., the only three techniques that treat post-consumer wastes) are highlighted in Table 1.

Energy recovery is the least profitable recycling technology, but it is extremely tolerant to contaminations and blended materials, thus it is not at all the best option when a clean and pure PET waste stream is available. On the contrary, mechanical recycling needs an extremely precise sorting procedure to eliminate contaminants and get a final pure stream made only by PET. Indeed, usually, the waste is collected and sorted to separate the

Table 1 Pathways for the treatment of the municipal mixed plastic wastes

Technique	Advantages	Challenges
Mechanical Recycling	Cost-Effective Efficient Well-Known and largely applied	Deterioration of product properties Very precise pre-treatments are necessary
Chemical Recycling	Simple Technology Saves fossil feedstocks	Mainly limited to condensation polymers Quite expensive
Energy Recovery	Generates considerable energy Extremely tolerant to contaminants	Not environmentally acceptable

most valuable compounds, while the rest is sent to energy recovery or even to landfills. In 2004 the European Union legislated on this, stating that each country had to start the collection of post-consumer plastic wastes, but let the country themselves decide independently on how to implement a method for this collection (Ragaert et al., 2017). Thus, currently, many countries have implemented slightly different methods for post-consumer waste collection and sorting. For example, some countries request the citizens to separate at home all the plastic wastes (LDPE, HDPE, PP, PET, PU, etc.) from paper, glass, and compost, and the plastic sorting is completed in specific centralized facilities. On the other side, countries like Switzerland, Germany, and Belgium, do even request a separate and individual collection for PET wastes, which are therefore separated directly by the consumers from all the other plastics types.

The pre-treatment suitable to manage municipal plastic waste and to obtain a pure PET waste fraction involves several steps. First, a sorting step to separate PET from the other plastics, is carried out automatically by floatation (density separation), electrostatic separation, magnetic separation, melt filtration, X-ray detection, or infrared detection (Ruj et al., 2015; Schyns and Shaver, 2021). Second one or more washing phases to remove dirt and other contaminants, and finally the grinding to reduce the polymer particle size and to form flakes (Kenny et al., 1992) which are ready to be re-extruded and then, post-processed (Ragaert et al., 2017).

1.3 Food contact polymers: Virgin and recycled PET

Speaking of food-contact polymers, three phenomena could potentially affect the integrity of food (and also of the polymer itself). One of these

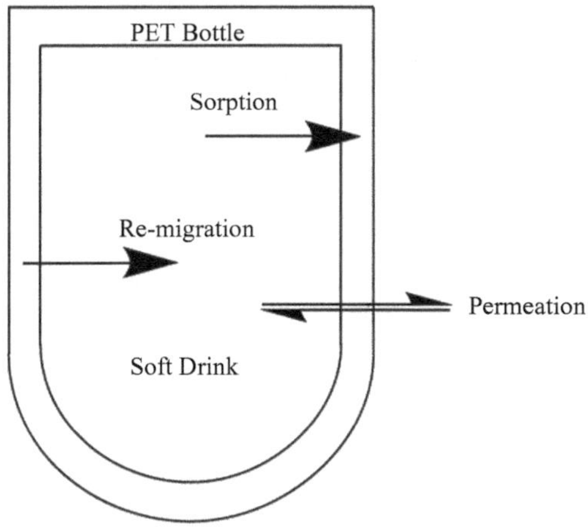

Fig. 7 Three main interactions between food and polymer in packaging.

phenomena is the permeation of gasses (carbon dioxide and oxygen) through polymeric packaging. The other two, schematically illustrated in Fig. 7, are the migration of impurities or additives from the package into the food, and the absorption of impurities or additives from the food into the package.

In the EU, a law enacted in 2008 stated the criteria to be used for safety evaluation of a recycling process for the production of recycled PET intended for food packaging (EuropeanUnion, 2008). According to this legislation, the reference post-consumer contaminated PET is chosen, conservatively, as if it has $3\,mg\,kg_{PET}^{-1}$ contaminants. The resulting residual concentration of each contaminant in recycled PET (Cres) is then compared to a predicted concentration value (C_{mod}). This C_{mod} is calculated using generally recognized conservative migration models. The migration calculated by such diffusion models overestimates the real migration from around 5 to 100 times (EuropeanCommission, 2012). The default scenario considers an infant of 5 kg who consumes every day 0.75 L of water. According to this scenario, the highest concentration of a substance in water is smaller than or equal to $0.017\,\mu g\,kg_{food}^{-1}$. Taking into account a conservative diffusion model that overestimates 5 times the surrogate diffusion, the limit is 0.1, 0.15, and $0.75\,\mu g\,kg_{food}^{-1}$ for infants, toddlers, and adults respectively. A scheme summarizing this evaluation method is shown in Fig. 8.

1.4 Recycling: Orders and loops

Under the name of recycling, we group a bunch of concepts that have in common the idea of giving a second life to treated wastes. Each material

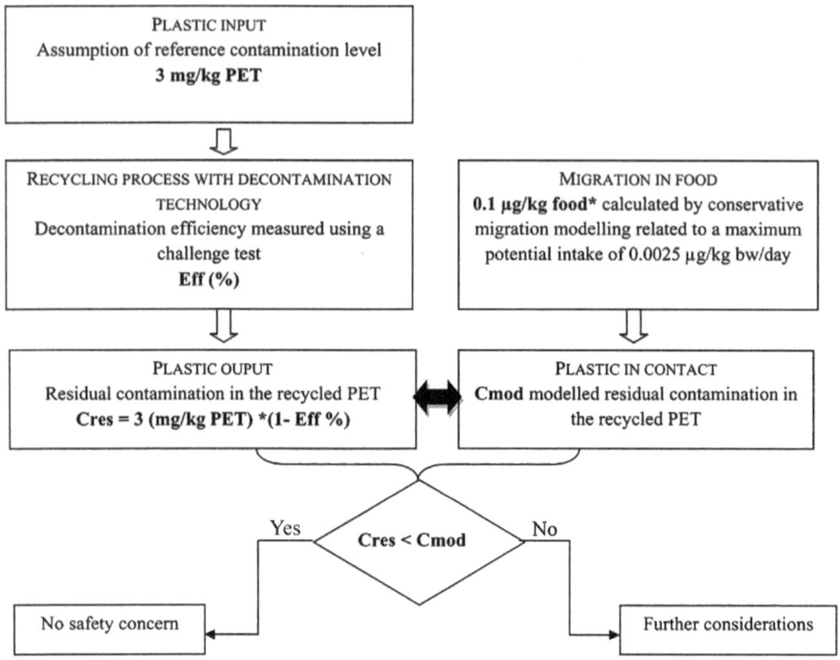

Fig. 8 Evaluation method for safety requirements of recycled PET.

has some preferred recycling methods, and condensation polymers like PET can be recycled in many different ways, universally listed as zero, first, second, third, and fourth order recycling (Ignatyev et al., 2014).

The **zero order recycling** is the simplest method to apply the concept of "reuse." Refilling directly the PET bottles, transforming them into a vase, or taking advantage of its thermoplastic properties to create an enclosure, and many other uses, make PET an easily reusable material. On the other hand, this great variety of uses makes it difficult to detect the impurities and contaminants present after the service life. For this reason, the PET recycling industry makes many efforts to recover this material often categorized as a recycling method, and is listed together with the others. This practice is called reuse, and it is even more sustainable compared with the other recycling processes, since it doesn't require any energy or raw materials for being carried out, and it directly prevents the consumer from buying a new item. This is of course a good practice in everyday life, but clearly, it can't solve the problem of plastic accumulation. Indeed, plastic products like those made of polyethylene terephthalate are persisting in the environment for centuries and probably one can refill a PET bottle up to 10–15 times before a hole appears, it starts leaking, and it necessarily needs to be thrown into the trash.

The **first order recycling** is also called "in-plant recycling" because it is the most established method to recycle the polymer during the production lines. This method is based on the concept of "reduce" because it is possible to reduce raw material consumption avoiding wastage. Indeed, this practice is straightforward and cheap, especially for plastics since the scrap materials are pure, uncontaminated, and can be easily reprocessed in loco. For instance, in the case of PET, the industrial (pre-consumer) scrap material is pelletized, mixed with some virgin material, re-extruded, and then post-processed. This kind of procedure only employs clean and uncontaminated wastes produced in loco. Therefore, it can be seen as a method to reduce the amount of lost raw material and not correctly as a real recycling technique.

The **second order recycling**, also referred to as "mechanical recycling," is the first real recycling technique, and it is the most common and diffused recycling method for plastics and PET in particular (Al-Salem et al., 2009). This process is already well established at the industrial level, and it typically includes the steps of collection, sorting, washing, and grinding of the waste materials, where these steps can also occur in a different order or can take place also more than once, depending on the quality and the composition of the treated waste materials. For instance, the mechanical recycling of post-consumer PET bottles usually requires the collection of the wastes, the sorting procedure to separate the non-PET fraction, one or more washing to remove any food contamination or traces of paper and glue, and then one or two grinding steps to get a size compatible with the employed extruder (Schyns and Shaver, 2021; Ragaert et al., 2017).

The **third order recycling** is also referred to as "chemical recycling." Chemical recycling processes for PET were implemented in parallel to the manufacture of the polymer on a commercial scale as confirmed by numerous patents coming from the early 50s (Han, 2019). This technology is a less advanced recycling technique that has not been vastly employed at an industrial level yet but has been gaining much attention during the last years (Ghosal and Nayak, 2022). Chemical recycling is aimed to fully or partly depolymerize the plastic waste back to the initial monomers or a recycled pre-polymer that can be mixed and used with virgin pre-polymers (Al-Sabagh et al., 2015). This technique has two main advantages: first, it allows the treatment of highly contaminated and heterogeneous wastes (i.e., multi-layered food containers or textiles made of blended fibers) that cannot be mechanically recycled. Second, even when starting from contaminated and low-quality wastes, it allows monomers that are precisely equal to those obtained from fossil feedstocks, which can be used to manufacture high-quality polymers. On the other hand, the main drawback of chemical recycling stands in the step of monomer purification, which may become tricky and energy expensive (Barnard et al., 2021). The most common techniques are divided as follows: glycolysis, methanolysis, hydrolysis, ammonolysis, aminolysis, and combined methods. The main available routes for the PET depolymerization and the product they lead are schematized in Fig. 9.

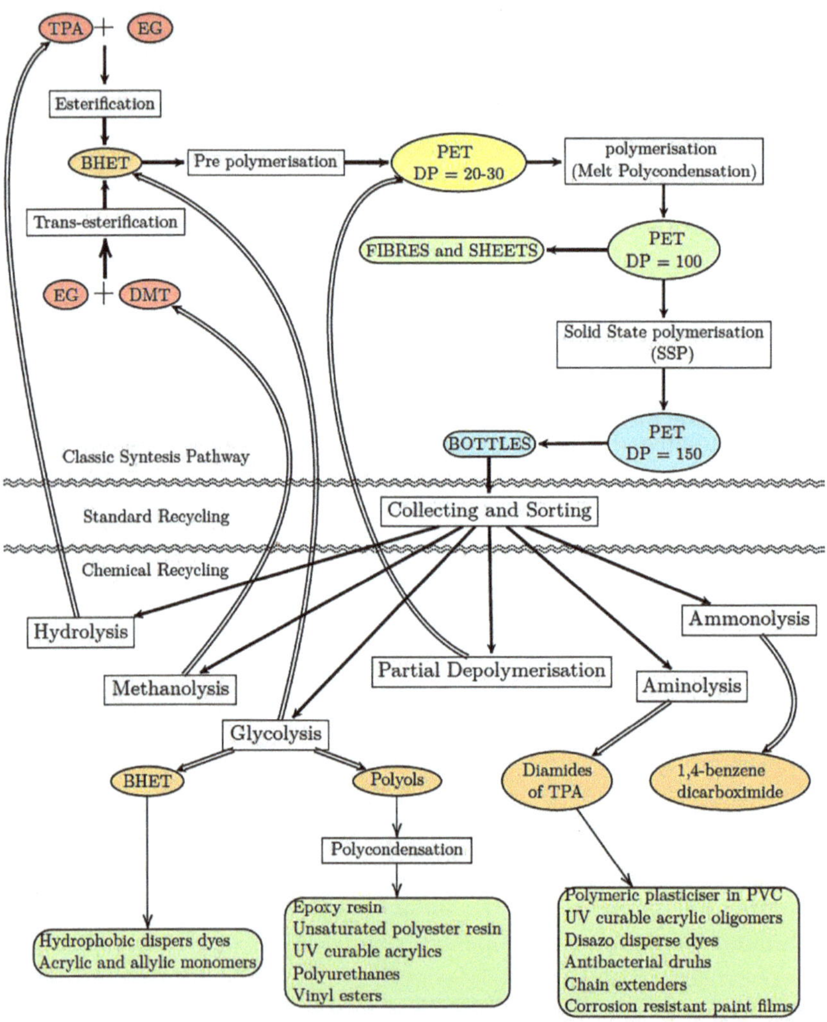

Fig. 9 Available routes for PET chemical recycling.

In addition, under the name of chemical recycling, other methods can be included: the thermochemical approaches that are usually called "*from waste to fuel.*" Those solutions have roots in common with the plants for the valorization of heavy petroleum fractions, and they are exploited to recycle plastic waste forming new energy vectors (i.e., fuels) or also other petrochemical feedstoks. This kind of process involves gasification, thermal de-polymerization and thermal cracking, catalytic cracking, and pyrolysis, where the latter is the most employed and promising thermochemical approach especially for PET materials and mixed plastic wastes in general. Moreover, some novel chemical approaches have been proposed in the literature recently, for example, cyclodepolymerisation and recycling via ring-chain equilibria (Hodge, 2015), photodegradation (Pires et al., 2015), ultrasound degradation (George and Kurian, 2014), combined ultrasound-enzymatic hydrolysis, and degradation in a microwave reactor (Yang et al., 2010).

Finally, the **Fourth order recycling**, also referred to as "energy recovery," is the last option available to avoid landfilling of the wastes, and it consists in burning them as if they were fossil fuels and recovering the energy stored. Energy recovery can be employed for almost any kind of waste, and in particular, it is preferred to treat mixed and not separable waste streams such as multilayered items made for example by PET and polyolefines. This last "recycling" option is less friendly from an environmental point of view if compared with all the other ones, and it has nothing to do with the goal of the circular economy. However, it is still preferable compared to landfilling since the power plants which are currently recovering energy from wastes are strictly regulated in terms of pollution. These power plants are much less efficient compared to those employing fossil fuels since the calorific value of the most common plastics is around 30 [MJ/kg] while that of methane is almost double, being around 53 [MJ/kg] (Panda et al., 2010). However, since this method is extremely cheap and tolerant to mixed wastes and any kind of contamination, many countries are betting on energy recovery by waste incineration to reduce the amount of landfilled wastes.

1.5 Polyethylene terephthalate

The chemical structure of the PET repeating unit is illustrated in Fig. 10, and as already discussed, PET is one of the most diffused plastics worldwide, and due to its excellent chemical and physical properties, PET is extensively used

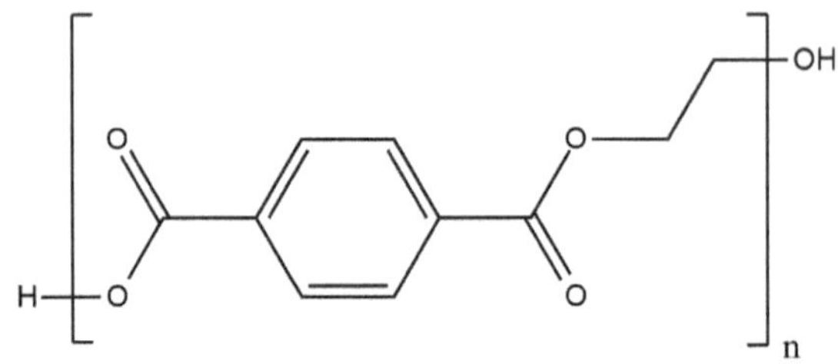

Fig. 10 Chemical structure of the PET repeating unit.

in a wide range of applications. Indeed, it is inert, clear, transparent, and lightweight, it has low gas diffusivity and is waterproof, it has excellent mechanical properties, and resists high temperatures as well (Omnexus, 2022). Moreover, PET is quite easy to be treated and post-processed. Due to all these features, PET is mainly employed in the manufacture of items that have to be in contact with food and drinks, and the lower quality quote is employed to manufacture synthetic cheap garments. In other words, polyethylene terephthalate is mostly employed in the production of items that have a very short lifetime or even which are designed for single use (Fortune Business Insights, 2022). In particular, each application requires different mechanical properties and thus, different polymer chain lengths: for example in bottles an average molecular weight of 50 kDa is required, while for textiles 30 kDa is well enough.

In contrast, polymers such as PVC or PU are widely employed also for manufacturing items that are designed to last much longer, therefore it's not surprising that PET has the highest ratio between produced plastic and generated wastes among the top 5 commercial polymers (Geyer et al., 2017; Hondo, 2022). Indeed, when one thinks about plastic wastes in general, the first image that comes to mind is probably that of a cumulus of PET bottles landfilled somewhere far. Indeed, PET has been the leading actor in the narration about ocean plastics, pollution, and unsustainability in general.

To be fair, on one hand, this is understandable, since the vast majority of PET products are single-use items, and nowadays a large quote of them do end up in landfills or the oceans (for a total of 8 million tons of plastics per year) (Jambeck et al., 2015). However, on the other hand, there's something more that deserves to be considered: as a matter of fact, only a few are aware that PET is by far the easiest polymer to be recycled among the top 5 polymers produced by volume on the planet (PE, PP, PVC, PU, PET). Therefore, the bad reputation that PET is gaining in the last years is not completely justified: PET has a great recycling potential, which is looking forward to being exploited (ZeroWasteEurope, 2022).

This potential has not been exploited so far, since the only waste fraction which is industrially recycled nowadays is the one coming from clean and transparent bottles, and every year the actual volumes of plastic items made of recycled PET are way smaller than those which are produced from virgin PET (therefore, derived directly from fossil feedstocks). In any case, the recycling chain of PET is by far the most developed one among the other main produced plastics, which are much more problematic to be recycled (PlasticsForChange, 2021): for example, the chemical recycling technologies available for the olefines (PE and PP) are nothing else than just pyrolysis (Kaminsky et al., 1995; Sivagami et al., 2021).

On the contrary, for PET there is a much broader spectrum of available recycling methods (which will be extensively discussed in the next paragraph), and a bunch of industries, as well as many universities, have already started focusing on this topic and investing in research. Of course, these three players together are not enough to achieve any consistent results, since the recycling chain has to work properly from the first step toward the last one to be fully efficient, therefore from the waste collection to the actual recycling plant.

Thus, people consciousness is crucial for the success of the whole process, and during the last decade, the feeling about the "plastic issue" has gained a great impact on the public debate. Thanks to this increasing awareness of sustainability issues, the practice of waste collection and sorting is experiencing a vigorous boost in the last years (PetcoreEurope, 2021), and laws on these topics are currently being enacted in several countries (Jamison, 2021). The result is that the great potential of PET in terms of recycling is being more and more exploited every year, while many of the other most diffused plastics are not experiencing such a boost in their recycling chain, since they are not suited to be chemically recycled to monomers (besides PS and PMMA). The large majority of common plastics (for instance, polyolefins) can be treated only by thermochemical approaches, by landfilling or they can be burnt for energy recovery purposes. At best, they can be downcycled once to some lower quality products and then, in any case, landfilled or incinerated for energy recovery.

1.6 Different approaches to PET recycling

PET wastes can be recycled according to any of the 5 different types of processes previously introduced. Energy recovery is the easiest and most tolerant to contaminations among the three options for the recycling of PET, but on the other hand, it is the most pollutant and less convenient from an economical point of view, since the recovered energy is much less valuable than the original PET material. Moreover, it is not an option to achieve a circular economy and to operate closed-loop recycling, thus mechanical or chemical recycling has always been preferred, if possible.

The mechanical recycling of PET is already widespread at the industrial level, since it is extremely cheap, easy to operate, and it allows to obtain high-value products (i.e., recycled PET, referred also as rPET) starting from clean waste materials. The prices of rPET are even higher than those of virgin PET (vPET) as the market demand for the "green rPET" is being extremely high in the last years (Kahlert and Bening, 2022). However, mechanical recycling cannot handle streams of PET that are contaminated or blended, thus it cannot intercept a huge fraction of waste PET (for instance, layered food containers and clothes). This waste fraction, which currently has no economic value, can be treated by chemical recycling instead, which could be the keystone in the achievement of a circular lifecycle for PET, allowing the closing of the recycling loop (PackagingEurope, 2022).

2. PET mechanical recycling

Industrially, mechanical recycling is the most employed recycling method for the treatment of PET wastes, and since it is cheaper than chemical recycling and leads to products with higher economical value compared to energy recovery, it has always been preferred over the other recycling methods, when possible. Being such cheap, mechanical recycling turns out to be economically profitable also at medium-small scales (Mastellone, 2019). The extreme simplicity and low cost of this kind of process made it the most widespread technology for the recycling of plastics, and especially if the initial waste stream consists only of clean PET, the automatic sorting procedure can be skipped and the process becomes in principle nearly inexpensive. However, the market prices for these kinds of clean and pure post-consumer PET wastes have hugely increased, so the prices of the mechanically recycled PET pellets are currently even higher than those of the virgin ones (Kahlert and Bening, 2022). Indeed, many companies are ready to pay more for high-quality recycled plastics for the sake of their image, since nowadays consumers are willing to spend more to buy a fancy garment made out of recycled PET rather than a similar one made out of virgin PET. However, mechanical recycling has two big issues that strongly limit its applicability: firstly degradation, in the form of thermal-mechanical and lifetime degradation, and secondly the fact that this technology can just recycle the highest quality PET fraction (i.e., in the form of clear and pure PET bottles and containers).

2.1 Degradation

The most problematic form of degradation for PET is thermal-mechanical degradation, which occurs mainly during the melting and extrusion procedures due to the high temperatures and shear (Nait-Ali et al., 2011). The consequence is the breaking of the polymer chains, the reduction of the molecular weights, and thus the lowering of the polymer mechanical properties. For PET it was reported that just after two mechanical recycling steps, the molecular weight nearly halves (Schyns and Shaver, 2021). This reduction of the PET molecular weight can be compensated by the addition of additives and stabilizers, and since PET is synthetized through a polycondensation, it is also possible to operate a final post-condensation of the mechanically recycled PET (a solid-state polycondensation step) to increase the molecular weight back to the original value (EREMA, 2022; Forrest, 2019). The other type of degradation which is responsible for the loss in the quality and the properties of the mechanically recycled PET is what is referred to as lifetime degradation. The latter is the effect that heat, light, moisture, oxygen, and mechanical stress have on the molecular structure of the polymer, thus on its chain length and therefore the mechanical properties it may achieve. This kind of degradation is less problematic than thermal-mechanical degradation since it can be almost totally avoided by the addition of well-known stabilizers in the polymer formulation (Chang et al., 1982).

2.2 Quality of the recyclable fraction

The other main limitation of all the mechanical recycling processes is related to the quality of the wastes that it can treat. Indeed, since they do not include a deep purification step and not even a decolorization procedure, mechanical recycling processes may at best produce pellets containing the same amount of contaminants and colors which were trapped inside the starting PET waste stream. Therefore, a PET mechanical recycling plant may only treat a very clean, pure, and preferably transparent fraction of PET wastes, which requests, as already introduced, very precise automatic sorting and manual sorting procedures (Damayanti and Wu, 2021). The fraction of PET waste that is mixed with other plastics (for instance, layered materials) is therefore cut out from the mechanical recycling plants and most importantly, the huge fraction of PET wastes in the form of textile fibers (for instance, clothes and carpets which are lower quality PET products and even mixed with many other compounds and non-PET fibers) cannot be recycled at all with any mechanical recycling technology.

Therefore, PET mechanical recycling facilities can only treat the highest value fraction of wastes and it can only reduce the quality of the treated material, performing what is typically called "downcycling" (Geyer et al., 2016). Indeed, PET usually undergoes just one mechanical recycling process, as after just one cycle its quality decreases to the point where it cannot be used anymore for food contact application and it is rather used to manufacture textile fibers, which then can't be recycled mechanically further (Assadi et al., 2004). In other words, nowadays just 7% of the recycled PET available on the market comes from two or more mechanical recycling loops, and this fraction is usually used for the very lowest quality products: mechanical recycling can only perform what we call "open loop recycling."

Summing up, on one hand, mechanical recycling is a good technology that can be employed to extend the lifetime of a PET product, since it can give it a new second life, but on the other hand, this technology has a huge limitation since it is not able to recycle any kind of PET waste for infinite times. Mechanical recycling turns out not to be a robust option for the achievement of the circular economy goal since it is not able to perform an upcycling and not even closed-loop recycling of the waste PET that it treats.

Even if mechanical recycling is currently the only technique available at the industrial level, and even if it is very simple and cheap, it can't be seen as a long-term solution to the plastics issue. Indeed, the only existing possibility for increasing the quote of mechanically recycled PET stands in the increase of the volumes of collected bottles, but once the bottles are downcycled, they irreversibly exit the loop.

2.3 Industrial solutions and PET super-clean recycling technologies

After the collection and sorting step, mechanical recycling involves several treatments and operations: stabilization, extrusion by heating, pelletization, and reprocessing into new PET products.

Mechanical recycling is commonly used to perform open-loop recycling: indeed it is finalized to the production of polyester sheets and fibers from waste bottles. This process offers the advantage to be very simple thanks to the possibility of directly re-granulate and re-process the plastic into new products, but on the other hand, this approach is strongly influenced by both the impurities and the discontinuity of the feed. Moreover, because of the thermal treatment and also because of the formation of free radicals, the polymer quality, and consequently its market price, decreases (López et al., 2014; Rieckmann et al., 2011). A useful parameter for describing the mechanical and physical properties of a polymer is the intrinsic viscosity (IV), and its change during the polymer life cycle and recycling processes is shown in Fig. 11, from which it is clear that post-processing of the mechanically recycled material is crucial to restore the intrinsic viscosity and to put back on the market such polymers.

In Table 2, the main advantages and disadvantages of the mechanical recycling processes are summarized.

Beyond the conventional mechanical recycling processes, it is worth to mention also the "super-clean" technologies, which are finalized to the closed-loop recycling (performing a bottle-to-bottle recycling loop) (Welle, 2011). This kind of mechanical recycling processes differ from the classical ones because of two pieces of equipment: firstly the *decontamination unit*, which is used in every recycling process that aims to achieve a

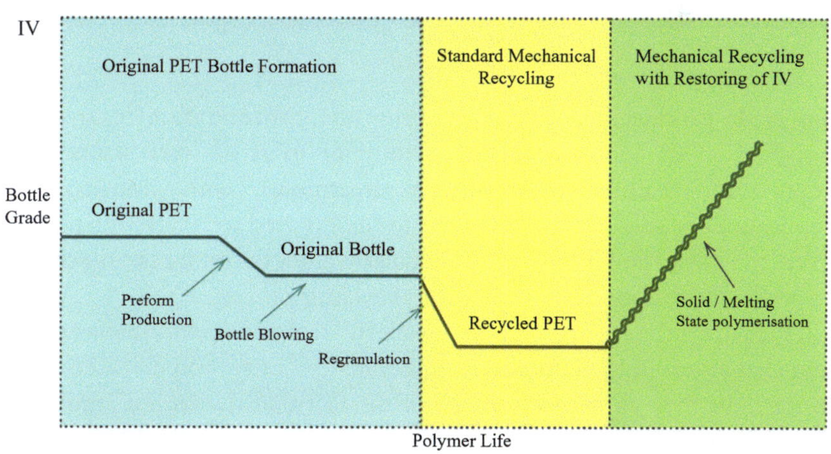

Fig. 11 The behavior of the PET intrinsic viscosity during the PET lifetime and the recycling process.

Table 2 Main advantages and disadvantages of the mechanical recycling processes.

Advantages	Disadvantages
Relatively simple	Paper labels and label adhesive (polyvinyl acetate/ ethylene vinyl acetate) cause the PET to discolor and lose clarity, and the presence of colored PET wastes imparts an undesirable gray color to the recycled PET.
Require low investments	PET containing residual moisture degrades readily when reprocessed, if not dried, and Thermal and oxidative degradation products cause yellowing and diminish the mechanical properties.
Make use of established equipment	Collection, sorting, and separation costs are high because of the low-bulk density of PET bottles and the stringent requirement to have well-sorted feedstock.
Flexible in terms of feedstock volume	PVC liners in bottle caps are problematic to the PET recycling process. PVC and PET have almost the same density and are difficult to separate from each other. PVC releases hydrochloric acid during PET reprocessing reducing the commercial value of recycled PET.
Have a minor environmental impact	Trace metals such as antimony, cobalt, and manganese (from the catalyst and additives) present in post-consumer PET waste promote transesterification and polycondensation reaction in the r-PET. This makes the r-PET chemically heterogeneous and may affect the melt rheology behavior from batch to batch.

food-grade product, in order to decrease the concentration of unwanted substances in the polymer using vacuum or inert gas treatments, high-temperature treatments, or surface treatments with non-hazardous chemicals. Secondly, the *re-polymerization unit* is used to restore the polymer intrinsic viscosity and its average molecular weight to restore the mechanical performances that got lost during the recycling process.

There are different kinds of "super-clean" technologies available, each of them characterized by different decontamination or re-polymerization procedures, but they all take advantage of the fact that the largest amounts of impurities are absorbed in the first 10 μm surface layer of the PET polymer layer, as shown in Fig. 12. Therefore, the decontamination step can be focused on the very thin external layer of the PET bottle itself, and how

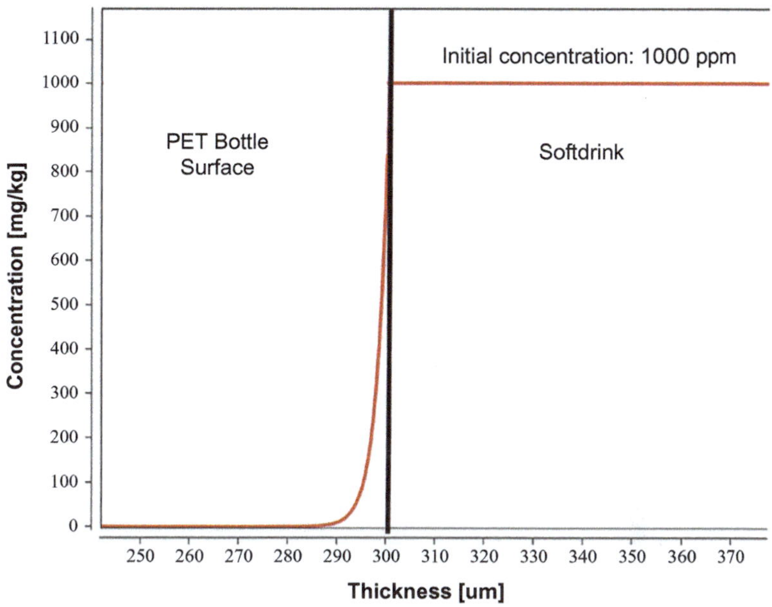

Fig. 12 Distribution of absorbed contaminants in a PET bottle as a function of its thickness.

this is done, it is the differentiation between different "super-clean" technologies from the other:

The three most diffused "super-clean" technologies for PET mechanical recycling are the one based on pellets, the one based on flakes, and the one based on the depolymerization of the flake surface, and their main advantages and disadvantages are highlighted in Table 3.

In the processes **based on pellets** (Starlinger, 2022), the flakes are carefully sorted and they are degassed and dried in a first unit to remove all the oxygenated compounds and to prevent the formation of free radicals responsible for the polymer deterioration during the extrusion and re-pelletization processes. Then, the pellets are transferred into a solid-state polymerization reactor (SSP) that works under a vacuum and below the polymer melting point (at temperatures ranging from 180 °C and 230 °C) for 6 to 14h. Inside the SSP unit, it is possible not only to increase the IV but the volatile organic compounds (VOCs) are also removed from the pellets by diffusion.

Table 3 Main advantages and disadvantages of the mechanical recycling processes.

Technology	Advantages	Disadvantages
Pellet Technology	a. Relatively simple process b. Homogeneity of pellet feed after extrusion c. Pellet easily moved	a. High diffusion limitation b. High reaction time
Flake Technology	a. Reduce the reaction time and the energy consumption	a. Flakes inhomogeneity b. Require one more unit
Surface Depolymerisation Technology	a. Decontamination unit improved	a. Flakes inhomogeneity b. Require one more unit c. Cost of the solvent

In the processes **based on flakes** (EREMA, 2022), there's an important improvement compared to the previous one: indeed, differently from the technology based on pellets, the decontamination and re-polymerization are carried out in two different units. Then, since the major amount of impurities is located in the first layer of the flake surface (as shown in Fig. 12), this process directly decontaminates the flakes both from the oxygenated compounds and the volatile impurities, taking advantage of the higher specific area of the flakes (compared to the pellets). Finally, the clean flakes are directly re-polymerized in a solid or molten state. This final re-polymerization step requires less time (around 3 to 4h) thanks to the absence of the decontamination step. This process ends up with the usual extrusion and pelletization.

In the processes **based on the depolymerization of the flake surface**, the decontamination and re-polymerization steps are still performed in two different units, but the decontamination step is carried out differently. Indeed, in this process, the content of the contaminants inside the final recycled PET is reduced by depolymerizing the flake surface, and this is usually done using a caustic solution (often with NaOH) (Welle, 2011).

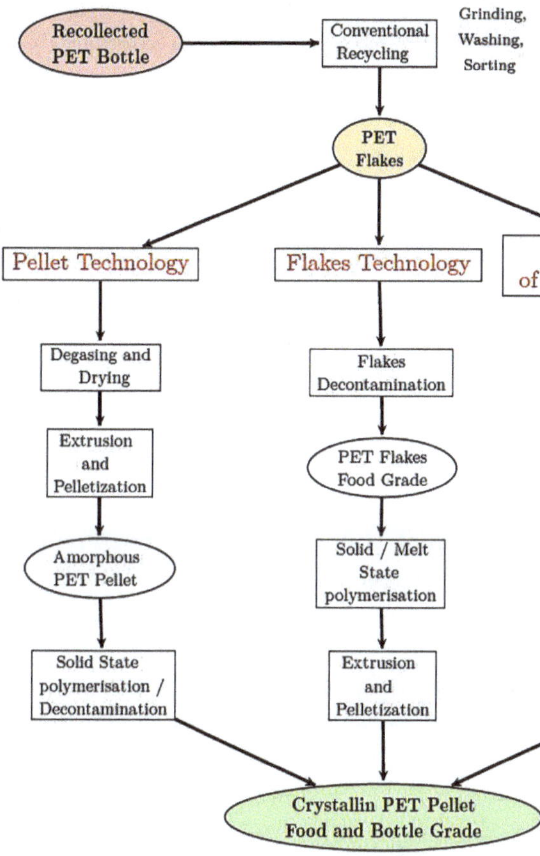

Fig. 13 Flow diagram of the three presented "super-clean" technologies.

In Fig. 13, the flow diagram for all the presented "super-clean" PET mechanical recycling technologies is schematized, from the collected post-consumer bottle waste to the final food-grade polymer.

3. PET chemical recycling

Chemical recycling could be the next big thing about PET recycling, allowing the production of monomers or other petrochemical feedstocks starting from any post-consumer waste stream (Shojaei et al., 2020). In particular, a first distinction can be made between methods that lead to the production of fuels (i.e., pyrolysis), and methods that lead to the production of monomers which then can be used to get back a polymer having the same quality as the starting one (i.e., monomer recycling).

Pyrolysis can be applied to any of the synthetic polymers found in post-consumer waste streams (and on biomasses as well), and it is a much easier and tolerant technique for the treatment of mixed and contaminated wastes (Qureshi et al., 2020). On the other hand, only monomer recycling may give the possibility to perform closed-loop recycling. Therefore, this chapter will mainly be focused on the latter, as it represents the ideal method for the preservation of the limited fossil resources and the protection of the environment from the production of large volumes of non-degradable virgin plastics.

3.1 Monomer recycling

Monomer recycling cannot be applied to all of the most diffused polymers on the planet, since it is usually exploitable only for condensation polymers (such as PET and PU) and not on step-growth polymers (such as PE and PP). Indeed, PET comes from the polycondensation of terephthalic acid (TPA) with ethylene glycol (EG), and it can be quite easily depolymerized by the cleavage of the ester bonds. In other words, monomer recycling is surely a feasible option for polyethylene terephthalate (Thiyagarajan et al., 2022).

Different technologies are currently available for carrying out PET monomer recycling, such as hydrolysis, glycolysis, methanolysis (which are schematized in Fig. 14), and many others. These are indeed depolymerization reactions, aiming to selectively depolymerize PET straight to its

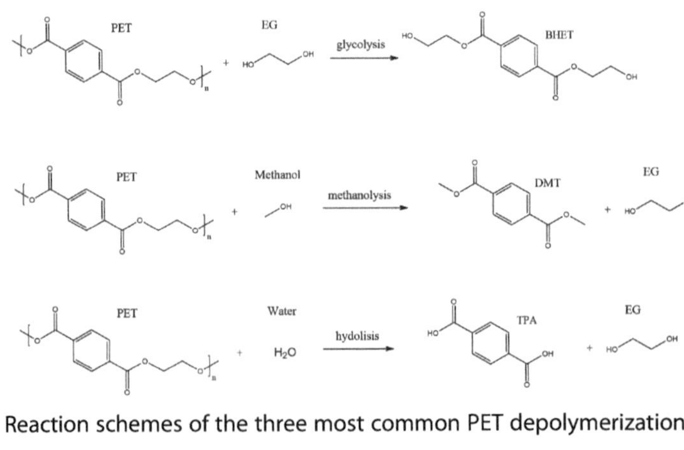

Fig. 14 Reaction schemes of the three most common PET depolymerization reactions.

monomer building blocks (namely terephthalic acid—TPA, or bis hydroxyethyl terephthalate—BHET, or dimethyl terephthalate—DMT) (Shojaei et al., 2020). In any case, the depolymerization step is followed by downstream processing which allows to firstly separate the PET monomers from any of the other materials, such as residual oligomers, other blended polymers, colorants, dyes, and contaminants, and secondly to purify the obtained monomers. The final goal is to obtain monomers of the same purity grade as the virgin ones, i.e., the ones obtained directly from fossil feedstocks (in the case of TPA, the virgin monomer is produced from the para-xylene obtained by catalytic reforming of petroleum naphtha) (Tomás et al., 2013).

Therefore, PET monomer recycling can overcome the two main limitations of mechanical recycling: it can treat any kind of PET wastes (even textile fibers) and separate monomers that are exactly equal (i.e., having the same molecular structure) to those obtained starting from fossil resources. Hence, the polymer obtained from the chemical recycled monomers is expected to have the same quality as the virgin PET.

However, the main reason why monomer recycling is currently not diffused at all in the industry is the higher cost and complexity of such a technique when compared to mechanical recycling, especially concerning the steps of monomer recovery and purification. Indeed, monomer recycling processes may be quite simple and relatively cheap when they have to treat the purest fraction of the PET wastes (which however can be treated also by mechanical recycling), but they gain significance when they start treating the lowest quality fraction of the wastes (Nikiema and Asiedu, 2022). As a matter of fact, in the actual market situation, mechanical recycling is cheaper and it has to be preferred over monomer recycling when possible.

Indeed, as already introduced, the price of this high-quality waste fraction is increasing such that the PET pellets obtained from mechanical recycling of the cleanest waste fraction have reached prices even higher than those of virgin PET pellets. Of course, this price trend makes sense only when considering the increasing social awareness about sustainability, as also many governments are enacting laws and incentives on this issue (for instance, forcing companies to add a certain variable amount of recycled material in their plastic containers) (PlasticsForChange, 2020).

Therefore, during the last years, industrial and academic research has been focused on new technologies able to treat lower quality PET wastes or to maintain the quality level of the treated wastes, but currently, only the treatment of the cleanest and purest waste fraction is economically feasible. No solution able to treat mixed and low-quality wastes is indeed currently widespread at the industrial level. Chemical recycling is the only method that has the potential to treat the lowest quality wastes such as blended layered plastic products or textile fibers, but even if some technologies do already exist, they are still at the development stage and they are mainly operated on very small-scale plants (NextChem, 2022). Moreover, they mostly use quite clean PET wastes rather than textile fibers or layered materials as raw input materials, even if the great advantage of such technologies is the possibility to use very low-quality wastes which have low or even zero economic value.

This evidence, together with the increasing prices of both mechanical recycled and virgin PET products, may make some of the chemical recycling processes that are nowadays being developed economically feasible in the close future. Indeed, academic research as well as many companies have started focusing on the chemical recycling of PET in the last years, and in particular on the treatment of low-quality PET wastes, i.e., mixed textiles made of polyethylene terephthalate blended with other non-polyester fibers (Jeplan, 2022).

3.2 Different approaches to the PET monomer recycling

Polyethylene terephthalate is the most diffused condensation polymer worldwide, and its depolymerization reaction had been one of the first studied polymer chemolysis and is currently being studied extensively.

The three main technologies employed in the monomer recycling of PET are glycolysis, methanolysis, and hydrolysis, and they are the most present both in the scientific literature and in industry practice even if mainly just on prototypes or pilot plants. Lately, some works on other technologies such as aminolysis, ammonolysis, enzymatic depolymerization, and UV-assisted depolymerization have also been published, but since they seem very far from being economically feasible, none of these has been explored industrially, at the best of our knowledge. Moreover, apart from enzymatic depolymerization which is currently being largely studied, their contribution is quite marginal also in academic research, Indeed, the scientific publications are still mostly focused on glycolysis, methanolysis, and hydrolysis as feasible and most promising technologies for carrying out the PET depolymerization and thus, for performing PET monomer recycling (Damayanti and Wu, 2021).

In any case, PET chemolysis methods can produce monomers (for instance BHET, DMT, or TPA) that are employed to re-polymerize

PET, or they may lead to a mixture of oligomers (polyols) which can be employed in the synthesis of other polymers such as polyurethanes or polyesters. The main strategies to depolymerize PET, and the products which are generated in each case, are depicted in Fig. 9.

3.2.1 *Glycolysis*

Glycolysis is the most established method used to depolymerize PET since it is extremely versatile and quite tolerant to contaminations present in the waste materials. The glycolysis reaction, employing glycols such as ethylene glycol (EG), diethylene glycol (DEG), or triethylene glycol (TEG), leads to depolymerization of PET and to the formation of the monomer BHET together with minor yields of small oligomers (mostly dimers and trimers). Then, the obtained BHET can be used as the starting material to produce PET via polycondensation, or if the glycolysis is operated in order not to produce the monomers but longer oligomers, then those oligomers can be employed for the synthesis of other polymers such as unsaturated polyester resins, polyurethanes, epoxy resins, and many others (Sheel and Pant, 2019).

Usually, the glycolysis process is carried out in a range of temperatures between 170 and 260 [°C] at atmospheric pressure, and for reaction times between 20 min to 12 h. Conventional glycolysis processes make use of a catalyst to speed up the reaction and achieve complete depolymerization (Chaudhary et al., 2013; Chen et al., 1999). Usually, common transesterification catalysts such as Zinc Acetate are preferred, but many others have been tested. On the other hand, concerning the raw materials, in most publications, shredded small PET flakes are chosen (López-Fonseca et al., 2010).

The glycolysis reaction proceeds in three steps: firstly, the heterogeneous depolymerization of the starting PET polymer into small oligomeric chains, secondly the further depolymerization of these oligomers into dimers and very small oligomers of BHET, and lastly the equilibrium reaction in a homogeneous phase between the BHET dimer and the monomer. Therefore, the final product distribution will account for a quote of the BHET dimer together with the monomer (Sheel and Pant, 2019). Moreover, the rate-determining step of the conventional glycolysis reaction is the first one, i.e., the heterogeneous reaction (Viana et al., 2011) which is diffusion controlled, since it requires that EG is absorbed in the PET pellet which then swells and increases the diffusion rate of the ethylene glycol. Then the glycol reacts with an ester bond present in the PET chain and degrades the polymer into shorter chain oligomers. Since the reaction operates under a diffusion regime, Lòpez-Fonseca et al. demonstrated that the apparent depolymerization reaction rate was strongly dependent on the pellet surface area (López-Fonseca et al., 2011). Moreover, it was reported that an optimum for PET conversion after 90 min of reaction time was achieved for pellets as small as 0.1275 mm, thus for powdery-like pellets (El-Toufaili et al., 2006; Goje and Mishra, 2003). Summing up, the reaction occurs in a heterogeneous phase in the initial stages, and when a sufficient amount of BHET and oligomers are formed, the reaction proceeds in a homogeneous liquid phase until equilibrium is reached.

Mono ethylene glycol (MEG, also referred as EG) is the most common glycol used for the glycolysis reaction, and since it boils at 197 [°C], most of the scientific publications report 197 °C as the maximum reaction temperature. Higher temperatures were also explored (i.e., up to 240 °C) when using DEG or TEG instead of MEG (Hoang et al., 2019).

In the case of EG, temperatures below 180 °C are scarcely reported since both kinetics and thermodynamics allow low yields, while just by increasing the temperatures between 190 and 197 °C, the yields in BHET are reported to reach values up to 87% - 90%.

3.2.1.1 Catalyzed glycolysis

As anticipated, in common glycolysis reactions, these yields are only achievable when adding a catalyst to the reaction system, the most common ones are: metal-derived catalysts, zeolites, and ionic liquids.

Among the metal-derived catalysts several sub-categories can be identified: metal acetates (manganese, zinc, cobalt, lead) are the most employed (Khoonkari et al., 2015), as they allow yields in BHET as high as 90%, and the activity trend was reported to be as $Zn^{2+} > Mn^{2+} > Co^{2+} > Pb^{2+}$. On the other hand, also carbonates and bicarbonates were reported to be effective catalysts for PET glycolysis, reaching conversions as high as 79% after comparable reaction times, and these are also much less detrimental to the environment compared to the metal acetates (López-Fonseca et al., 2011). Also, other metal-derived catalysts such as sodium and potassium sulfate or lithium and magnesium chlorides have been tested, but the results in terms of depolymerization efficiency turn out to be lower (Bartolome et al., 2012; Pingale et al., 2009). Zeolites have been tested as well, and according to Shukla et al., they were able to achieve maximum yields of BHET around 68% at 196 °C (Shukla et al., 2008), while metal oxide catalysts impregnated on silica nanoparticles managed to improve the catalytic activity reaching BHET yields equal to 90% (Imran et al., 2011). Lastly, ionic liquids (salts with a melting point below 100 °C) were also tested for degrading PET via glycolysis, and they allowed an easy removal and repetitive utilization but required high volumes and higher reaction times (4 to 8 h at 180 °C) compared to the common transesterification catalysts for achieving nearly 100% PET conversion (Yue et al., 2011).

Together with the conventional methods used to carry out PET glycolysis (using a catalyst), during the last few years other 3 interesting techniques have emerged: supercritical glycolysis, microwave-assisted glycolysis, and solvent-assisted glycolysis.

3.2.1.2 Supercritical glycolysis

Supercritical glycolysis was carried out at 450 °C and 15.3 MPa in an autoclave reactor with a 10 mL volume by Imran et al., leading to yields in BHET of 93% after just 30 min of reaction time, while for longer reaction times the yields in BHET decreased sharply (Imran et al., 2010). Supercritical glycolysis was not studied extensively since it requires very harsh reaction conditions and it leads to comparable results with those of catalyzed glycolysis in terms of BHET yields, even if the reaction times are reported to be reduced from 120 min to only 30 min.

3.2.1.3 Microwave-assisted glycolysis

Microwave-assisted glycolysis is another option for PET depolymerization and it is being studied as well since it may help reduce the reaction time without compromising the BHET yields and keeping milder reaction conditions compared with supercritical glycolysis. In particular, Chaudhary et al. reported yields in BHET above 95% after just 30 min of reaction time and using the transesterification catalysts typically used for conventional PET glycolysis, such as zinc acetate and sodium carbonate (Chaudhary et al., 2013).

3.2.1.4 Solvent assisted glycolysis

On the other hand, in solvent-assisted glycolysis, an external solvent is introduced into the reaction environment together with ethylene glycol. Wang et al. reported that the glycolysis of PET using EG as a solvent and a deep eutectic solvent (as Urea/ZnCl2 and Urea/MnAc2) managed to reach 100% PET conversion and selectivity in BHET above 80% (Wang et al., 2015). Many other studies report other solvents such as Urea (Wang et al., 2012), or ionic liquids as 1,3-diethylimidazolium acetate ([deim]ZnOAc) (Zhou et al., 2012), reaching selectivities in BHET above 65%.

3.2.1.5 Combined glycolysis methods

Furthermore, some works are also popping up combining different glycolysis strategies, such as using ionic liquids under microwave irradiation, improving the results obtained with ionic liquids and conventional heating (which led BHET selectivity up to 78% after 24h of reaction time) to 99% BHET selectivity after just 3h reaction time at 170 °C (Cot et al., 2019).

3.2.2 Methanolysis

Methanolysis is the alcoholysis reaction of PET with methanol. In principle, other alcohols may be used to perform the PET depolymerization, which in general is referred to as "alcoholysis." Methanol is by far the most employed alcohol for PET depolymerization, but also ethanol, propanol, isoctyl alcohol, and many others were tested (De Castro et al., 2006; Liu et al., 2013).

In the methanolysis reaction, methanol is used to break down the PET polymer chain into the monomer DMT (dimethyl terephthalate) and ethylene glycol, which are also raw materials required for the production of PET, even if recently the PET polycondensation plants usually prefer TPA as monomer raw material.

Differently from glycolysis, no equilibrium exists between the monomer DMT and its dimer or trimer, thus the methanolysis reaction leads only to the monomer DMT. Oligomers are not formed and thus the depolymerized products are not directly employed for the production of polyurethanes or other resins. PET methanolysis has been known for many years, and it is one of the most studied and promising technologies for the chemical recycling of polyethylene terephthalate since it can treat more contaminated and lower-quality feedstocks compared to glycolysis. Indeed, the subsequent step of DMT purification is easier with respect to the BHET purification. The drawback is that methanolysis is more expensive than glycolysis, and it may become economically feasible only on very large-scale plants.

The PET methanolysis reaction can be operated in a wide range of conditions, with temperatures between 180 °C and 280 °C, pressures between 20 bar and 110 bar, and also with or without catalyst. In particular, according to the employed reaction conditions, 3 different methods for the PET methanolysis reaction are distinguished: liquid, vapor, and supercritical methanolysis.

3.2.2.1 Liquid Methanolysis

Liquid methanolysis is usually carried out at pressures ranging from 20 bar to 40 bar and temperatures ranging from 140 °C to 280 °C. It is a catalytic reaction, where the employed catalyst is usually one of the common trans-esterification catalysts used also in catalyzed glycolysis (Paszun and Spychaj, 1997). The middle-high pressures are selected to maintain methanol in the liquid state at the selected temperature, and yields in DMT above 80% were reported for these conventional liquid methanolysis processes. In particular, Mishra and Goje studied the influence of particle size and reaction time on a conventional PET liquid methanolysis reaction at 120 °C using zinc acetate and lead acetate as catalysts, reporting the best results for 127.5 [um] particle size and for 120 min reaction time, which allowed a 97.8% PET conversion and 97.6% DMT yield, while complete PET depolymerization was reached at 130 °C (Mishra and Goje, 2003b). On the other hand, some recent studies also claim 93% yields in DMT operating the methanolysis reaction at 25 °C and atmospheric pressure using sodium carbonate as the catalyst and DCM as cosolvent, even if they required up to 24 h of reaction times (Pham and Cho, 2021).

Since methanolysis is the most suited PET chemical recycling method for the treatment of contaminated wastes, and since it can compete with glycolysis only when treating this lower-quality and cheaper wastes, then the DMT purification step gains great significance in the process economics. DMT is usually separated from the methanolysis products by crystallization and then further purified by standard procedures such as re-crystallization (as shown in Fig. 15), melt crystallization, or distillation (Robert Donald Slockett, 1972; Toot, 1995).

3.2.2.2 Vapor methanolysis

Vapor methanolysis allows milder reaction conditions as it is carried out at similar temperatures as liquid methanolysis (i.e., between 140 °C and 280 °C). High pressure is not required as the methanol is fed as of super-heated vapor. Methanol bubbles through the reacting mixture constituted of PET, oligomers, and ethylene glycol. PET vapor methanolysis is, therefore, a liquid-vapor reaction, thus intrinsically much more complex than homogeneous PET liquid methanolysis. Mass transfer limitations are much more relevant in vapor methanolysis thus requiring vigorous mixing. However, reaction rates remain smaller than those of liquid methanolysis, as reported in the literature. In addition, as typical for heterogeneous liquid-vapor reactions, the reaction rates depend on temperature and pressure. Kim and Han studied these dependencies as well as that from methanol feeding rate and catalyst amounts, identifying the optima operating condition in terms of DMT yield at 250 °C and 5 bar (Kim and Han, 2004).

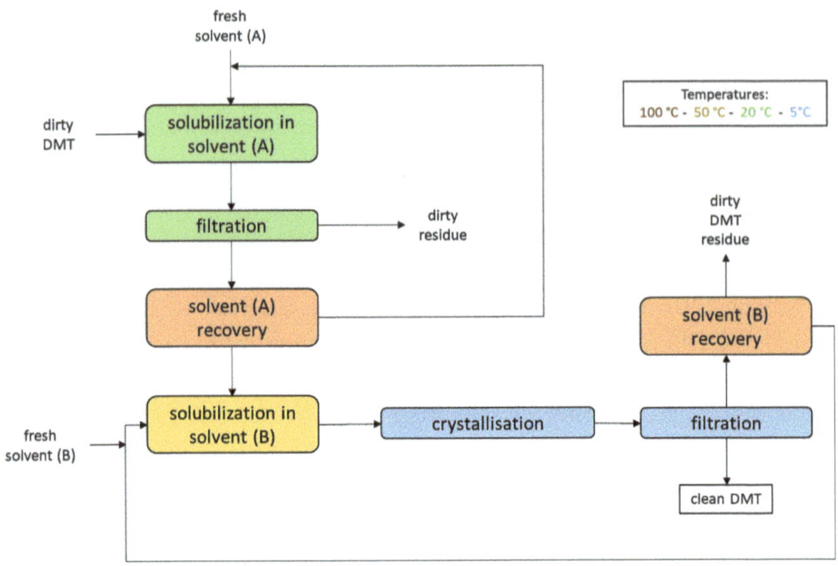

Fig. 15 Flow diagram for a DMT purification and re-crystallization process.

Indeed, higher pressures reduce the DMT removal rate, while higher temperatures excessively reduce methanol solubility in the liquid phase. However, the main advantage of vapor methanolysis is the possibility of removing part of the DMT in the vapor stream, since methanol acts as a carrier that manages to strip the monomer DMT leaving the oligomers in the liquid phase. This enhances PET conversions and yields in DMT. Another advantage of vapor methanolysis is the possibility to treat more contaminated PET wastes since it allows the instantaneous separation of DMT from the high boiling materials which may be present in the waste stream such as colorants and dyes (Gamble, 1993). However, when starting from highly contaminated wastes, DMT has to be further treated and purified before re-polymerization (Pudack et al., 2020). A patented technology was presented by Eastman Kodak (Eastman Kodak, 1999), even if currently most of the attention is on PET supercritical methanolysis.

3.2.2.3 Supercritical Methanolysis

Supercritical methanolysis requires quite severe reaction conditions to keep methanol in the supercritical state, i.e., temperatures above 300 °C and pressure above 80 bar. In this region, the reaction was reported to proceed much faster than both in the liquid and the vapor methanolysis cases (Han, 2019). This observation increased interest and after the first studies carried out in

1997 (Sako et al., 1997) supercritical methanolysis is now the most studied methanolysis reaction. As the supercritical fluids exhibit high densities (i.e., similar to liquids) and high kinetic energies (i.e., similar to gases), they constitute a convenient medium for any kind of reaction. Indeed they allow a faster reaction rate compared to their liquid or vapor counterparts. Supercritical methanolysis of PET is reported to have extremely high reaction rates, and the addition of a catalyst is not even necessary to achieve full depolymerization in reasonable times (i.e., around 10 min vs around 5 h for liquid methanolysis (Shojaei et al., 2020)). Goto et al. developed a reaction mechanism and performed a detailed kinetic study on PET supercritical methanolysis in batch conditions (Goto et al., 2002). Many studies focused on optimizing process conditions and on the identification of the most influential parameters on DMT yield. Yang et al. reported that supercritical methanolysis is mostly influenced by three main factors: the amount of methanol, temperature, and reaction time, while the process is not very sensitive to pressure (Yang et al., 2002). Further studies claimed that maximum yield in DMT can be obtained with methanol to PET ratio equal to 6, a temperature of 298 °C, and a reaction time of 112 min (Liu et al., 2015). Lower reaction times are claimed by Shojaei, Abtahi and Najafi, which report times as low as 5 min. Longer times (30 min) are reported by Sako et al.. The conclusions of Liu et al. and of Yang et al. differ also in relation to the influence of methanol to PET ratio, which is not found to impact PET conversion nor the DMT yield for the former and of relevance for the latter. Supercritical methanolysis is not been deeply investigated so far, and it is currently one of the main topics of research in the area of PET chemical recycling. Its main advantage is that of leading to a high-quality DMT matching the standards of virgin DMT. DMT is then separated from ethylene glycol and methanol together with the monomer within the reactor. The main disadvantage, motivating a long time to market of such technology, is related to the economics and the complexity of the process. The severe conditions, the complicated continuous operation and the high operating costs are sustainable only on a very large scale. Additional costs are caused by hydrolization of DMT to TPA which is the typical raw material in PET polycondensation plants (Achilias and Karayannidis, 2004).

3.2.3 *Hydrolysis*

In the hydrolysis reaction, water is used to break down the PET polymer chains into the monomer terephthalic acid (TPA) and ethylene glycol (EG). PET hydrolysis is a mature technology and it can be carried out in acid, neutral, or alkaline conditions at temperatures ranging from 100 °C to 300 °C and pressures up to 100 bar. Most of the published studies on hydrolysis report two major drawbacks: the high purification costs of TPA, and the long reaction times. However, hydrolysis attracts great attention as it requires TPA as a reactant rather than BHET or DMT.

To obtain pure TPA rather than BHET or DMT, hydrolysis was also investigated as a final step applicable to a purified DMT stream previously obtained from other processes (i.e., methanolysis). Combined approaches for PET depolymerization such as this methanolysis-hydrolysis approach will be discussed more in detail in the following.

3.2.3.1 Neutral hydrolysis

Hydrolysis under neutral conditions had been one of the first investigated methods for PET depolymerization, and it may be performed with liquid water or steam. Neutral hydrolysis is typically carried out at relatively severe conditions, i.e., temperatures ranging from 200 °C to 300 °C and pressures from 10 bar to 40 bar (Mandoki, 1986). Such conditions are needed to activate neutral water toward PET depolymerization. To speed up the kinetics, temperatures above the PET melting point (265 °C) are often preferred, and pressures to keep water in the liquid state are chosen accordingly. Studies on heterogeneous hydrolysis reaction was reported in (Mancini and Zanin, 2004), but reaction times of at least 6 h are needed for temperatures below 205 °C. This makes homogeneous neutral hydrolysis reaction preferable.

Campanelli et al. reported that neutral hydrolysis at 265 °C led to complete depolymerization within 2 h of reaction time, and claimed that transesterification catalysts such as zinc acetate and sodium acetate had a slight effect on the increase of the depolymerization rate (Campanelli et al., 1994).

In the hydrolysis reaction, since water is both the reactant and the solvent, despite its poor features as a solvent for PET, a large excess is required. Some researchers proposed the addition of a cosolvent, together with PET and water. This also allows adding water at stoichiometric conditions. High boiling xylenes are within the tested co-solvents and they allow a faster depolymerization rate even at lower temperatures and pressures compared to conventional neutral hydrolysis methods (Güçlü et al., 2003). The addition of a cosolvent to better solubilize PET flakes and to increase the reaction rates is gaining quite a lot of interest in the last years, also for the PET methanolysis reaction (Pham and Cho, 2021).

The main advantage of hydrolysis is the neutral and non-corrosive medium: no salts or wastes are generated and the equipment does not require much maintenance. This may endorse hydrolysis as a "greener" approach compared to acid or alkaline hydrolysis. On the other hand, the harsh conditions in terms of temperatures and pressures are responsible for high operation costs, and, most importantly, the TPA produced is not easily separable from the contaminants of the initial raw materials. Thus some complex and expensive purification units such as filtrations and multiple crystallizations are required (Karayannidis et al., 2002; Michalski, 1987).

As a matter of fact, neutral hydrolysis is not much promising for industrial applications, and studies on the topic in the last years are scarce. A recent publication investigated PET neutral hydrolysis in a metal salts mixture from ocean waters acting as catalysts, but the assessment of its possible industrial application was out of the scope of the study (Stanica-Ezeanu and Matei, 2021).

3.2.3.2 Acid hydrolysis

To soften the reaction conditions typical of neutral hydrolysis, acid hydrolysis makes use of strong acids, such as sulfuric acid or nitric acid in concentrated solutions ranging from 5 M to 15 M. The operating temperatures are much lower compared to neutral hydrolysis, and also to inhibit corrosion, and pressures are usually atmospheric.

Yoshioka et al. proved that even a 7 M sulfuric acid solution allowed to achieve complete PET depolymerization in 5 h at 150 °C (Yoshioka et al., 1994), while previous processes had been designed using much concentrated 15 M sulfuric acid solutions (A. et al., 1985).

The same researchers published a study testing the performances of a nitric acid solution with a similar concentration (between 7 M and 13 M), and they highlighted not very promising results: PET conversion values of 80% were achieved at 100 °C and atmospheric pressure only after 30 h reaction times.

Recently, following the same trend as for neutral hydrolysis and methanolysis, the use of a cosolvent able to speed up the reaction and to ease the downstream purification procedures was studied. For instance, the use of dimethyl sulfoxide turned out to have positive effects on both the aspects: conversions up to 87% were reached using a solution 8 M of nitric acid at 120 °C and atmospheric pressure, for 140 min of reaction time feeding PET in 127.5 um flake size (Mishra et al., 2003).

Moreover, in acid hydrolysis, it is well reported that the depolymerized products have then to be separated and since TPA is insoluble in water, usually it is neutralized in a strongly alkaline solution of sodium hydroxide, then acidified again, and finally reprecipitated. These operations are complex, and expensive, and they carry a consistent amount of wastes containing strong acids and salts. This problem, the stringent corrosion issues that emerge when working at high temperatures in the presence of highly concentrated acidic solutions and the long reaction times required to complete the depolymerization, have slowed down advances in acid hydrolysis in the last decades.

3.2.3.3 Alkaline hydrolysis

Typically, PET alkaline hydrolysis is carried out at temperatures between 150 °C and 250 °C, at pressures up to 20 bar, and using an aqueous solution of sodium hydroxide (4% to 20% weight). Reaction times are mostly reported in the range of a few hours. Alkaline hydrolysis operating conditions are similar to neutral hydrolysis. As an advantage, the purification phase of TPA is easier and the process seems to be applicable at the industrial level. Polk et al. reported a 93% yield in TPA using an aqueous sodium hydroxide solution at 5% by weight after 60 min of reaction time (Polk et al., 1999). Mishra and Goje reported yields in TPA higher than 90% after 90 min of reaction time at 90 °C and atmospheric pressure (Mishra and Goje, 2003a). Several other studies were performed to test the performances of various catalysts. In addition to common transesterification catalysts, some phosphonium and ammonium salts showed promising results in terms of reaction times and TPA yields (López-Fonseca et al., 2008; López-Fonseca et al., 2009).

Moreover, as with many other depolymerization methods discussed above, the effect of cosolvents is being investigated also for alkaline hydrolysis. Cosolvents speed up the reaction rate, increase the yields in TPA, and ease the final purification of the produced TPA. Many examples using ethers (Hu et al., 1997), 2-Methoxyethanol (Karayannidis et al., 2002), and also combining catalysts as magnesium carbonate and cosolvents as dimethyl sulfoxide (Sharma et al., 2013) are published in the literature. Very promising results have been claimed: Sharma et al. later managed to reach complete depolymerization in 10 min at 190 °C.

The easier purification step of the obtained monomer (i.e., TPA) is a great advantage of alkaline hydrolysis, therefore, decreasing the costs. Alkaline hydrolysis does not present corrosion issues, and milder conditions reduce operating costs. Pilot-scale plants are being developed thanks to the great research efforts in recent years and the process is approaching the industrial scale level.. Karayannidis and Achilias report that alkaline hydrolysis, together with methanolysis, are currently the only technologies for PET depolymerization which have the potential to scale up (Karayannidis and Achilias, 2007).

3.2.4 Other techniques (aminolysis, ammonolysis, and enzymatic depolymerization)

Beyond the three most diffused methods for PET depolymerization, other methods such as aminolysis, ammonolysis, and enzymatic depolymerization are currently being investigated, even if the higher costs compared to glycolysis, methanolysis and hydrolysis currently inhibit the pilot-scale and industrial-scale perspective.

3.2.4.1 Aminolysis

Aminolysis of PET is the reaction between the polyester and an amine, leading to the formation of the diamides of terephthalic acid (BHETA) and ethylene glycol, as shown in Fig. 16. Several amines have been reported in the literature to be effective in PET depolymerization. The most common are methylamine and ethylamine (Gupta and Bhandari, 2019). Usually, PET aminolysis is performed at temperatures ranging from 20 °C to 160 °C and at atmospheric pressures. A catalyst is also often employed since the kinetics is reported to be very slow in the absence of a catalyst. Organic catalysts such as 1,5,7-triazabicyclo[4.4.0]dec-5-ene (TBD) (Demarteau et al., 2020; Fukushima et al., 2013) or dibutyl tin oxide (DBTO) (Taw and Eskander, 2010) are preferable.

Cheaper and more common catalysts such as sodium acetate, glacial acetic acid, and potassium sulfate have been tested. The aminolysis reaction was performed using ethanolamine in a 1:6 M ratio with respect to PET, and less than 1% weight of catalyst was added with respect to PET. Reaction times of 8–9 h yielded 91% in BHETA at 80 °C using sodium acetate as the catalyst. Both fibers and bottle wastes were tested in such reactions. Fibers showed slightly better results in terms of BHETA yields in all the experimented conditions (Shukla and Harad, 2006).

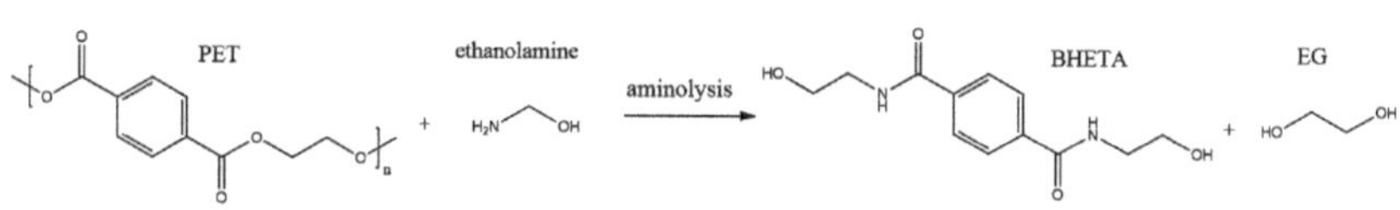

Fig. 16 Reaction scheme of the aminolysis PET depolymerization reaction.

As for the other depolymerization techniques, more advanced studies are reported in the literature, for instance testing the effect of cosolvents and that of advanced catalysts. Musale et al. managed to synthesize two deep eutectics (namely, choline chloride.x ZnCl2 and choline chloride.2 urea) and used those as a catalyst for the PET aminolysis. They report very high yields in BHETA, such as 95%, after 30 min of reaction time and using diethanolamine in a 6:1 M ratio with respect to PET and their catalyst in concentrations up to 8% by weight (Musale and Shukla, 2016).

Moreover, aminolysis was tested also in the presence of ionic liquids: in this case, the PET depolymerization was carried out at 196 °C and at atmospheric pressure using ethanolamine and hydrazine hydrate. Five different ionic liquids were tested, and 1-Hexyl-3-methylimidazolium trifluoromethanesulfonate (Hmim.TfO) turned out to be the best performing one, allowing BHETA yields as high as 89% after 60 min reaction time (Shukla et al., 2008). Some trials were also carried out using UV light to induce the PET aminolysis reaction (Yang et al., 2005).

Furthermore, Ghorbantabar et al. reported an investigation on different analytical methods to track the PET degradation during an aminolysis reaction. Both differential scanning calorimeter analysis (DSC) and thermogravimetric analysis (TGA) turned out to be reliable methods that can be used instead of the conventional gravimetric analysis, which required cumbersome prior product separation and purification procedures. They carried out an aminolysis reaction using ethanolamine in a 5:1 M ratio with respect to PET, and the reaction was kept running for 2 h at 160 °C and atmospheric pressure without any catalyst (Ghorbantabar et al., 2021).

The last technique which has been explored, trying to reduce aminolysis costs is microwave-assisted aminolysis since it allows to heat the reaction mixture faster, therefore reducing the reaction times and saving a considerable amount of energy in comparison to conventional heating methods. Pingale and Shukla tested the aminolysis of PET wastes (fibers and bottles) using very similar conditions to the ones previously chosen by Shukla and Harad (i.e. using cheap and nontoxic catalysts such as sodium acetate and sodium sulfate), but employing microwave heating. The reaction time decreased from 8 h to 4 min, obtaining similar BHETA yields (up to 90%) (Pingale and Shukla, 2008). Similar results are also reported in other studies. For instance Achilias et al. showed that under a 100 W microwave irradiation for 5 min the PET depolymerization through aminolysis is complete (Achilias et al., 2011).

3.2.4.2 Ammonolysis

Similar to aminolysis, the PET ammonolysis reaction makes use of ammonia to depolymerize the polyester, leading to terepthalamide (TPD), but it is much slower compared both to aminolysis and the other depolymerization processes described above. As usual reaction times in the order of days are necessary to achieve good monomer yields. Usually, this reaction is carried out in an aqueous solution of ammonia (for instance 1:10 w/v), at low pressures, temperatures ranging from 40 °C to 180 °C, and in the presence of a catalyst.

Ammonolysis has been less explored compared to other reactions. Jain and Soni reported complete depolymerization after 15 days carrying out the ammonolysis reaction at room temperature, and atmospheric pressure, and using zinc acetate as a catalyst (Jain and Soni, 2007). Mittal et al. reported similar results as they carried out the reaction for 30 days at 40 °C and atmospheric pressure, using a 1:10 w/v ammonia aqueous solution, an ammonium quaternary salt. 100% TPD selectivity was reached but not high conversions (up to 38%) (Mittal et al., 2010).

Processes for PET depolymerization based on ammonolysis have also been patented. For example, Fox and Shafer report yields up to 90% after 7 h of reaction time at 20 bar and 120 °C (Fox and Shafer, 1990).

3.2.4.3 Enzymatic depolymerization

Enzymatic depolymerization in the form of hydrolysis is gaining a lot of attention in the last years, as a greener and less pollutant route for depolymerizing PET. Indeed, enzymatic depolymerization does not generate the so-called "second pollutants" which on the contrary are generated by chemical treatments. Loads of enzymes and bacteria have been tested in the last few years, and this is one of the most active research fields concerning polyester depolymerization. However, several limitations exist in the use of such enzymes. Such limitations are not even considered in chemical depolymerization methods as they are not extremely relevant. For instance, even the degree of crystallinity does strongly influence depolymerization performances (Kawai, 2019). According to Maurya et al., the four main constraints regarding PET enzymatic depolymerization are the low catalytic performances due to the limited accessibility of active sites; the enzymes inhibition caused by MHET (a reaction intermediate product); the low stability of enzymes, and the loss in their performances above 70 °C; and finally the inactivation of the enzymes caused by the progressive acidification of the solution as the reaction proceeds (Maurya et al., 2020).

Two of the most promising microorganisms (and their corresponding enzymes) for carrying out the depolymerization of PET are the "Candida Cyclindracea" (Lipases) (Ma et al., 2012), and the Ideonella Sakaiensis'j (MHETase and PETase) (Ribitsch et al., 2012), but many other enzymes have been tested and showing promising results in terms of depolymerization efficiencies. Such enzymes include cutinase, esterase, and hydrolase (Maurya et al., 2020).

De Castro and Carniel published work on enzyme-catalyzed glycolysis which managed to overcome one of the main disadvantages of conventional enzymatic hydrolysis, i.e., the deactivation of the enzymes due to the acidification of the reaction medium) (Machado et al., 2017).

3.2.5 Combined approaches

All the available methods discussed in the previous paragraphs present both advantages and disadvantages, which are often complementary one to the other. This is the reason why, in principle, two or more methods could be operated in series to get the advantages of both, performing what is called a combined chemolysis process. When using this kind of approach, the main gains are the reduction of the reaction time, the possibility of treating lower quality feedstocks, and the possibility to make reaction conditions milder. In other words, the combined chemolysis methods may allow to reduce the cost of a PET monomer recycling process and make it economically feasible.

The combined approaches which have been more reported are glycolysis-methanolysis, glycolysis-hydrolysis, and methanolysis-hydrolysis.

The two latter combined processes are characterized by a final hydrolysis step, therefore they yield TPA which is the preferred monomer in most PET polycondensation plants. This is the first advantage. Moreover, these two combined approaches use either glycolysis (which is way faster than hydrolysis) or methanolysis (which is much more tolerant to contaminations than hydrolysis) and leave the hydrolysis step at the end. This shortens reaction times (up to one-third) in the case of the glycolysis-hydrolysis process (Saadet and Orbay, 2003) and allows to treat lower quality feedstocks in the case of the methanolysis-hydrolysis process (Genta, 2003).

The third approach (i.e., glycolysis-methanolysis) is also very interesting since it combines the two fastest methods to carry out PET depolymerization. PET is initially dissolved in ethylene glycol, and the first phase of the glycolysis reaction starts, leading to a mixture of PET oligomers; then, the stream of reactants are mixed with methanol and the depolymerization ends. The final product is DMT (Gamble et al., 1994), and it is reported that the second step of the process (i.e., methanolysis) is the most important one, while the "pre-glycolysis" step is just needed to dissolve the polymer to make the subsequent methanolysis much faster and efficient (Bo-Kyung et al., 2008).

Combined chemolysis methods represent a very promising perspective for scaling up PET monomer recycling at the industrial level (Padhan and Sreeram, 2019). Their main advantages are depicted in Fig. 17.

3.2.6 Monomer recycling in the industry

Mechanical recycling is already widespread at the industrial level, energy recovery plants are widely operated as well (Cewep, 2022), and also some large-scale pyrolysis plants started operations in the last years. Conversely, monomer recycling is not mature enough to be profitably employed in industry, but during the very last years, many investments are being funded on several of the different available technologies for PET depolymerization discussed in this chapter.

As already pointed out, the majority of PET polycondensation plants are currently making use of TPA as the starting monomer, while those starting from BHET and DMT are being progressively abandoned or revamped. Due to this reason, hydrolysis gained much attention, while if the PET depolymerization is performed by glycolysis, methanolysis, or any other method, subsequent hydrolysis step to finally yield TPA may have to be considered in the economic analysis of the integrated process (Han, 2019).

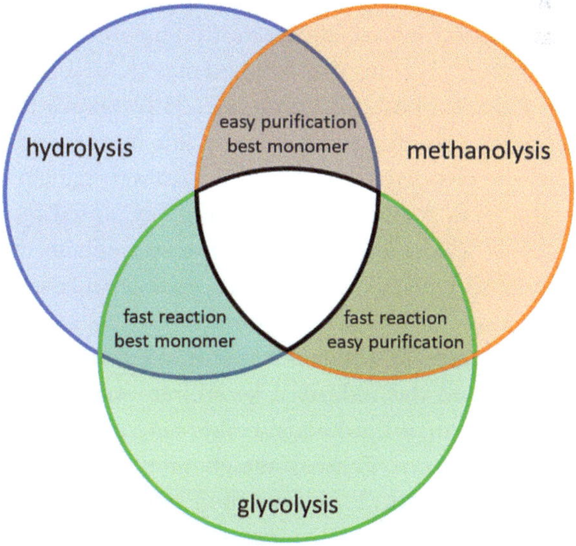

Fig. 17 Main advantages of each of the presented combined chemolysis approach.

Thus hydrolysis is preferable since it produces directly TPA, but on the other side, it is not much suited for the treatment of highly contaminated wastes since TPA purification is quite complex and expensive. Indeed, the hydrolysis reaction is usually reported as feasible on clean wastes, so that the subsequent purification step is not much demanding and the obtained products may be used as such. Therefore, hydrolysis has not been widely explored at an industrial level, not even on the pilot scale. The only two companies which are known to be pursuing hydrolysis to reach the industrial scale are FENC from Taiwan (FENC, 2019), RITTEC from Germany (RITTEC, 2022), and the start-up Gr3n from Switzerland (Gr3n, 2022), with the latter trying to combine the advantages of alkaline hydrolysis (the only type of hydrolysis which is reported to be, potentially, economically feasible) with the advantages of microwave heating.

Conversely, methanolysis offers the advantage of an easier monomer purification step but is only marginally explored at the industrial level due to high costs requiring large-scale plants. Only two big companies are betting on methanolysis: Eastman and Loop Industries. In particular, in 2021 Eastman invested 250 million dollars in building a large-scale molecular recycling facility in Kingsport, Tennessee (Eastman, 2021), while in 2022 they invested 1 billion dollars for another project in France (Eastman, 2022), both based on their methanolysis technology. Together with Eastman, also Loop Industries is investing in France (Goldsberry, 2020).

On the other hand, glycolysis is currently the most trusted technique for a potential scale-up in the industry, and many companies are investing in it. For instance, Jeplan in Japan (Jeplan, 2022), PerPETual in the UK (PerPETual, 2022), Ioniqua and CuRe in the Netherlands (CuRe, 2022; Ioniqua, 2022), Garbo in Italy (Garbo, 2022), and IFPEN in France (IFPEN, 2022). The majority of these technologies claim to be able to treat also the low quality, contaminated and blended fraction of PET wastes (i.e., even textiles), even if in principle glycolysis is not the best technology suited for such wastes. Indeed, they include a customized purification step which, for instance in the case of Garbo and Jeplan, is reported to be able to reach a food-grade purity.

Among the other presented techniques, only enzymatic depolymerization has already reached the industrial level, but very few companies are active in this area. The major industrial reality active in enzymatic depolymerization is Carbios from France, which after having validated their technology at a small scale, 2021 launched a first demonstration plant (Carbios, 2022).

3.2.7 Recycling via ring-chain equilibria

Polymer recycling via ring-chain equilibria is one of the most promising alternatives and possible approaches for the chemical recycling of polycondensation polymers, yielding macrocyclic oligomers instead of the canonic monomers. For example, polybutylene terephthalate, polyethylene terephthalate (PET), and polyethylene furanoate (PEF), can be degraded by a process known as ring-closing depolymerization or cyclodepolymerization (CDP). Usually, this technique is often applied without solvent and under vacuum, and the macrocyclic oligomers are distilled away from the reacting mixture during the reaction, as for example the poly lactic acid (PLA) (Ehsani et al., 2014). CDP of PET has been reported only with the use of di-n-butyltin oxide, stannoxane and tetraisopropyl ortho titanate as catalysts (Bryant and Semlyen, 1997; Hodge et al., 2000; Matsumura et al., 2008). All these three catalysts show a distribution of macrocyclic oligomers, with the highest contribution being the PET trimer. The macrocyclic oligomers can be re-polymerized, in the presence of a catalyst, to reform PET by Ring-Opening Polymerization (ROP). This technique has great advantages compared to the normal polycondensation reaction since during the ROP no by-products are produced leading to a faster polymerization reaction with a characteristic time of only 30 min instead of the hours (or even days) of the normal polycondensation which starts from linear monomers and oligomers. In Fig. 18, all the presented recycling techniques are schematized. In particular, compared to classic mechanical recycling, ring-chain recycling can treat chemically more complex plastic sources with simple polymerization systems, while when compared to the chemical complete depolymerization techniques, the ring-chain recycling needs lower investements and requires less operative costs (since the repolymerisation reaction is much simpler and faster).

3.3 Pyrolysis

In parallel to the already presented chemical depolymerization reactions (which fall into the category of monomer recycling), PET can be also chemically recycled through a completely different methodology, i.e., pyrolysis.

A pyrolysis is a thermochemical approach, where PET is treated in the absence of oxygen at high pressures and temperatures usually ranging from 400 °C to 900 °C (studies up to 3000 °C are also reported for the treatment of polymers), leading to small hydrocarbons to be used, for instance, as fuels or chemical building blocks. Pyrolysis is classified as a chemical recycling

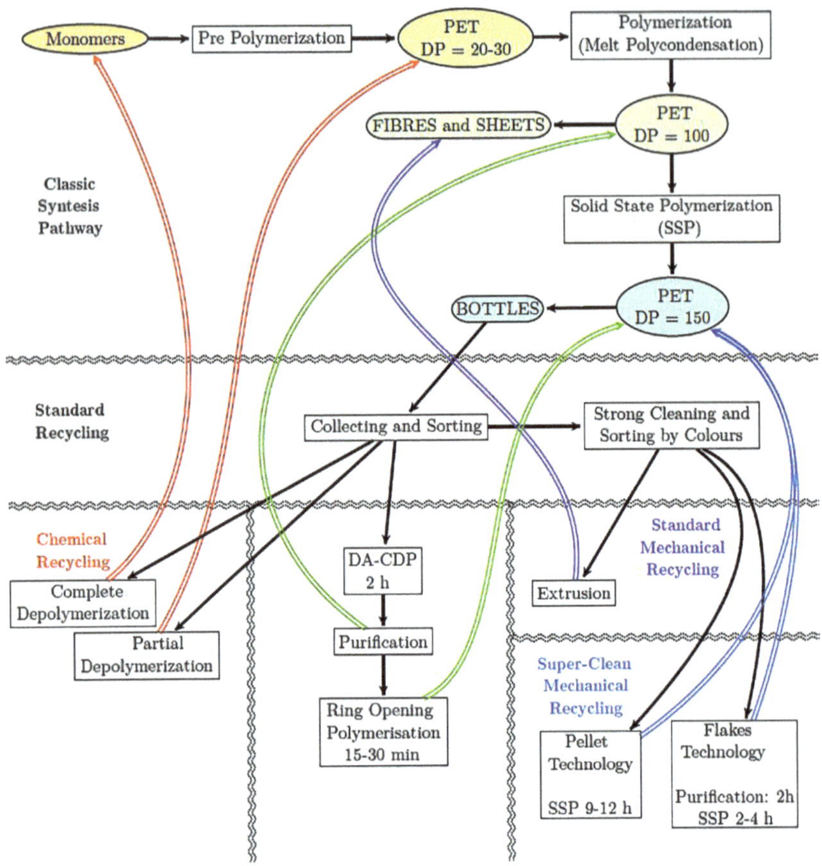

Fig. 18 Comparison of Total Depolymerization, Partial depolymerization and mechanical recycling by Polycondensation and the Ring-chain recycling by Ring-Opening Polymerization.

approach, even if it does not lead to any of the PET monomers (such as BHET, DMT, or TPA) but it rather degrades the polymer chain directly into smaller oxygenated hydrocarbons molecules (1-propanone, benzoic acid, biphenyl, diphenylmethane, 4-ethylbenzoic acid, 4-vinylbenzoic acid, fluorene, benzophenone, 4-acetylbenzoic acid, anthracene, biphenyl-4-carboxylic acid, 1-butanone, *m*-terphenyl) and gases such as CO and CO_2. The pyrolysis products can then be further treated to yield fuels, which are valuable products when the alternative is waste incineration to yield energy or landfill. However, from the perspective of a circular economy, the pyrolysis of pure PET wastes is a much less attractive option with respect to any of the monomer recycling processes presented in this chapter.

However, due to its extremely high tolerance to any contaminants and to the possibility to treat blends of any materials present in post-consumer waste streams, pyrolysis has already reached the industrial scale, even if it is not widespread as mechanical recycling or energy recovery plants. The main drawback of pyrolysis is that plants are quite complex, consists of several units, and the reaction conditions are quite demanding, thus high investments are required as well as quite frequent maintenance. In any case, some plants are currently being funded, constructed, and operated (Johnson, 2022).

More details on many aspects of thermal chemical recycling processes (i.e. pyrolysis and gasification) are provided in previous chapters of this volume.

References

A. LR, Barna BA, Johnsrud DR: *Process for recovering terephthalic acid from waste polyethylene terephthalate*, vol. 19, 1985, pp 1–7.

Achilias DS, Karayannidis GP: The chemical recycling of PET in the framework of sustainable development, *Water, Air and Soil Pollution* 4(4–5):385–396, 2004. https://doi.org/10.1023/B:WAFO.0000044812.47185.0f.

Achilias D, et al.: *Aminolytic depolymerization of poly (ethylene terephthalate) waste in a microwave reactor Dimitris S Achilias*, 2011, Georgia P Tsintzou, Alexandros K Nikolaidis, pp 500–506. https://doi.org/10.1002/pi.2976 (July 2010).

Al-Sabagh AM, et al.: Ionic liquid-coordinated ferrous acetate complex immobilized on bentonite as a novel separable catalyst for PET glycolysis, *Ind Eng Chem Res* 54(50):12474–12481, 2015. https://doi.org/10.1021/acs.iecr.5b03857.

Al-Salem SM, Lettieri P, Baeyens J: Recycling and recovery routes of plastic solid waste (PSW): A review, *Waste Manag* 29(10):2625–2643, 2009. https://doi.org/10.1016/j.wasman.2009.06.004.

Assadi R, Colin X, Verdu J: Irreversible structural changes during PET recycling by extrusion, *Polymer* 45(13):4403–4412, 2004. https://doi.org/10.1016/j.polymer.2004.04.029.

Barnard E, Rubio Arias JJ, Thielemans W: Chemolytic depolymerisation of PET: A review, *Green Chem* 23(11):3765–3789, 2021. https://doi.org/10.1039/d1gc00887k.

Bartolome L, et al.: Recent developments in the chemical recycling of PET, In *Material Recycling - Trends and Perspectives*, 2012. https://doi.org/10.5772/33800 (Preprint).

Bo-Kyung K, et al.: Chemical recycling of poly (ethylene terephthalate) Using a New Hybrid Process, *J Chem Eng Japan* 41(9):923–928, 2008.

Bryant JJL, Semlyen JA: Cyclic polyesters: 6. Preparation and characterization of two series of cyclic oligomers from solution ring-chain reactions of poly(ethylene terephthalate), *Polymer* 38(10):2475–2482, 1997. https://doi.org/10.1016/S0032-3861(96)00801-4.

Campanelli JR, Cooper DG, Kamal MR: Catalyzed hydrolysis of polyethylene terephthalate melts, *J Appl Polym Sci* 53(8):985–991, 1994. https://doi.org/10.1002/app.1994.070530801.

Carbios: *Carbios enzymatic recycling*, 2022, Available at: https://www.carbios.com/en/enzymatic-recycling/.

Cewep: *A voice of waste-to-energy*, 2022, Available at: https://www.cewep.eu/what-cewep-does/.

Chang S, Sheu MF, Chang NH: Effect of stabilizers on the preparation of poly(ethylene terephthalate), *J Polym Sci, Part A-1: Polym Chem* 20(8):2053–2061, 1982. https://doi.org/10.1002/pol.1982.170200808.

Chaudhary S, et al.: Microwave assisted glycolysis of poly(ethylene terepthalate) for preparation of polyester polyols, *J Appl Polym Sci* 129(5):2779–2788, 2013. https://doi.org/10.1002/app.38970.

Chen J, Chen L, Cheng W: Kinetics of glycolysis of polyethylene terephthalate with zinc catalyst, *Polym Int* 48(9):885–888, 1999. https://doi.org/10.1002/(sici)1097-0126(199909)48:9<885::aid-pi216>3.3.co;2-k.

Cot S, et al.: An oxalate-bridged binuclear iron(III) ionic liquid for the highly efficient glycolysis of polyethylene terephthalate under microwave irradiation, *ChemPlusChem* 84(7):786–793, 2019. https://doi.org/10.1002/cplu.201900075.

CuRe: *CuRe technology*, 2022, Available at: https://curetechnology.com/.

Damayanti, Wu HS: Strategic possibility routes of recycled pet, *Polymers* 13(9):1–37, 2021. https://doi.org/10.3390/polym13091475.

De Castro REN, et al.: Depolymerization of poly(ethylene terephthalate) wastes using ethanol and ethanol/water in supercritical conditions, *J Appl Polym Sci* 101(3): 2009–2016, 2006. https://doi.org/10.1002/app.23748.

Demarteau J, Olazabal I, Jehanno C: Aminolytic upcycling of poly(ethylene terephthalate) wastes using a thermally-stable organocatalyst, *Polymer Chem* 2, 2020. https://doi.org/10.1039/d0py00067a.

Eastman: *Eastman and governor lee announce world-scale plastic-to-plastic molecular recycling facility to be built in Kingsport*, 2021, Tenn. Available at: https://www.eastman.com/Company/News_Center/2021/Pages/Eastman-and-Governor-Lee-Announce-Plastic-to-Plastic-Recycling-Facility.aspx.

Eastman: *Eastman to invest up to $1 billion to accelerate circular economy through building world's largest molecular plastics recycling facility in France*, 2022, Available at: https://www.eastman.com/Company/News_Center/2022/Pages/Eastman-to-invest-to-accelerate-circular-economy.aspx.

Eastman Kodak: *Depolymerization process for recycling polyesters*, 1999.

Ehsani M, Khodabakhshi K, Asgari M: Lactide synthesis optimization: Investigation of the temperature, catalyst and pressure effects, *E-Polymers* 14(5):353–361, 2014. https://doi.org/10.1515/epoly-2014-0055.

El-Toufaili FA, Feix G, Reichert KH: Mechanistic investigations of antimony-catalyzed polycondensation in the synthesis of poly(ethylene terephthalate), *J Polym Sci A Polym Chem* 44(3):1049–1059, 2006. https://doi.org/10.1002/pola.21200.

EREMA: *EREMA plastic recycling system*, 2022, Available at: https://www.erema.com/en/applications_bottle-to-bottle.

EuropeanCommission: *Communication from the Commission to the European Parliament*, 2012, The Council, The European Economic And Social Committee And The Committee Of The Regions. Available at: https://eur-lex.europa.eu/LexUriServ/LexUriServ.do?uri=COM:2012:0673:FIN:EN:PDF.

EuropeanUnion: EC/282/2008 on recycled plastic materials and articles intended to come into contact with foods, *Official J European Commun* L86(282):9–18, 2008.

FNEC: *Method for manufacturing terephthalic acid and system thereof*, 2019, Available at: https://patents.justia.com/patent/10604634.

Forrest MJ: 6. Recycling technologies for polyethylene terephthalate, In *Recycling of polyethylene terephthalate*, 2019, pp 101–120. https://doi.org/10.1515/9783110640304-006.

Fortune Business Insights: *PET market size, share & COVID-19 impact analysis*, 2022, Available at: https://www.fortunebusinessinsights.com/industry-reports/polyethylene-terephthalate-pet-market-101743.

Fox DW, Shafer SJ: *United States patent (19)*, 1990 (19).

Fukushima K, et al.: Advanced chemical recycling of poly(ethylene terephthalate) through organocatalytic aminolysis†, *Polym Chem*:1610–1616, 2013. https://doi.org/10.1039/c2py20793a.

Gamble WJ: *A process of recovering components from scrap polyester*, 1993, Available at: https://patentimages.storage.googleapis.com/74/6e/b5/d97de56c478cec/EP0599309A2.pdf.

Gamble IWJ, et al.: *'United States patent 5,298,530', (19)*, 1994.

Garbo: *Garbo ChemPET*, 2022, Available at: https://garbo.it/en/chempet/.

Genta M: Development of chemical recycling process for post-consumer PET bottles by Methanolysis in supercritical methanol, 40(1):1–4, 2003.

George N, Kurian T: Recent developments in the chemical recycling of postconsumer poly(ethylene terephthalate) waste, *Ind Eng Chem Res* 53(37):14185–14198, 2014. https://doi.org/10.1021/ie501995m.

Geyer B, Lorenz G, Kandelbauer A: Recycling of poly(ethylene terephthalate) – A review focusing on chemical methods, *Express Polym Lett* 10(7):559–586, 2016. https://doi.org/10.3144/expresspolymlett.2016.53.

Geyer R, Jambeck JR, Law KL: Production, use, and fate of all plastics ever made, *Sci Adv* 3(7):25–29, 2017. https://doi.org/10.1126/sciadv.1700782.

Ghorbantabar S, et al.: Investigation of conventional analytical methods for determining conversion of polyethylene terephthalate waste degradation via aminolysis process, *J Mater Cycles Waste Manag* 23(2):526–536, 2021. https://doi.org/10.1007/s10163-020-01149-5.

Ghosal K, Nayak C: Recent advances in chemical recycling of polyethylene terephthalate waste into value added products for sustainable coating solutions-hope vs. hype, *Materials Advances* 3(4):1974–1992, 2022. https://doi.org/10.1039/d1ma01112j.

Goje AS, Mishra S: Chemical kinetics, simulation, and thermodynamics of glycolytic depolymerization of poly(ethylene terephthalate) waste with catalyst optimization for recycling of value added monomeric products, *Macromol Mater Eng* 288(4):326–336, 2003. https://doi.org/10.1002/mame.200390034.

Goldsberry C: *Loop industries: Pay no attention to the man behind the curtain*, 2020, Available at: https://www.plasticstoday.com/advanced-recycling/loop-industries-pay-no-attention-man-behind-curtain.

Gopalakrishna KG, Reddy N: *Regulations on recycling PET bottles, recycling of polyethylene terephthalate bottles*, 2019, Elsevier Inc. https://doi.org/10.1016/b978-0-12-811361-5.00002-x.

Goto M, et al.: Depolymerization of polyethylene terephthalate in supercritical methanol, *J Phys Condens Matter* 14(44 SPEC ISS):11427–11430, 2002. https://doi.org/10.1088/0953-8984/14/44/494.

Gr3n: *Gr3n recycling*, 2022, Available at: https://gr3n-recycling.com/.

Güçlü G, et al.: Hydrolysis of waste polyethylene terephthalate and characterization of products by differential scanning calorimetry, *Thermochim Acta* 404(1–2):193–205, 2003. https://doi.org/10.1016/S0040-6031(03)00160-6.

Gupta P, Bhandari S: *Chemical depolymerization of PET bottles via ammonolysis and Aminolysis, recycling of polyethylene terephthalate bottles*, 2019, Elsevier Inc. https://doi.org/10.1016/b978-0-12-811361-5.00006-7.

Han M: *Depolymerization of PET bottle via Methanolysis and hydrolysis, recycling of polyethylene terephthalate bottles*, 2019, Elsevier Inc. https://doi.org/10.1016/b978-0-12-811361-5.00005-5.

Hoang CN, Le TTN, Hoang QD: Glycolysis of poly(ethylene terephthalate) waste with diethyleneglycol under microwave irradiation and ZnSO4 ·7H2O catalyst, *Polymer Bull* 76(1):23–34, 2019. https://doi.org/10.1007/s00289-018-2369-z.

Hodge P: Recycling of condensation polymers via ring-chain equilibria, *Polym Adv Technol* 26(7):797–803, 2015. https://doi.org/10.1002/pat.3469.

Hodge P, et al.: Cyclo-depolymerisation of poly(ethylene naphthalene-2,6-dicarboxylate) and ring-opening polymerisations of the cyclic oligomers obtained, *J Mater Chem* 10(7):1533–1537, 2000. https://doi.org/10.1039/b002786n.

Hondo D: *Plastic waste and the circular economy*, Available, 2022, MarshMcLennan. at: https://www.brinknews.com/plastic-waste-and-the-circular-economy/.

Hu LC, et al.: Alkali-decomposition of poly(ethylene terephthalate) in mixed media of non-aqueous alcohol and ether. Study on recycling of poly(ethylene terephthalate), *Polymer J* 29(9):708–712, 1997. https://doi.org/10.1295/polymj.29.708.

IFPEN: *IFPEN - our solutions for plastics recycling*, 2022, Available at: https://www.ifpenergiesnouvelles.com/innovation-and-industry/our-expertise/climate-environment-and-circular-economy/plastics-recycling/our-solutions.

Ignatyev IA, Thielemans W, Vander Beke B: Recycling of polymers: A review, *ChemSusChem* 7(6):1579–1593, 2014. https://doi.org/10.1002/cssc.201300898.

Imran M, et al.: Sub-and supercritical glycolysis of polyethylene terephthalate (PET) into the monomer bis(2-hydroxyethyl) terephthalate (BHET), *Polym Degrad Stab* 95(9):1686–1693, 2010. https://doi.org/10.1016/j.polymdegradstab.2010.05.026.

Imran M, et al.: Metal-oxide-doped silica nanoparticles for the catalytic glycolysis of polyethylene terephthalate, *J Nanosci Nanotechnol* 11(1):824–828, 2011. https://doi.org/10.1166/jnn.2011.3201.

Ioniqua: *Ioniqa's Circular Solution*, 2022, Available at: https://ioniqa.com/applications/.

Jain A, Soni RK: *Spectroscopic investigation of end products obtained by ammonolysis of poly (ethylene terephthalate) waste in the presence of zinc acetate as a catalyst*, 2007, pp 475–481. https://doi.org/10.1007/s10965-007-9131-9.

Jambeck J, et al.: Plastic waste inputs from land into the ocean, *Mar Pollut* 347(6223):768, 2015. Available at: https://science.sciencemag.org/CONTENT/347/6223/768.abstract.

Jamison O: *Plastics and packaging laws in the european union*, 2021, Available at: https://cms.law/en/int/expert-guides/plastics-and-packaging-laws/european-union.

Jeplan: *Jeplan BRING technology*, 2022, Available at: https://www.jeplan.co.jp/en/technology/.

Johnson J: *New Hope building world's largest pyrolysis plant*, 2022, Available at: https://www.plasticsnews.com/news/new-hope-building-worlds-largest-pyrolysis-plant.

Kahlert S, Bening CR: Why pledges alone will not get plastics recycled: Comparing recyclate production and anticipated demand, *Resour Conserv Recycl* 181(April), 2022, 106279. https://doi.org/10.1016/j.resconrec.2022.106279.

Kaminsky W, Schlesselmann B, Simon C: Olefins from polyolefins and mixed plastics by pyrolysis, *J Anal Appl Pyrolysis* 32:19–27, 1995. https://doi.org/10.1016/0165-2370(94)00830-T.

Karayannidis GP, Achilias DS: Chemical recycling of poly(ethylene terephthalate), *Macromol Mater Eng* 292(2):128–146, 2007. https://doi.org/10.1002/mame.200600341.

Karayannidis GP, Chatziavgoustis AP, Achilias DS: Poly(ethylene terephthalate) recycling and recovery of pure terephthalic acid by alkaline hydrolysis, *Adv Polym Technol* 21(4):250–259, 2002. https://doi.org/10.1002/adv.10029.

Kawai F: *Current knowledge on enzymatic PET degradation and its possible application to waste stream management and other fields*, 2019, pp 4253–4268.

Kenny GR, Morgan DR, Stephens AH: *Automated separation of commingled plastic containers*, 1992.

Khoonkari M, et al.: Chemical recycling of PET wastes with different catalysts, *Int J Polym Sci* 2015, 2015. https://doi.org/10.1155/2015/124524.

Kim S, Han M: A Methanolysis process for recycling polyester oligomer waste, *J Korean Institute Chem Engin* 42:53–58, 2004.

Liu F, et al.: Alcoholysis of poly(ethylene terephthalate) to produce dioctyl terephthalate with sub- and super-critical isooctyl alcohol, *J Anal Appl Pyrolysis* 99:16–22, 2013. https://doi.org/10.1016/j.jaap.2012.11.008.

Liu Q, Li R, Fang T: Investigating and modeling PET methanolysis under supercritical conditions by response surface methodology approach, *Chem Eng J* 270(2015):535–541, 2015. https://doi.org/10.1016/j.cej.2015.02.039.

López DMC, M., et al.: Assessing changes on poly(ethylene terephthalate) properties after recycling: Mechanical recycling in laboratory versus postconsumer recycled material, *Mater Chem Phys* 147(3):884–894, 2014. https://doi.org/10.1016/j.matchemphys.2014.06.034.

López-Fonseca R, et al.: Chemical recycling of PET by alkaline hydrolysis in the presence of quaternary phosphonium and ammonium salts as phase transfer catalysts, *WIT Trans Ecol Environ* 109:511–520, 2008. https://doi.org/10.2495/WM080521.

López-Fonseca R, González-Velasco JR, Gutiérrez-Ortiz JI: A shrinking core model for the alkaline hydrolysis of PET assisted by tributylhexadecylphosphonium bromide, *Chem Eng J* 146(2):287–294, 2009. https://doi.org/10.1016/j.cej.2008.09.039.

López-Fonseca R, et al.: Chemical recycling of post-consumer PET wastes by glycolysis in the presence of metal salts, *Polym Degrad Stab* 95(6):1022–1028, 2010. https://doi.org/10.1016/j.polymdegradstab.2010.03.007.

López-Fonseca R, et al.: Kinetics of catalytic glycolysis of PET wastes with sodium carbonate, *Chem Eng J* 168(1):312–320, 2011. https://doi.org/10.1016/j.cej.2011.01.031.

Ma M, Wang L, Zhu H: Enzymatic degradation of polyester-nanoparticles by lipases and adsorption of lipases on the polyester-nanoparticles, *Adv Mat Res* 420:2302–2307, 2012. https://doi.org/10.4028/www.scientific.net/AMR.418-420.2302.

Machado A, Castro D, Carniel A: Short communication A novel process for poly (ethylene terephthalate) depolymerization via enzyme-catalyzed glycolysis, *Biochem Eng J* 124:64–68, 2017. https://doi.org/10.1016/j.bej.2017.04.011.

Mancini SD, Zanin M: Optimization of neutral hydrolysis reaction of post-consumer PET for chemical recycling, *Prog Rubber Plast Recycl Technol* 20(2):117–132, 2004. https://doi.org/10.1177/147776060402000202.

Mandoki, J.W. (1986) 'Depolymerization of condensation polymers', United States Patent [Preprint], (19).

Marathe KV, Chavan KR, Nakhate P: *Life cycle assessment (LCA) of PET bottles, recycling of polyethylene terephthalate bottles*, 2019, Elsevier Inc. https://doi.org/10.1016/b978-0-12-811361-5.00008-0.

Mastellone ML: A feasibility assessment of an integrated plastic waste system adopting mechanical and thermochemical conversion processes, *Resources, Conservation and Recycling: X* 4, 2019, 100017. https://doi.org/10.1016/j.rcrx.2019.100017.

Matsumura S, et al.: Stability and utility of pyridyl disulfide functionality in RAFT and conventional radical polymerizations, *J Polym Sci A Polym Chem* 46(April):7207–7224, 2008. https://doi.org/10.1002/pola.

Maurya A, Bhattacharya A, Khare SK: Enzymatic remediation of polyethylene terephthalate (PET)– Based polymers for effective Management of Plastic Wastes : An Overview, *Front Bioeng Biotechnol* 8:1–13, 2020. https://doi.org/10.3389/fbioe.2020.602325.

Michalski A: Hydrolysis of poly (ethylene terephthalate) waste to obtain terephthalic acid, *Wl Chem* 49:144, 1987.

Mishra S, Goje AS: Chemical recycling, kinetics, and thermodynamics of alkaline depolymerization of waste poly(ethylene terephthalate) (PET), *Polym React Eng* 11(4 Spec. Iss):963–987, 2003a. https://doi.org/10.1081/PRE-120026382.

Mishra S, Goje AS: Kinetic and thermodynamic study of methanolysis of poly(ethylene terephthalate) waste powder, *Polym Int* 52(3):337–342, 2003b. https://doi.org/10.1002/pi.1147.

Mishra S, Goje AS, Zope VS: Chemical recycling, kinetics, and thermodynamics of poly (ethylene terephthalate) (PET) waste powder by nitric acid hydrolysis, *Polym React Eng* 11(1):79–99, 2003. https://doi.org/10.1081/PRE-120018586.

Mittal A, et al.: Scanning electron microscopic study of hazardous waste flakes of polyethylene terephthalate (PET) by aminolysis and ammonolysis, *J Hazard Mater* 178(1–3):390–396, 2010. https://doi.org/10.1016/j.jhazmat.2010.01.092.

Musale RM, Shukla SR: Deep eutectic solvent as effective catalyst for aminolysis of polyethylene terephthalate (PET) waste, *Int J Plast Technol* 20(1):106–120, 2016. https://doi.org/10.1007/s12588-016-9134-7.

Nait-Ali LK, Colin X, Bergeret A: Kinetic analysis and modelling of PET macromolecular changes during its mechanical recycling by extrusion, *Polym Degrad Stab* 96(2):236–246, 2011. https://doi.org/10.1016/j.polymdegradstab.2010.11.004.

NextChem: *Nextchem completes italy's 1° demonstration plant for pet and textiles polyester chemical recycling*, 2022, Available at: https://nextchem.it/news/nextchem-completes-italys-1deg-demonstration-plant-pet-and-textiles-polyester-chemical.

Nikiema J, Asiedu Z: A review of the cost and effectiveness of solutions to address plastic pollution, *Environ Sci Pollut Res* 29(17):24547–24573, 2022. https://doi.org/10.1007/s11356-021-18038-5.

Nye JVC: *Standards of living and modern economic growth*, 2018, Available at: https://www.econlib.org/library/Enc/StandardsofLivingandModernEcon omicGrowth.html.

Omnexus: *Polyethylene terephthalate (PET): A comprehensive review*, 2022, Available at: https://omnexus.specialchem.com/selection-guide/polyethylene-terephthalate-pet-plastic.

PackagingEurope: *Closing the loop for PET*, 2022, Available at: https://packagingeurope.com/comment/closing-the-loop-for-pet/7952.article.

Padhan RK, Sreeram A: *Chemical depolymerization of PET bottles via combined Chemolysis methods, recycling of polyethylene terephthalate bottles*, 2019, Elsevier Inc. https://doi.org/10.1016/b978-0-12-811361-5.00007-9.

Panda AK, Singh RK, Mishra DK: Thermolysis of waste plastics to liquid fuel. A suitable method for plastic waste management and manufacture of value added products-A world prospective, *Renew Sustain Energy Rev* 14(1):233–248, 2010. https://doi.org/10.1016/j.rser.2009.07.005.

Paszun D, Spychaj T: Chemical recycling of poly(ethylene terephthalate), *Ind Eng Chem Res* 36(4):1373–1383, 1997. https://doi.org/10.1021/ie960563c.

PerPETual: *PerPETual technology*, 2022, Available at: https://www.perpetual-global.com/manufacturing/#technology.

PetcoreEurope: *The road to 90%. How to achieve the eu's pet bottle collection bottle collection targets by 2029*, 2021, Available at: https://www.petcore-europe.org/legislation/334-how-achieve-eu-pet-bottle-collection-bottle-collection-targets-2029.html.

Pham DD, Cho J: Low-energy catalytic methanolysis of poly(ethyleneterephthalate), *Green Chem* 23(1):511–525, 2021. https://doi.org/10.1039/d0gc03536j.

Pingale ND, Shukla SR: Microwave assisted ecofriendly recycling of poly (ethylene terephthalate) bottle waste, *Eur Polym J* 44(12):4167–4172, 2008. https://doi.org/10.1016/j.eurpolymj.2008.09.019.

Pingale ND, et al.: Glycolysis of postconsumer polyethylene terephthalate waste, *J Appl Polym Sci* 116(5):2658–2667, 2009. https://doi.org/10.1002/app.

Pires HM, et al.: Effect of weathering and accelerated photoaging on pet/pc (80/20 wt/wt%) melt extruded blend, *Mater Res* 18(4):763–768, 2015. https://doi.org/10.1590/1516-1439.010115.

Plastics For Change: *Which plastic can be recycled?*, 2021, Available at: https://www.plasticsforchange.org/blog/which-plastic-can-be-recycled.

PlasticsEurope: *'Plastics the fact 2006', plastics Europe market research group (PEMRG) and Conversio Market & Strategy GmbH*, 2006 (Preprint).

PlasticsEurope: *Plastics – The facts 2019*, 2019, Available at: https://plasticseurope.org/wp-content/uploads/2021/10/2019-Plastics-the-facts.pdf.

PlasticsEurope: *'Plastics the fact 2021', plastics Europe market research group (PEMRG) and Conversio Market & Strategy GmbH*, 2021, [Preprint]. Available at: https://plasticseurope.org/.

PlasticsForChange: *What are the regulations for the use of recycled plastic?*, 2020, Available at: https://www.plasticsforchange.org/blog/category/what-are-the-current-regulations-for-the-use-of-recycled-plastic.

Polk MB, et al.: Nylon 66, nylon 46, and pet phase-transfer-catalyzed alkaline depolymerization at atmospheric pressure, *Polym-Plast Technol Eng* 38(3):459–470, 1999. https://doi.org/10.1080/03602559909351593.

Pudack C, Stepanski M, Fässler P: PET recycling – Contributions of crystallization to sustainability, *Chemie-Ingenieur-Technik* 92(4):452–458, 2020. https://doi.org/10.1002/cite.201900085.

Qureshi MS, et al.: Pyrolysis of plastic waste: Opportunities and challenges, *J Anal Appl Pyrolysis* 152(March), 2020. https://doi.org/10.1016/j.jaap.2020.104804.

Ragaert K, Delva L, Van Geem K: Mechanical and chemical recycling of solid plastic waste, *Waste Manag* 69:24–58, 2017. https://doi.org/10.1016/j.wasman.2017.07.044.

Recycling Magazine: *Human beings cannot live without plastics for one hour*, 2019, Available at: https://www.recycling-magazine.com/2019/11/18/human-beings-cannot-live-without-plastics-for-one-hour/.

Reike D, Vermeulen WJV, Witjes S: The circular economy: New or refurbished as CE 3.0?—Exploring controversies in the conceptualization of the circular economy through a focus on history and resource value retention options, *Resour Conserv Recycl* 135(August 2017):246–264, 2018. https://doi.org/10.1016/j.resconrec.2017.08.027.

Ribitsch D, et al.: Characterization of a new cutinase from Thermobifi da alba for PET-surface hydrolysis, *Biocatal Biotransformation* 30(February):2–9, 2012. https://doi.org/10.3109/10242422.2012.644435.

Rieckmann T, Frei F, Völker S: Modelling of PET quality parameters for a closed-loop recycling system for food contact, *Macromol Symp* 302(1):34–45, 2011. https://doi.org/10.1002/masy.201000069.

RITTEC: *RITTEC RevolPET technology*, 2022, Available at: https://www.rittec.eu/solutions/revolpet-r.html.

Robert Donald Slockett: *United States patent: 3871965*, 1972.

Ruj B, et al.: Sorting of plastic waste for effective recycling, *Int J Appl Sci Eng Sci* 4(4):564–571, 2015. Available at: www.ijaser.com.

Saadet O, Orbay M: Simultaneous glycolysis and hydrolysis of polyethylene terephthalate and characterization of products by differential scanning calorimetry, *Polymer* 44:7609–7616, 2003. https://doi.org/10.1016/j.polymer.2003.09.062.

Sako T, et al.: Depolymerization of polyethylene terephthalate to monomers with supercritical methanol, *J Chem Eng Japan*:342–349, 1997. https://doi.org/10.1252/jcej.30.342.

Schyns ZOG, Shaver MP: Mechanical recycling of packaging plastics: A review, *Macromol Rapid Commun* 42(3):1–27, 2021. https://doi.org/10.1002/marc.202000415.

Sharma V, et al.: Recycling of waste PET-bottles using dimethyl sulfoxide and hydrotalcite catalyst, *J Appl Polym Sci* 129(3):1513–1519, 2013. https://doi.org/10.1002/app.38829.

Sheel A, Pant D: *Chemical depolymerization of PET bottles via glycolysis, recycling of polyethylene terephthalate bottles*, 2019, Elsevier Inc. https://doi.org/10.1016/b978-0-12-811361-5.00004-3.

Shojaei B, Abtahi M, Najafi M: Chemical recycling of PET: A stepping-stone toward sustainability, *Polym Adv Technol* 31(12):2912–2938, 2020. https://doi.org/10.1002/pat.5023.

Shukla SR, Harad AM: Aminolysis of polyethylene terephthalate waste, *Polym Degrad Stab* 91(8):1850–1854, 2006. https://doi.org/10.1016/j.polymdegradstab.2005.11.005.

Shukla SR, Palekar V, Pingale N: *Zeolite catalyzed glycolysis of poly(ethylene terephthalate) bottle waste*, 2008. https://doi.org/10.1002/app.

Sivagami K, et al.: Catalytic pyrolysis of polyolefin and multilayer packaging based waste plastics: A pilot scale study, *Process Saf Environ Prot* 149:497–506, 2021. https://doi.org/10.1016/j.psep.2020.10.038.

Stanica-Ezeanu D, Matei D: Natural depolymerization of waste poly(ethylene terephthalate) by neutral hydrolysis in marine water, *Sci Rep* 11(1):1–7, 2021. https://doi.org/10.1038/s41598-021-83659-2.

Stark M: Letter to the editor regarding 'are we speaking the same language? Recommendations for a definition and categorization framework for plastic debris, *Environ Sci Tech*, 2019. https://doi.org/10.1021/acs.est.9b01360.

Starlinger: *Starlinger recycling technology*, 2022, Available at: https://www.starlinger.com/en/recycling/.

Taw ME, Eskander SB: Chemical recycling of poly (ethylene terephthalate) waste using ethanolamine. Sorting of the end products -C-X-C- + YZ, *Polym Degrad Stab* 95:187–194, 2010. https://doi.org/10.1016/j.polymdegradstab.2009.11.026.

Thiyagarajan S, et al.: Back-to-monomer recycling of polycondensation polymers: Opportunities for chemicals and enzymes, *RSC Adv* 12(2):947–970, 2022. https://doi.org/10.1039/d1ra08217e.

Tomás RAF, Bordado JCM, Gomes JFP: P-xylene oxidation to terephthalic acid: A literature review oriented toward process optimization and development, *Chem Rev* 113(10):7421–7469, 2013. https://doi.org/10.1021/cr300298j.

Toot WEJ: *Removal of dimethylterephthalate from a methanolysis vapor stream*, 1995.

ZeroWasteEurope: *PET, the most circular of all plastics, is far from real circularity*, 2022, Available at: https://zerowasteeurope.eu/press-release/new-report-pet-the-most-circular-of-all-plastics-is-far-from-real-circularity/.

Viana ME, et al.: Chemical recycling of PET by catalyzed glycolysis: Kinetics of the heterogeneous reaction, *Chem Eng J* 173(1):210–219, 2011. https://doi.org/10.1016/j.cej.2011.07.031.

Wang Q, et al.: Urea as an efficient and reusable catalyst for the glycolysis of poly(ethylene terephthalate) wastes and the role of hydrogen bond in this process, *Green Chem* 14(9):2559–2566, 2012. https://doi.org/10.1039/c2gc35696a.

Wang Q, et al.: Deep eutectic solvents as highly active catalysts for the fast and mild glycolysis of poly(ethylene terephthalate)(PET), *Green Chem* 17(4):2473–2479, 2015. https://doi.org/10.1039/c4gc02401j.

Welle F: Twenty years of PET bottle to bottle recycling - an overview, *Resour Conserv Recycl* 55(11):865–875, 2011. https://doi.org/10.1016/j.resconrec.2011.04.009.

Yang Y, et al.: Study on methanolytic depolymerization of PET with supercritical methanol for chemical recycling, *Polym Degrad Stab* 75(1):185–191, 2002. https://doi.org/10.1016/S0141-3910(01)00217-8.

Yang BP, et al.: *Facile preparation of a patterned, aminated polymer surface by UV-light-induced surface Aminolysis***, 2005, pp 1415–1425. https://doi.org/10.1002/adfm.200400335.

Yang Z, et al.: Crystallization behavior of poly(ε-caprolactone)/layered double hydroxide nanocomposites, *J Appl Polym Sci* 116(5):2658–2667, 2010. https://doi.org/10.1002/app.

Yoshioka T, Sato T, Okuwaki A: Hydrolysis of waste PET by sulfuric acid at 150°C for a chemical recycling, *J Appl Polym Sci* 52(9):1353–1355, 1994. https://doi.org/10.1002/app.1994.070520919.

Yue QF, et al.: Glycolysis of poly(ethylene terephthalate) (PET) using basic ionic liquids as catalysts, *Polym Degrad Stab* 96(4):399–403, 2011. https://doi.org/10.1016/j.polymdegradstab.2010.12.020.

Zhou X, et al.: Effective catalysis of poly(ethylene terephthalate) (PET) degradation by metallic acetate ionic liquids, *Pure Appl Chem* 84(3):789–801, 2012. https://doi.org/10.1351/PAC-CON-11-06-10.

Further reading

Bujak JW: *Thermal utilization (treatment) of plastic waste*, vol. 90, 2015, pp 1468–1477. https://doi.org/10.1016/j.energy.2015.06.106.

European Council: *Directive (Eu) 2019/904 of the European Parliament and of the Council of 5 June 2019 on the reduction of the impact of certain plastic products on the environment*, 2019, pp 1–19. Available at: https://www.plasticseurope.org/en/resources/publications/1689-working-together-towards-more-sustainable-plastics%0Ahttps://www.plasticseurope.org/en/resources/publications%0Ahttps://eur-lex.europa.eu/legal-content/EN/TXT/PDF/?uri=CELEX:32019L0904&from=EN.

Gradus RHJM, et al.: A cost-effectiveness analysis for incineration or recycling of Dutch household plastic waste ☆, *Ecol Econ* 135:22–28, 2017. https://doi.org/10.1016/j.ecolecon.2016.12.021.

Honus S, et al.: Pyrolysis gases produced from individual and mixed PE, PP, PS, PVC, and PET—Part I : Production and physical properties, *Fuel* 221:346–360, 2018. https://doi.org/10.1016/j.fuel.2018.02.074 (September 2017).

Jouhara H, et al.: (2018) 'pyrolysis of domestic based feedstock at temperatures up to 300 ° C, *Therm Sci Eng Prog* 5:117–143, 2017. https://doi.org/10.1016/j.tsep.2017.11.007.

Nguyen T: *Sources of economic growth : Physical capital, human capital, natural resources, and TFP Tu-Anh Nguyen to cite this version : HAL id : Tel-00402443 sources of economic growth : Physical capital, human capital*, 2009, Natural Resources and TFP.

OECD: *Plastic pollution is growing relentlessly as waste management and recycling fall short, says OECD*, 2022, Available at: https://www.oecd.org/environment/plastic-pollution-is-growing-relentlessly-as-waste-management-and-recycling-fall-short.htm#:~:text=Globalplasticsproductiondoubledfrom,ofglobalgreenhousegasemissions.

Ozsin G: *A comparative study on co-pyrolysis of lignocellulosic biomass with polyethylene terephthalate, polystyrene, and polyvinyl chloride : Synergistic effects and product characteristics*, vol. 205, 2018, pp 1127–1138. https://doi.org/10.1016/j.jclepro.2018.09.134.

CHAPTER SEVEN

Poly(methyl methacrylate): Market trends and recycling

Mattia Sponchioni[a,*] and Sibel Altinok[b]

[a]Department of Chemistry, Materials and Chemical Engineering, Politecnico di Milano, Milano, Italy
[b]Organik Kimya, Kemerburgaz Cd. 146, Istanbul, Turkey
*Corresponding author: e-mail address: mattia.sponchioni@polimi.it

Contents

Abstract

Poly(methyl methacrylate) (PMMA) is currently produced at 3.9 million metric tons per year, being involved in different sectors, due to its high transparency, lightness and barrier properties that make this material particularly suitable for optical and electronic applications. However, it is estimated that nowadays only 10% of this production is recycled. Landfilling and incineration are still the dominant fates for waste PMMA. Since its market is expected to expand in the coming 7 years at a compound annual growth rate of 3.4%, this poses urgent concerns about the disposal of end-of-life PMMA. In this chapter, the PMMA market trends are first introduced to better define the boundaries of the topic. Then, the available technologies for PMMA recycling are discussed. Given the possibility of quantitatively depolymerize this polymer back to methyl methacrylate, the thermal decomposition is by far the most convenient and explored recycling route. The main focus is then on this process. Even though the knowledge on the reaction mechanism is still limited, the fundamentals accepted by the scientific community are presented. Then, the discussion is moved to the reactor configurations developed over the years for the thermolysis of PMMA, extensively referring to examples available in the patent literature. From this analysis, fluidized bed reactors and reactive extruders result the most appealing solutions in terms of environmental footprint and efficiency. However, additional research effort is required to properly design a process capable of treating not only high-quality feedstocks, but also end-of-life PMMA, possibly containing additives and comonomers.

Advances in Chemical Engineering, Volume 60
ISSN 0065-2377
https://doi.org/10.1016/bs.ache.2022.09.004

1. Introduction

The global production of plastic has reached 368 million metric tons in 2019, according to the analysis conducted by PlasticsEurope (PlasticsEurope, 2019). This market volume is rapidly increasing over the years, as a result of the unbeaten thermomechanical properties that polymers display. Particularly remarkable are their light weight, flexibility, excellent insulation properties, impermeability to gas and humidity, and durability, just to mention a few. These undoubted advantages compared to other materials make polymers a key component in most everyday commodities. The packaging field is the one absorbing most of the industrial production of plastics. It has been estimated that almost 40% of the annual production of polymers go to packaging. This sector is followed by building, automotive, electronics and clothes & textile. The rapid expansion of the plastic production is producing a turnover of more than 350 billion euros per year, with the plastic supply chain counting on more than 55,000 companies, mostly small and medium enterprises and giving job to more than 1.5 million people (Plastics Europe - Association of Plastic Manufacturers (Organization), 2020).

Despite the annual production of polymers looks dangerous in terms of lacking raw materials for their manufacturing, it is worth underlying that the most urgent concern related to their usage is still not related to the scarcity of fossil resources. As a matter of fact, the production of plastics contributes for <10% to the global carbon footprint. The majority of crude oil (>85%) is still dedicated to the production of energy (Sasse and Emig, 1998). What is critical related to the plastic market is rather the consideration that most of the goods and polymeric products have a shelf-life shorter than 50 years. This scenario is even worse if the analysis is limited to the packaging sector, where the material is designed for a single use. It turns out that the most concerning impact of the polymer industry is related to the waste generation. As a matter of fact, the great majority of the polymers populating the market are not degradable, at least under environmental conditions, which poses many concerns related to the plastic handling after usage. The leading polymers in the plastic market are polyolefins, with polyethylene (both high density and low density) and polypropylene covering almost 50% of the world annual production of plastic materials. The second and third place in this ranking are covered by polyvinylchloride and polyethylene terephthalate. These extremely durable materials spend centuries in the environment, which makes a correct disposal particularly critical. This is even more so considering that only 30% of the plastics ever produced is still in use. This highlights the necessity of polymer recycling, whose aims are manifold: reducing the waste generation on one side, reducing the pressure on landfills on the other, with the additional advantage of generating added value out of a waste material (Hundertmark et al., 2018).

Despite the benefits brought about by recycling, this is still scarcely used in the common waste management practice. As a matter of fact, it was estimated that <30% of the plastic collected every year is recycled, but only 0.9% is recycled more than once (Geyer et al., 2017). This is ascribed to two main factors: the relative infancy of most recycling technologies, still requiring harsh conditions in terms of temperatures, pressures, organic solvents and equipment complexity to make them affordable and appealing, as well as the specificity of the recycling processes to the different materials. Since most of the plastic goods are actually a mixture of different materials, the latter aspect requires a careful separation of the constituting components before more specific processes can be put in place. This is the concern related to the so-called Plasmix.

In the following, the issue related to the polymer recycling will be specifically investigated for poly(methyl methacrylate) (PMMA) which, despite being produced at a much lower volume than other plastics, approximately 3.9 million metric tons per year, plays an important role in many everyday applications (De Tommaso and Dubois, 2021). This chapter in particular encompasses briefly the market and applications of PMMA, in order to highlight its importance worldwide. This analysis is followed by the different processes that have been specifically designed or successfully applied to the recycling of PMMA, concluding with the reactor technologies developed so far for the chemical recycling of PMMA, the most important process for this polymer. Extensive reference to examples taken from the patent literature will be made in order to highlight advantages and criticalities of the reactor configurations proposed, as well as the most relevant option from an efficiency and an environmental point of view.

2. PMMA: History and market

Unlike polyolefins, PET and PVC, PMMA cannot be defined as a bulk polymer. PMMA belongs indeed to the family of acrylic resins that, due to their high transparency and appreciable optical properties, are also referred to as acrylic glasses (SRI International, 1994). As a matter of fact,

PMMA made its appearance on the market for the first time in 1933, when Röhm & Haas launched a transparent and hard plastic under the brand name of Plexiglas®. After an initial blossoming for civilian applications, mainly instrument covers and watch glass, the real affirmation of this material was during World War II, when it was used for military aircraft windows. This allowed a dramatic expansion of the company, which increased the personnel by 10 times. After World War II, the civilian use of Plexiglas® was reestablished and today, it is still in use in many everyday applications, including optical elements in the automotive sector, furniture, electronics, optical applications, glasses for aquariums and many others. The reason for this success is imputable to the outstanding waterproofing, durability, breaking strength, simple manufacturing, and lightness that add to the notorious transparency (Albrecht et al., 2013; Ali et al., 2015).

Nowadays, the production of Plexiglas® is carried out by Evonik, which expanded the market and scopes of PMMA by introducing new acrylic glasses, in particular Deglas® and Acrylite® (Evonik, n.d.). Depending upon the expected applications in fact, slight modifications to the pure PMMA are required, which justifies the advent of these variants. Additives or acrylic comonomers, for example, can be added to MMA with the aim of increasing the flexibility of the final polymer, its strength or, when extrusion is carried for the manufacturing of the final goods, its thermal resistance.

Regardless of the presence a comonomer, PMMA is synthesized, in most cases, via free-radical polymerization (FRP) of methyl methacrylate. The reaction is conducted at temperatures between 333 and 393 K but inevitably leads to pretty high dispersity, which compromises the homogeneity of the polymer chains, which is important for advanced applications in the optic and electronic sectors (Christian et al., 2000; Scorah et al., 2004). Another important consideration related to the FRP of methyl methacrylate, which strongly impacts the recycling of PMMA, is the possibility of having four different propagation reactions according to the binding sites between the growing radical and the monomer. For methacrylates, it is common practice to refer to the most substituted side of the vinyl group as the "head" and to the less substituted side of the vinyl group as the "tail." Accordingly, as schematically shown in Fig. 1, four combinations are possible, namely, head-to-tail, the most likely, head-to-head, tail-to-head and tail-to-tail. The strength of these bonds is not the same. As it will be discussed in more details later in this chapter, the head-to-tail bond is the most resistant and therefore hampers the thermal recycling of PMMA, which is typically initiated from the defects in the polymer chain, represented by the other three possible combinations (Albrecht et al., 2013).

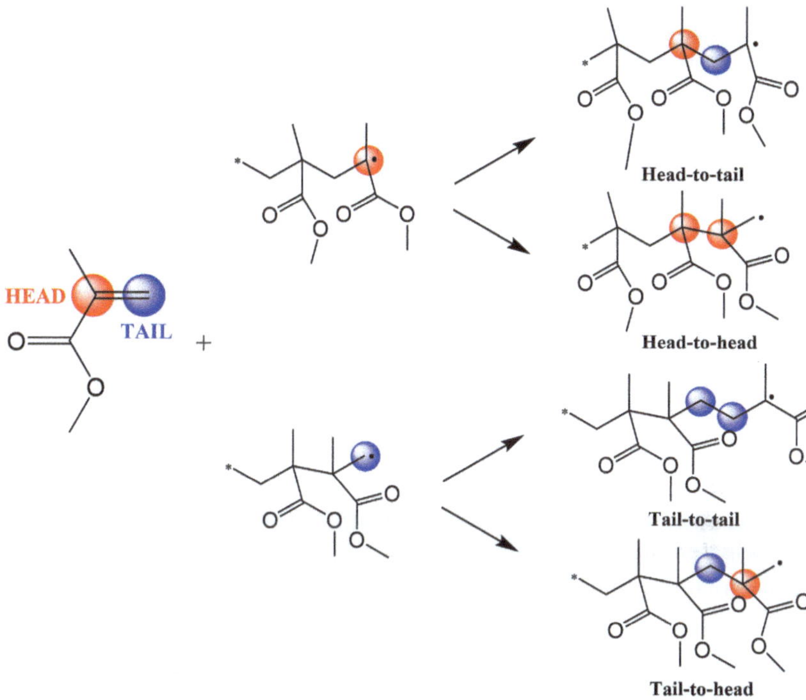

Fig. 1 Four possible combinations for the chain propagation mechanism of a PMMA radical.

Another more likely possibility of forming head-to-head defects, which is important in the thermal recycling of PMMA, is through chain recombination. Since the formation of head-like radicals is more thermodynamically favorable, a large population of these macroradicals is expected, which can then recombine through head-to-head addition.

When low interchain drift is required, the anionic polymerization (AP) of methyl methacrylate is the solution of choice. Following the ideal absence of termination reactions as well as a faster initiation compared to FRP, much lower dispersities can be obtained with AP. The drawback in this case is represented by a limitation in the maximum molecular weights achievable, with an upper threshold of few tens of thousands Dalton. In addition, through AP, it is possible to almost completely exclude structural defects, with the head-to-tail propagation being far more probable than the other possible combinations due to a strong dipole in methyl methacrylate (Glusker et al., 1961; Goode et al., 1960). In addition, ideally excluding any termination, the formation of head-to-head defects by combination is prevented in AP. Both these aspects give PMMA synthesized via AP higher thermal resistance compared to a similar material synthesized via FRP.

3. PMMA recycling processes

As mentioned in Section 1, PMMA is produced at approximately 300,000 million metric tons per year considering Europe only (Plastics Europe - Association of Plastic Manufacturers (Organization), 2020). However, only 10% of the produced PMMA is currently recycled, with the remaining fraction being landfilled or incinerated (van der Heijden et al., 2018). The reason for this being the difficulty in obtaining similar purities as the virgin PMMA from end-of-life scraps.

Different recycling technologies for PMMA have nowadays been proposed. However, it should be underlined that all of them require either high energy inputs or the use of organic solvents. It results that a univocally optimal solution to the plastic recycling has not been identified yet, at least from an environmental point of view. The most convenient process should be then selected on a case-by-case approach, considering the specificities of the waste material under consideration and the process performances that could be achieved on that specific material.

The current technologies for the recycling of PMMA can be broadly divided in two families: mechanical recycling, consisting in the processing of the polymer waste without intervening on its chemical structure, and feedstock or chemical recycling, which first degrades the waste polymer into lower molecular weight compounds that are then repolymerized leading to the recycled product. In the following, a detailed discussion of these two pathways is proposed reporting the currently available processes and highlighting their unique features.

3.1 PMMA mechanical recycling

Mechanical recycling relies on the recovery of solid plastic waste (SPW) and its reutilization after physical and non-reactive unit operations, most often melt extrusion. It is evident that this strategy is suitable mainly for thermoplastic materials, while the internal crosslinking of thermosets intrinsically prevents this form of recycling. The easiest implementation of the mechanical recycling is the primary recycling, which applies to industrial scraps with a known history and composition. Because of the lack of impurities, controlled processes can be put in place to restore the original polymer quality. However, it is evident that this primary recycling can involve only a limited amount of plastic waste. The core of the mechanical recovery is instead the secondary recycling, addressed to the treatment of civilian plastic waste.

In this case, the composition is broader and preliminary processes aimed at the removal of contaminants from the SPW are required before moving to the extrusion (Achilias and Karayannidis, 2004; Scheirs, 1998). Despite waste PMMA is typically a clean and well-defined feedstock, only few literature examples are available about its mechanical recycling for further formulation. These examples refer to the recovery of highly pure PMMA from LCD screens (Prado et al., 2017; Suresh et al., 2017a). The workflow includes a washing step, whose intensity depends on the amount of additives and impurities expected in the SPW, followed by extrusion (Fig. 2). In the case of a highly pure feedstock, the cleaning procedure is simply obtained with a mixture of surfactants, followed by rinsing with water. Whenever a higher content of contaminants is expected, more delicate procedures should applied. Fig. 2, for example, shows the case of the dissolution/reprecipitation technique (Achilias, 2006). In these cases, PMMA is first solubilized in an appropriate solvent like toluene or xylene and then mixed with a non-solvent, causing the precipitation of the polymer while keeping impurities and eventual additives in the solution. A filtration stage is then typically applied for the recovery of the wet PMMA that is dried and finally extruded. The extruder chamber temperature in this case is kept in between 190 °C and 230 °C, to ensure melting but with an appropriate safety margin to avoid the risk of thermal decomposition of the polymer.

The reason for the scarce use of this recycling technology in the case of PMMA is twofold. On one side, to produce high quality recycled polymer (r-PMMA), expensive unit operations are necessary, together with a proper management of the organic solvents involved in the washing phases. The second reason is ascribed to the deterioration of the polymer quality after each recycling cycle. Suresh et al. systematically investigated the mechanical and thermal properties of PMMA recovered from end-of-life LCD screens

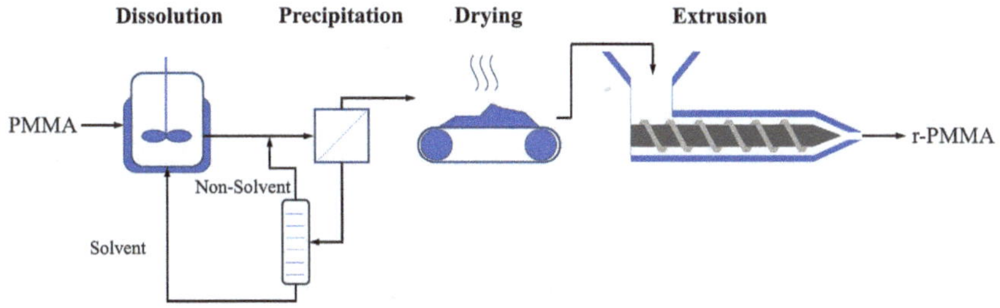

Fig. 2 Schematic representation of a typical process for PMMA mechanical recycling, including dissolution and precipitation to remove additives and impurities, drying and melt extrusion.

and mechanically processed and compared them with those of the virgin polymer (18). While certain mechanical properties like tensile and flexural strength could actually be improved, the r-PMMA showed a reduced impact resistance and elongation at rupture compared to virgin PMMA. This was attributed by the authors to random chain breakages experienced by the polymer during its life cycle as well as during the recycling process. An even more critical drawback for a polymer like PMMA is the loss of transparency following extrusion, which prevents the reutilization of this recycled material for optical elements, like the same screens from which it was recovered. Instead, an interesting research stream is currently focused on the formulation and utilization of poly(vinyl chloride) (PVC)/ PMMA blends. In fact, waste PVC can be collected from data cable waste and combined with PMMA recovered from LCD/LED screens to form a recycled blend through mechanical recycling and in particular exploiting the melt co-extrusion of these plastics. The resulting blend showed superior mechanical properties compared to the parent materials and, the co-extrusion with PMMA, allowed to increase the stability of PVC, reducing the autocatalytic formation of hydrochloric acid. Given the large application of both PVC and PMMA in computer and electronic sectors, this strategy is therefore appealing for the reduction of the waste from electric and electronic equipment (WEEE), which is expanding exponentially due to the advancements in the technology (Sasikala Suresh et al., 2018; Suresh et al., 2017b).

Despite this niche applications, most of the research effort was dedicated to the chemical recycling, which even if more complex as it typically requires a repolymerization step, can lead to an upcycling of the polymer, thus showing the actual potential of closing the loop.

3.2 PMMA chemical recycling

The term chemical recycling, also known as feedstock recycling or tertiary recycling, actually comprises manifold processes that can be even very diverse in terms of unit operations involved and process conditions but that have the common feature of reducing the treated polymer waste into low molecular weight molecules. Typically, these processes aim at producing the pure monomer, which could then be repolymerized to achieve a high-quality recycled polymer. In practice, all these chemical processes end up with a mixture of monomer, short oligomers and gases, which can be used as fuels or as raw materials for chemical syntheses (Al-Salem et al., 2009).

The most important chemical processes for the recycling of polymers are:

- *Gasification processes*: High reaction temperatures and pressures are applied in order to produce syngas. At the scope, oxygen in substoichiometric amounts is used for the partial polymer oxidation, in order to reach the proper O/C/H ratio. Among them, particularly relevant are the fixed-bed pressure gasification, which converts a mixture of brown coal and plastic waste into syngas used for electricity generation, and the Winkler Process, using air and steam in an autothermal fluidized-bed reactor to produce syngas (Lopez et al., 2018)
- *Reduction in the blast furnace*: SPW can be used in replacement of traditional fuels to operate blast furnaces. The polymer scraps are almost instantaneously gasified at the temperatures of approximately 2000 °C reached in the furnace. A critical point here is the generation of chloride-containing gases. Therefore, the chlorine content should be accurately limited in the SPW fed to the furnace (Sahajwalla et al., 2012)
- *Polymer hydrogenation*: Given the similarity of the majority of polymers with heavy oils, this process operates concomitantly with the oil refining operations. In particular, a partial depolymerization can be obtained after the visbreaking of the polymers, eventually in a mixture with heavy oil, followed by hydrocracking at conditions of 200 bar and 400–500 °C (McManus and Rempel, 1995)
- *Solvolysis*: Particularly suitable for condensation polymers like polyurethanes and polyethylene terephthalate, processes exploiting the amenability for rupture of the chemical bonds connecting the repeating units in the polymer chain by reaction with a specific solvent have been developed. These processes rely on solvents like water (i.e., hydrolysis), ethylene glycol (i.e., glycolysis), a low molecular weight alcohol (i.e., methanolysis or ethanolysis) or an acidic solvent (i.e., acidolysis) to operate the depolymerization up to the individual monomer. They are particularly effective on pre-cleaned feedstocks (Vollmer et al., 2020)
- *Pyrolysis*: This process degrades the polymer into products with a high-added value, typically the monomer itself. It is conducted under mild conditions, at temperatures in the range 350–600 °C and in an inert atmosphere. One of the brightest examples of pyrolysis process is the BASF thermolysis. Integrated in the Ludwigshafen plant, it can treat more than 15,000 tons of mixed plastic waste per year. In a first step, the polymer melting and concomitant dehalogenation is achieved in a train of continuous stirred tank reactors operated at 350 °C. The liquid is then passed to a tubular cracker producing a mixture with a similar composition as the naphtha (Stabel et al., 1995)

It is evident that the most convenient process for the generation of products with an important market value is the pyrolysis (Senthil Kumar et al., 2017). This is indeed the most exploited process for the chemical recycling of PMMA, which, due to its relatively low ceiling temperature (Tc = 198 °C), can be quantitatively depolymerized to its monomer (Albrecht et al., 2013). Already before the massive diffusion of PMMA, back in the '50s a plethora of thermal processes for the pyrolytical depolymerization to MMA were introduced. The first process, introduced by Degussa AG, exploited a molten metal bath as the depolymerization reactor (Domingo Segui and Cabanero, 1958). Suitable metals used for this scope are lead (melting point, $T_m = 327\,°C$) and tin ($T_m = 232\,°C$). This solution is appreciated for the good heat transfer, incremented by a stirring device operating directly in the metal bath. The depolymerization products can be recovered in the gas phase leaving the reactor. Through a proper adjustment of the reaction temperature, it is possible to induce PMMA to depolymerize spontaneously, with high yields in the monomer. Despite the good potential of this process, some drawbacks prevented its application on a large scale, mainly the use of significant amount of metals that could contaminate the product, the possibility for carbonaceous species to accumulate on the metal deteriorating the heat transfer and the possible formation of toxic organometallic compounds (Popescu et al., 2009). For these reasons, alternative process configurations, replacing the metal bath with cheaper and less hazardous solutions were explored. This led to the definition by Kaminsky et al. of the Hamburg Process, which allows the pyrolysis of PMMA in a fluidized bed reactor in which plastic flakes are directly fed and mixed with sand or super-heated steam to avoid the uncontrolled pyrolysis (Kaminsky, 2006).

Before describing the currently available and scaled-up technologies for the pyrolysis of PMMA in detail, it is worth highlighting the key reaction mechanisms underlying its thermochemical recycling.

The analysis of the PMMA thermal degradation has been carried out during the years mainly through thermogravimetric analysis (TGA), which allowed to unravel the underlying radical mechanism. In particular, Kashiwagi et al. for the first time realized that this process occurs in three steps, similarly to what occurs during the synthesis via FRP, as judged by the three different stages of weight loss observed during TGA experiments (Kashiwagi et al., 1986). In particular, a preliminary initiation stage is required to generate active radical centers out of unreactive dead polymers. Once the first radicals are formed, the reaction proceeds through the depropagation step until termination reactions come into play. Ideally, the depropagation is desired to occur until completion, in order to increase as much as possible the yield in monomer, avoiding residual oligomers.

The mechanism of the initiation step, which is an endothermic, homolytic bond rupture involving a dead polymer chain and leading to the formation of smaller macroradicals, is dependent on the type of chain where the radical is generated. In particular, the FRP of MMA generates mainly two types of polymer chain, unsaturated chains coming from termination by disproportionation and saturated chains produced by combinations. In the former case, the initiation occurs by chain-end fission, with the bond cleavage occurring at the chain-end where the unsaturation is located. This requires temperatures in the order of 230–290 °C and leads to the formation of a tertiary and an allyl radical, as shown in Fig. 3A. Alternatively, in the case of both saturated and unsaturated chains, the initiation can occur by homolytic cleavage of a mid-chain bond. The most favorable reaction in this case is the head-head fission (Fig. 3B). Head-to-head bonds come from terminations by combination or by structural defects in the polymer chains as discussed previously. Since this reaction leads to the formation of two tertiary macroradicals, particularly stable, it can occur at relatively mild temperatures, in the range 160–200 °C, and hence justifies the first stage of weight

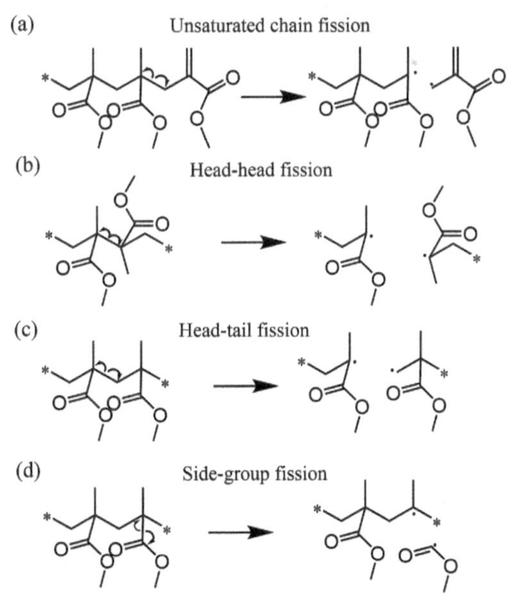

Fig. 3 Postulated mechanisms for the initiation of the thermal degradation of PMMA.

loss observed during the thermogravimetric analysis. Also head-tail fissions can occur (Fig. 3C). If from one side head-to-tail linkages are the most abundant in the PMMA chain, making their cleavage kinetically favored, on the other the bond rupture requires more energy than head-head fissions as it leads to the formation of a tertiary and a primary macroradical, a less stable configuration. As a result, this initiation mechanism is typically appreciated at higher temperatures than head-head fissions, in the range 290–360 °C (Bahadur et al., 2022; Berdie et al., 2021; Chen et al., 2021). As a matter of fact, it was hypothesized that these head-tail fissions can take place even at the early stages of the thermal degradation of the PMMA. However, given the limited mobility of the macroradicals generated and their intrinsic reactivity, the overall trend is toward rapid recombination, which makes this mechanism hardly observable at low temperatures, despite being kinetically favored (Manring et al., 1989). Finally, an additional possibility is represented by the side-group fissions, depicted in Fig. 3D.

Once two macroradicals are formed, the depropagation mainly occurs through an unzipping process that generates MMA and progressively reduces the chain length of the macroradical until termination reactions intervene. These are mostly disproportionations, generating again unsaturated polymer chains that can eventually be reactivated through a successive initiation reaction.

This ideal mechanism strictly applies to PMMA homopolymers. The situation in the case of waste PMMA is complicated by the possible presence of additives and comonomers. As an example, mercaptans added to the system can saturate the polymer chains, stabilizing the plastic. A similar stabilization effect can be introduced by comonomers, in particular acrylates, that are less prone to unzipping due to the formation of a secondary radical instead of a tertiary one. In this situation, the comonomer can block the unzipping and rather lead to disproportionation or cyclization reactions. On the other hand, other methacrylates used as comonomers with MMA have a lower influence on the thermal depolymerization, as the unzipping can still take place in this case (Bekanova et al., 2019; Chong et al., 2006; Galka et al., 2014; McNeill, 1968).

4. Reactor configuration for the PMMA thermal recycling

The choice of the proper reactor for the chemical recycling of PMMA is crucial to maximize the yield in monomer. It should indeed guarantee a strict control of the reaction temperature and efficient heat transfer while ensuring an inert atmosphere.

As discussed in the previous section, the first ever used reactor is a molten metal bath, where PMMA flakes can be directly fed. As a matter of fact, this solution, using molten lead as heat transfer medium, is still the most applied one at the industrial level. The produced MMA is recovered in the gas phase and then condensed. Purities up to 95% can be reached using this technology. However, it poses the important environmental issue of the disposal of the molten lead bath at the end of its lifetime.

For this reason, the research effort moved toward less environmentally impactful solutions. Among them, the fluidized bed reactor is one of the most studied technology for the depolymerization of plastics. In fact, it offers optimal heat transfer, operated through the addition of an inert material, like sand, silica or alumina. In particular, Vaughan et al. patented a peculiar geometry for the reactor (Vaughan and Highgate Donald, 1997). The cross-section of the main reactor chamber increases upwards, thus allowing the fluidization to be realized by the gas recirculated through a fan. PMMA scraps are fed in the middle of this truncated cone reactor, which is operated at 350–400 °C with a nitrogen atmosphere. Higher temperatures around 500 °C are used to conduct the depolymerization in the fluidized bed reactor patented by Sasaki et al (Sasaki et al., 2012). In this technology, sand is used as the heat transfer medium and the solid particles falling on the bottom of the reactor are collected, passed through a furnace in order to remove carbonaceous deposits formed over the sand particles and finally re-fed to the system through a screw feeder. This cleaning of the sand, cycle after cycle, is claimed to increase the duration of the process, reducing the preventive shut-downs for maintenance. The fluidization gas is represented by nitrogen as well as by a fraction of the depolymerization products, which are recirculated to the reactor. The patent reports a depolymerization yield of 95%, with a purity of the produced MMA of 96%. More recently, Dubois from Arkema proposed an alternative configuration of fluidized bed reactor for the chemical recycling of PMMA-fiber composites (Dubois, 2019). The composite material is fed to the reactor in the form of 25 mm particles directly through a hopper. No specific heat transfer medium is reported for this application, allowing to assume that any inert material could actually be exploited for this purpose. The fluidization is obtained through an inert gas fed from the bottom of the reactor. No spillages of the solid particles are performed with this configuration. The main issue in this case is the fate of the fibers composing the composite, which do not degrade during the process. The reactor is designed so that these fibers are dragged by the produced gas mixture into a cyclone, which separates them from the lighter components, including the produced MMA.

An alternative to the fluidized bed reactor configuration is a tubular reactor, exploiting inert material not only as fluidization and heat transfer medium but also to transport the PMMA along the reactor axis. The scraps on the other hand can be fed with a screw feeder. As a matter of fact, the earliest technologies combined this concept of tubular reactor with the extrusion. Indeed, the reactive extrusion process has been extensively reported in the patent literature. The conduction of the thermal depolymerization of PMMA in an extruder carries several advantages compared to the fluidized bed reactors. First, no inert material like sand, quartz or alumina is required, which facilitates the operation and avoids the risk of formation of deposits (e.g., soot and carbonaceous material) over the grain surface, with a negative impact on the heat transfer efficiency. An additional advantage is the easier scalability of the extrusion, which already showed examples applied at the industrial scale (De Tommaso and Dubois, 2021). Finally, the extrusion is intrinsically a continuous process, enabling higher throughputs and homogeneous heat transfer. PMMA flakes can be directly loaded through a hopper and the thermal degradation is initiated at the high temperatures reached inside the extruder, as a result of convection through the extruder wall and shear associated to the rotation of the screws and movement of the reaction mass along the longitudinal coordinate. This configuration was patented by Saito et al. (Saito et al., 1997). The molten PMMA moves along the extruder being progressively depolymerized to MMA at temperatures in the range 400–600 °C. The low boiling point of the monomer allows its recovery in the form of vapor, which is pushed toward the die of the extruder by the molten plastic. Therefore, at the reactor outlet a condenser is required to recover the MMA produced as a liquid. This basic configuration suffers a limited possibility for scale-up. In fact, when increasing the size of the extruder, the specific surface to volume ratio decreases, thus requiring higher temperatures at the wall to efficiently reach the conditions required by the thermal degradation. However, this increased temperature at the wall introduces the risk of forming hotspots and the consequent uncontrolled pyrolysis of a fraction of reactant (Moens et al., 2020). To overcome this limitation, the addition of a heat transfer medium to the extruder was proposed by Weiss et al. (Weiss et al., 2002). This configuration approaches the one of a fluidized bed, as this heat transfer medium acts as a fluidization agent and, at the same time, allows removing in continuous deposits and particulate formed on the screws, with a cleaning action that is reported to preserve the efficiency of the process for longer time.

A further type of reactor has been patented for the thermolysis of PMMA, which is the paddle reactor (Schola and Ruzicka, 2006). Here, the PMMA is kept stirred by a series of paddles and by an inert material that, together with a mechanical action devoted to the grinding of the PMMA scraps fed to the reactor, operates as a heat transfer medium. The produced monomer is recovered through a gas sampling port directly from the reactor. Despite the possibility of obtaining MMA with high purity, given the mechanical stresses at which the paddles and intern of the reactor are subjected, this technology is still far from being competitive.

Therefore, together with the fluidized bed reactor, the reactive extrusion shows the greatest promises for the development of a large-scale chemical recycling plant for PMMA. Most importantly, the two configurations represent desirable alternatives to the more traditional molten lead bath, which is still the most exploited technology for the depolymerization of PMMA. This holds particularly true for low-quality feedstocks, which causes the formation of solid complexes between the contaminants and the metal used in the reactor, bringing about important environmental concerns.

5. Conclusions

PMMA finds unbeaten applications in optical and electronic devices due to its transparency, light weight and insulation properties. Starting from the 1950s, different recovery and recycling technologies have been proposed for PMMA, whose use is expected to steadily grow in the coming years. Herein, the two main families of recycling processes, namely mechanical and chemical recycling, have been presented and discussed. While the former causes the deterioration of the material after each recycling cycle, only the chemical processes can actually preserve or even increase the polymer quality, with the potential of really closing the loop.

Among the significantly different processes proposed for the chemical recycling of plastic materials, the most convenient one is still dictated by the material itself and should therefore be selected on a case-by-case basis. In the peculiar case of PMMA, given its relatively low ceiling temperature, the thermal depolymerization is for sure the most relevant one. In fact, this radical process leads to high depolymerization yields producing the monomer MMA with, in most of the reported cases, high purity. MMA can be then repolymerized, thus allowing an upcycling of the end-of-life PMMA and obtain optimal properties for the optical and electronic market.

During the years, different reactor technologies for the thermolysis of PMMA have been patented. The configuration currently in operation at medium to large scales is constituted by a molten lead bath, ensuring optimal heat transfer and possibility of recovering the monomer in the gas phase with good purities. However, this technology is well suited for pure feedstocks. In fact, contaminants or additives present in the polymer scraps can easily form deposits and complexes with the metal, posing the important environmental issue of properly disposing the wastes. Therefore, more environmentally friendly solutions are desirable. Among them, the most appealing technologies, with scale-up capabilities, are the fluidized bed reactor and the reactive extrusion. These already showed good results on small to medium scale and more research effort is expected in this direction to bring them to full maturity. This effort, however, should be supported by advancements in the understanding of the PMMA depolymerization chemistry, which is still obscure at the single reaction level. Advances in this direction and in the modeling of the process, extended to the case of PMMA containing exogenous compounds and eventually comonomers, is highly desirable to identify the optimal operating conditions to be applied to the waste material collected from the different sources, which is crucial to achieve high depolymerization efficiencies and monomer purity.

References

Achilias DS: Chemical recycling of polymers. The case of poly(methyl methacrylate), *Proc 2006 IASME/WSEAS Int Conf Energy Environ Syst*:271–276, 2006.

Achilias DS, Karayannidis GP: The chemical recycling of PET in the framework of sustainable development, *Water, Air Soil Pollut Focus* 4(4):385–396, 2004. Available from https://doi.org/10.1023/B:WAFO.0000044812.47185.0f.

Albrecht K, Stickler M, Rhein T: *Polymethacrylates [Internet]. Ullmann's encyclopedia of industrial chemistry*, 2013, (Major Reference Works). Available from https://doi.org/10.1002/14356007.a21_473.pub2.

Ali U, Karim KJBA, Buang NA: A review of the properties and applications of poly(methyl methacrylate) (PMMA), *Polym Rev* 55(4):678–705, 2015. Available from https://doi.org/10.1080/15583724.2015.1031377.

Al-Salem SM, Lettieri P, Baeyens J: Recycling and recovery routes of plastic solid waste (PSW): A review, *Waste Manag* 29(10):2625–2643, 2009. Available from: https://www.sciencedirect.com/science/article/pii/S0956053X09002190.

Bahadur A, Iqbal S, Alsaab HO, Awwad NS, Ibrahium HA: Thermal degradation study of polymethylmethacrylate with AlI3 nanoadditive, *Microsc Res Tech* 85(4):1494–1501, 2022.

Bekanova MZ, Neumolotov NK, Jablanovic AD, Plutalova AV, Chernikova EV, Kudryavtsev YV: Thermal stability of RAFT-based poly(methyl methacrylate): A kinetic study of the dithiobenzoate and trithiocarbonate end-group effect, *Polym Degrad Stab* 164:18–27, 2019.

Berdie AD, Berdie AA, Jitian S: The degradation of thin poly(methyl methacrylate) films subjected to different destructive treatments, *J Polym Res* 28(2), 2021.

Chen RY, Pan RM, Li QW: Thermal degradation characteristics, kinetics and thermodynamics of micron-sized PMMA in oxygenous atmosphere using thermogravimetry and deconvolution method based on Gauss function, *J LOSS Prev Process Ind* 71, 2021.

Chong B, Moad G, Rizzardo E, Skidmore M, Thang SH: Thermolysis of RAFT-synthesized poly(methyl methacrylate), *Aust J Chem* 59(10):755–762, 2006.

Christian P, Giles MR, Griffiths RMT, Irvine DJ, Major RC, Howdle SM: Free radical polymerization of methyl methacrylate in supercritical carbon dioxide using a pseudo-graft stabilizer: effect of monomer, initiator, and stabilizer concentrations, *Macromolecules* 33(25):9222–9227, 2000. Available from https://doi.org/10.1021/ma0008948.

De Tommaso J, Dubois JL: Risk analysis on PMMA recycling economics, *Polymers (Basel)* 13(16), 2021.

Domingo Segui E, Cabanero Alarcon B. Process and device for the regeneration of monomers starting from polymethacrylate and, more especially, methyl polymethacrylatetle. US2858255A, 1958.

Dubois J-L. Method for Recycling Composite Materials with an Improved Energy Balance. WO2019207259A1, 2019.

Evonik: *Polymers for a clear, transparent view: Plexiglas® [Internet]*, Available from https://history.evonik.com/en/inventions/plexiglas.

Galka P, Kowalonek J, Kaczmarek H: Thermogravimetric analysis of thermal stability of poly(methyl methacrylate) films modified with photoinitiators, *J Therm Anal Calorim* 115(2):1387–1394, 2014.

Geyer R, Jambeck JR, Law KL: Production, use, and fate of all plastics ever made, *Sci Adv* 3(7):25–29, 2017.

Glusker DL, Lysloff I, Stiles E: The mechanism of the anionic polymerization of methyl methacrylate. II. The use of molecular weight distributions to establish a mechanism, *J Polym Sci* 49(152):315–334, 1961.

Goode WE, Owens FH, Fellmann RP, Snyder WH, Moore JE: Crystalline acrylic polymers. I. Stereospecific anionic polymerization of methyl methacrylate, *J Polym Sci* 46(148): 317–331, 1960.

Hundertmark T, Mayer M, McNally C, Simons TH, Witte C: *How plastics waste recycling could transform the chemical industry [Internet]*, 2018, Mckinsey & Company. Available from: https://www.mckinsey.com/industries/chemicals/our-insights/how-plastics-waste-recycling-could-transform-the-chemical-industry.

Kaminsky W: The Hamburg Fluidized-bed Pyrolysis Process to Recycle Polymer Wastes and Tires [Internet], In *Feedstock recycling and pyrolysis of waste plastics*, 2006, pp 475–491. (Wiley Online Books). Available from https://doi.org/10.1002/0470021543.ch17.

Kashiwagi T, Inaba A, Brown JE, Hatada K, Kitayama T, Masuda E: Effects of weak linkages on the thermal and oxidative degradation of poly(methyl methacrylates), *Macromolecules* 19(8):2160–2168, 1986. Available from https://doi.org/10.1021/ma00162a010.

Lopez G, Artetxe M, Amutio M, Alvarez J, Bilbao J, Olazar M: Recent advances in the gasification of waste plastics. A critical overview, *Renew Sustain Energy Rev* 82:576–596, 2018. Available from https://www.sciencedirect.com/science/article/pii/S1364032117312832.

Manring LE, Sogah DY, Cohen GM: Thermal degradation of poly(methyl methacrylate). 3. Polymer with head-to-head linkages, *Macromolecules* 22(12):4652–4654, 1989. Available from https://doi.org/10.1021/ma00202a048.
McManus NT, Rempel GL: Chemical modification of polymers: catalytic hydrogenation and related reactions, *J Macromol Sci Part C* 35(2):239–285, 1995. Available from https://doi.org/10.1080/15321799508009638.
McNeill IC: A study of the thermal degradation of methyl methacrylate polymers and copolymers by thermal volatilization analysis, *Eur Polym J* 4(1):21–30, 1968. Available from: https://www.sciencedirect.com/science/article/pii/0014305768900049.
Moens EKC, De Smit K, Marien YW, et al.: Progress in reaction mechanisms and reactor technologies for thermochemical recycling of poly(methyl methacrylate), *Polymers* 12, 2020.
Plastics Europe—Association of Plastic Manufacturers (Organization): *Plastics—the facts 2020*, 2020, PlasticEurope, p 16.
PlasticsEurope: *Plastics–the facts 2019 [Internet]*, 2019. Available from: https://www.plasticseurope.org/application/files/1115/7236/4388/FINAL_web_version_Plastics_the_facts2019_14102019.pdf.
Popescu V, Vasile C, Brebu M, et al.: The characterization of recycled PMMA, *J Alloys Compd* 483(1):432–436, 2009. Available from: https://www.sciencedirect.com/science/article/pii/S0925838808018252.
Prado AR, Leal-Junior AG, Marques C, et al.: Polymethyl methacrylate (PMMA) recycling for the production of optical fiber sensor systems, *Opt Express* 25(24):30051–30060, 2017. Available from: http://opg.optica.org/oe/abstract.cfm?URI=oe-25-24-30051.
Sahajwalla V, Zaharia M, Kongkarat S, et al.: Recycling end-of-life polymers in an electric arc furnace steelmaking process: fundamentals of polymer reactions with slag and metal, *Energy Fuel* 26(1):58–66, 2012. Available from https://doi.org/10.1021/ef201175t.
Saito H, Masaki T, Morooka S, Koyanagi K. Continuous Regeneration of Acrylic Resin and Device Therefor. JP3410343B2, 1997.
Sasaki A, Kikuya N, OOkubo T, Hayashida M. Recovery method of pyrolysis product of resin. US8304573B2, 2012.
Sasikala Suresh S, Mohanty S, Kumar NS: Preparation of poly(vinyl chloride)/poly (methyl methacrylate) recycled blends: effect of varied concentration of PVC and PMMA in stability of PVC phase on the recycled blends, *Mater Today Proc* 5(2, Part 3): 8899–8907, 2018. Available from: https://www.sciencedirect.com/science/article/pii/S2214785317333266.
Sasse F, Emig G: Chemical recycling of polymer materials, *Chem Eng Technol Ind Chem Equipment-Process Eng* 21(10):777–789, 1998.
Scheirs J: *Polymer recycling: science, technology and applications*, 1998, John Wiley & Sons Inc., p 616.
Schola E, Ruzicka M. Depolymerization Method and Device. US20060205845A1, 2006.
Scorah MJ, Dhib R, Penlidis A: Free-radical polymerization of methyl methacrylate with a tetrafunctional peroxide initiator, *J Polym Sci Part A Polym Chem* 42(22):5647–5661, 2004. Available from https://doi.org/10.1002/pola.20389.
Senthil Kumar P, Bharathikumar M, Prabhakaran C, Vijayan S, Ramakrishnan K: Conversion of waste plastics into low-emissive hydrocarbon fuels through catalytic depolymerization in a new laboratory scale batch reactor, *Int J Energy Environ Eng* 8(2):167–173, 2017.
SRI International: N.N. Marketing Research Report: Acrylic Resins and Plastics, In *Chemical Economic Handbook*, Zurich, 1994, SRI International.
Stabel U, Woerz H, Kotkamp R, Fried A. Process for Recycling Plastics in a Steam Cracker. WO9503375A1, 1995.

Suresh SS, Mohanty S, Nayak SK: Investigation into the mechanical and thermal properties of poly(methyl methacrylate) recovered from light guidance panels with a focus on future remanufacturing and sustainable waste management, *J Remanufacturing* 7(2):217–233, 2017a. Available from https://doi.org/10.1007/s13243-017-0041-7.

Suresh SS, Mohanty S, Nayak SK: Preparation and characterization of recycled blends using poly (vinyl chloride) and poly(methyl methacrylate) recovered from waste electrical and electronic equipments, *J Clean Prod* 149:863–873, 2017b. Available from: https://www.sciencedirect.com/science/article/pii/S0959652617302718.

van der Heijden S, Dubois J-L, Bossard M: *New innovative process by MMAtwo for recycling end-of-life PMMA waste [Internet]*, 2018, Available from: https://www.mmatwo.eu/wp-content/uploads/2019/01/Press-release-MMAtwo_ENG_VF.pdf.

Vaughan PW, Highgate Donald J. Depolymerisation. US5663420A, 1997.

Vollmer I, Jenks MJF, Roelands MCP, et al.: Beyond mechanical recycling: giving new life to plastic waste, *Angew Chem Int Ed* 59(36):15402–15423, 2020. Available from https://doi.org/10.1002/anie.201915651.

Weiss H-H, Schmalfeld J, Zentner U, Groschang T, Gropp U, Fuss W, et al. Method for depolymerizing polymethylmethacrylate. US6469203B1, 2002.

CHAPTER EIGHT

Systemic approaches to model plastics circularity

Meltem Urgun-Demirtas[a,*], Pahola Thathiana Benavides[b], Ulises R. Gracida-Alvarez[b], and Sean Riggio[a]
[a]Applied Materials Division, Argonne National Laboratory, Lemont, IL, United States
[b]Energy Systems and Infrastructure Analysis Division, Argonne National Laboratory, Lemont, IL, United States
*Corresponding author: e-mail address: demirtasmu@anl.gov

Contents

Abstract

As industry is moving from a linear to a circular economy (CE), analysis becomes crucial to develop strategies for plastic sustainability, identify opportunities for technology improvement, and provide research and development guidance to plastic manufactures and stakeholders. Life-cycle analysis (LCA) has been used to evaluate the environmental impact of plastic production in its first and post-use life. Material flow analysis (MFA) is another excellent tool for understanding how the new circular economy approach influences the supply chain and downstream markets of new products generated from plastic wastes. This assessment becomes necessary for large-scale changes in material circularity. There is a recent trend for integrating LCA and MFA methodologies to evaluate the degree of circularity and assess the effects of circularity at different application levels. The outcomes of this combined LCA and MFA analysis can provide a guiding framework for a consistent and transparent evaluation of increased plastics recycling and identification of an appropriate plastics recycling approach with large-scale implications across value and supply chains. Multiple driving forces will accelerate the circular economy around plastic materials, including but not limited to cost, environmental benefits, availability of collection/logistics systems and chemical recycling technologies, market conditions, policies and regulations, and social changes and adoption. Recently, there has been a growing interest in integrating different

Advances in Chemical Engineering, Volume 60
ISSN 0065-2377
https://doi.org/10.1016/bs.ache.2022.09.005

assessment tools to model the circularity transition and its impacts. Integrating a variety of modeling tools promotes an effective transition from a linear to a circular economy by identifying relevant opportunities, challenges, and research gaps under a systemic and dynamic framework.

1. Introduction

Plastic sustainability is a major global concern and has become a priority in the plastic industry. Increases in the global demand and production of plastic combined with <20% of plastic recycled around the world necessitates sustainable solutions to treat plastic waste. Ineffective plastic waste management around the world and its impact on human life and wildlife (Jambeck et al., 2015; Law et al., 2010; Lebreton et al., 2018) have driven attention toward expanding research and development in recycling technologies. A lot of plastic waste reaches the end of its useful life within a short period of time without leveraging its full potential (i.e., high energy and carbon content). Emerging advanced recycling technologies allow opportunities to recover this material, which can be used as an attractive feedstock for valuable fuels and chemicals. Plastic can be recycled via mechanical and chemical technologies. Although mechanical recycling is very popular, it can degrade the mechanical properties of plastic materials, which limits their application in the production of secondary and less valuable products (Jeswani et al., 2021). Conversely, chemical recycling generates new chemical feedstocks (monomers or intermediates) that can be upgraded into virgin-quality products. Thermochemical, hydrocracking, catalytic-based, and depolymerization processes are the most common methods used to produce intermediate chemicals or fuels from plastic waste (Kunwar et al., 2016; Phanisankar et al., 2020). Thermochemical processes use high temperatures to decompose plastic waste into combustible gases. Hydrocracking uses hydrogen as the main agent at high temperature to break larger hydrocarbons or polymers into fuel-range hydrocarbons. Catalytic conversion uses catalysts to improve conversion, product quality, increase selectivity, and improve operation conditions like temperature reduction and residence time. Depolymerization (e.g., hydrolysis, glycolysis or methanolysis) converts plastic waste into virgin-quality monomer precursors. This chapter focuses on pyrolysis conversion, as it is one of the most popular technologies. Pyrolysis uses high temperatures in the absence of oxygen to break down the macromolecule structures of polymers into smaller molecules and monomers (Khot and Basavarajappa, 2017). During

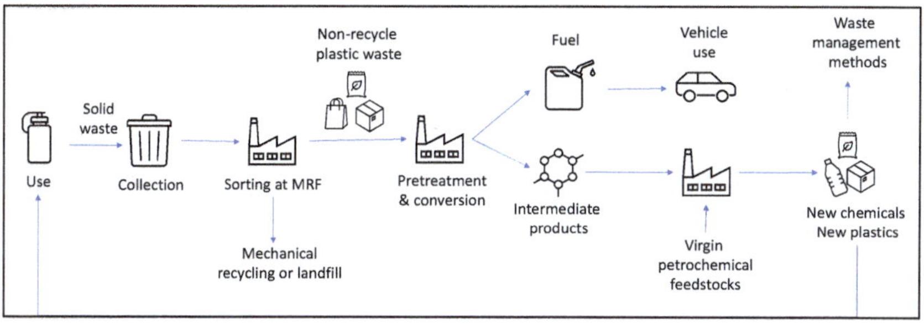

Fig. 1 Plastic waste pathways to fuels or chemicals. Collection and sorting at a material recovery facility (MFR) are the initial steps for plastic waste recovery. Plastic that cannot be mechanically recycled will undergo additional pretreatment to achieve the necessary composition and physical properties for conversion. Plastic waste is then converted into fuel or intermediate products such as pyrolysis oil. This latter can enter a petrochemical process and (in combination with virgin petrochemical feedstock) can be further synthetized into new chemicals or new plastics. Note: this figure only shows the flows of plastic waste being converted to fuel or new plastic, it does not include other input and outputs of the system such as materials for fuel or chemical production and emissions.

pyrolysis, char, fuel gas, and pyrolysis oil are produced. Char and fuel gas can be used to generate energy for internal use, while pyrolysis oil can be upgraded to fuel or rerouted for further processing of new plastic precursors. Fig. 1 presents two potential pathways of plastic waste conversion via pyrolysis to fuels or intermediate chemicals to new products or new plastics.

As industry is moving from a linear to a circular economy (CE) (Corona et al., 2019; Ellen MacArthur foundation, 2019), integrating multiple analysis methods such as LCA, and material flow analysis (MFA) within a circular economic framework is fundamental to determine the feasibility and potential benefits of chemical recycling technologies (Nicholson et al., 2022).

2. Life cycle analysis of recycling plastic wastes

Life-cycle analysis (LCA) can be used to evaluate the environmental impact of plastic production in its first and post-use life. This type of analysis would help develop strategies for plastic sustainability, identify opportunities for technology improvement, and provide research and development guidance to plastic manufactures and stakeholders. As this topic becomes increasingly relevant in the plastic industry, many LCA studies have been conducted to understand the environmental impact of plastic waste conversion technologies. Antelava et al. (2019) summarized LCA papers of pyrolysis of plastic waste and discussed its benefits and challenges. The authors provided a list of studies that discussed plastic waste management options (landfill, mechanical recycling and energy recovery) and compared conversion technologies to treat plastic solid waste. They concluded that technologies such as gasification, incineration or pyrolysis could decrease environmental effect in comparison to landfilling, and pyrolysis offers the advantage of producing oil and char with high calorific values that can be used as fuels. Jeswani et al. (2021) used LCA to assess the environmental impacts of pyrolysis of mixed plastic waste and compared the results with mechanical recycling and energy recovery. They found that 50% lower climate change impact and life cycle energy use could be achieved using pyrolysis of mixed plastic waste compared to the energy recovery while for the mechanical recycling the metrics were similar to pyrolysis technology. Building on previous work (Benavides et al., 2017), we conducted an updated LCA of producing ultra-low-sulfur diesel (ULSD) fuel and high-density polyethylene (HDPE) from plastic waste via pyrolysis. Fig. 2 shows the GHG emission comparison for the two pathways described in Fig. 1: (A) plastic-to-fuel and (B) plastic-to-plastic. To estimate the GHG emissions, we used the Greenhouse gases, Regulated Emissions,

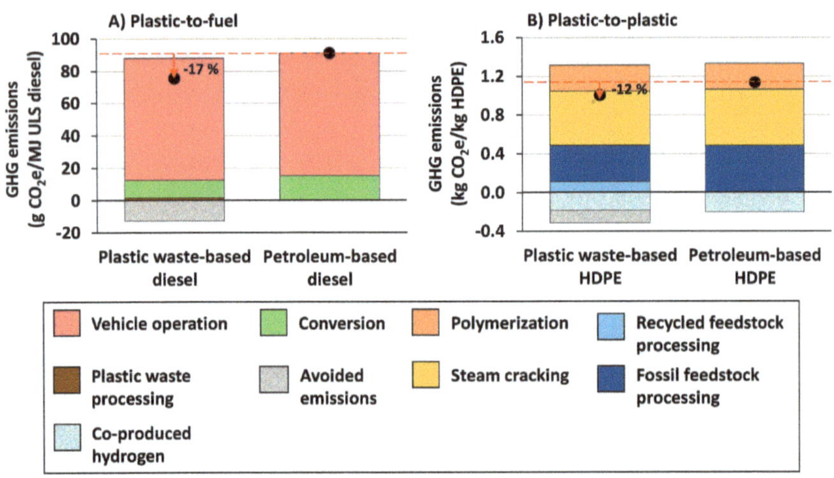

Fig. 2 GHG emission comparison of plastic waste via pyrolysis vs conventional petroleum-based technology. (A) Plastic-to-fuel, (B) Plastic-to-plastic. HDPE: high density polyethylene. Avoided emissions are evaded from traditional waste management, according to U.S. practices, by diversion of plastic waste to the production of new plastic. The results presented here were calculated using energy and mass allocation, for the plastic-to-fuel and plastic-to-plastic pathways, respectively, to allocate the emissions and energy between the products. For more detail on these results please refer to Benavides et al. (2022) and Gracida-Alvarez et al. (2022a).

and Energy use in Technologies (GREET®) model (Argonne, 2021). We leveraged industry data from eight pyrolysis companies with processing capacities ranging from 300 to 291,800 t of waste plastic per year. The results presented focus on plant capacities higher than 50,000 t, which were classified as the Nth-plant model. More details of the plastic-to-fuel pathway can be found at Benavides et al. (2022). The plastic-to-plastic pathway focused the analysis on the production of HDPE as the final product. A mass balance approach was used to determine the GHG emissions of co-processing pyrolysis oil feedstock at a dilution rate of 20%, with the remaining being virgin petrochemical feedstock. Under this dilution rate, the pyrolysis oil substituted the amount of naphtha and approximately 14% of the ethane, propane and butene used for steam cracking. For more details on this analysis, please refer to Gracida-Alvarez et al. (2022a). This analysis showed approximately 17% and 12% emission reductions for converting plastic waste into ULSD and HDPE, respectively, in comparison to their petroleum-based counterparts. Note that for the plastic-to-fuel pathway, the major contributor to the GHG emissions was the combustion emissions during vehicle use, regardless of the feedstock type and conversion technology used. For this analysis, it was assumed that the fuel gas (a co-product of the pyrolysis process) was used to produce internal energy, significantly reducing the energy demand. For the plastic-to-plastic pathway, the major contributor to emissions in the waste plastic- and petroleum-based HDPE pathway was the steam cracking process, accounting for approximately 50% of the total emissions in both products. Fig. 2B also shows that the reduction of GHG emissions of plastic waste-based HDPE compared to petroleum-based HDPE was driven by the inclusion of the avoided emissions from traditional waste management. Therefore, the diversion of plastic waste, used for the production of new plastic, from landfill and incineration resulted in GHG emissions savings of 0.13 kg CO_2 e/kg HDPE (Gracida-Alvarez et al., 2022a).

3. Material flow analysis of recycling plastics wastes: A case study focuses on lubricant production

MFA is an excellent tool for understanding how the new circular economy approach influences the supply chain and downstream markets of new products generated from plastic wastes. This assessment becomes necessary for large-scale changes in material circularity. The first MFA of polyurethane (PU) in the United States (US) was conducted recently

(Liang et al., 2021). The PU flow map provides essential information on how historical PU-related feedstock production, precursors, manufacturing, uses, and waste management data are crucial to provide a basis for decision making, improve the circularity of the economy and determine the challenges and opportunities for future scenarios on utilization of plastic waste streams. The developed PU flow map also pointed out limitations of the existing mechanical recycling plastics methods and identified a chemical recycling method to recycle the post-consumer PU waste streams. In another study, the MFA was used to evaluate data from material recovery facilities (MRFs) located in Europe to determine process efficiencies, materials flows and barriers from sorting and recycling of post-consumer plastic packaging waste streams (Antonopoulos et al., 2021). The modeling results showed that an end-of-life recycling rate of 49% would be possible in 2030 (compared to 14% in 2017) when best recycling practices and technologies are implemented. The results also showed the significant improvements need to be made at the MRFs, including plastics design (e.g., "design for recyclability," reduction in impurities and contaminants in the input), collection system (e.g., increase efficiency of operations) and market level (e.g., markets for low value products) to meet EU27's 55% recycling target in 2030 (Antonopoulos et al., 2021).

The application of the MFA illustrated in this chapter is for lubricant production in the U.S. since pyrolysis oil (aka lubricant) is one of the main products from pyrolysis of plastic waste streams. This MFA framework included precursors for the U.S. Lubricants Industry, U.S. base oil production and lubricant end-use, U.S. end-of-life and waste collection, and International Trade data. Quantities of lubricant precursors (crude oil and natural gas) were estimated using information from the Energy Information Administration (EIA, 2021a) as well as a yearly publication from Lubes 'N' Greases (LNG Publishing Company, Inc, 2021). The volume of oil and natural gas necessary to produce Group I–V base oils in the U.S. in 2021 was calculated using 2020 EIA data for crude oil volume flows into U.S. refineries per day combined with volume fraction estimates of a barrel of crude oil used to produce lubricants from the LNG factbook.

Volume fractions of each base oil group as a percentage of total base oil production in the U.S. were estimated from global lubricant demand data provided by the LNG Factbook. The volume fraction for each base oil was treated as a constant for each end-use application. See Table 1 for the volume fraction breakdown of each base oil. Total volume flows of base oils for each end-use application were estimated using data from the LNG

Table 1 Estimated volume fractions of base oils produced in the United States in 2021.

Base oil group	Volume fraction
Group 1	35.39%
Group 2	38.26%
Group 3	13.39%
PAO/Group 4	4.35%
Napthenic/Group 5	8.61%

factbook. Additive volumes for each end-use application were also calculated using estimates from Chevron Corporation, Acton Chemical, and the LNG factbook (2021).

Estimates for lubricant recycling and other end-of-life scenarios were estimated from a 2020 DOE report on lubricant recycling in the U.S. (DOE, 2020). Overall, five scenarios were considered, including lost-in-use, re-refining to base oils, re-refining to non-base oil products, lubricants collected but not re-refined, and accounted-for lubricants. In the Sankey diagrams provided in the results section of this paper, re-refining, re-refining to non-base oils, lubricants collected but not refined, and lubricants accounted for at the source all fall under the category of generated lubricants. According to the 2020 DOE report, approximately 44% by volume of U.S. lubricants are lost-in use, while the remaining lubricants are considered generated. These volume fractions were applied to all end-use applications to calculate the total volume of lubricants lost-in-use and generated. The volume flows for each of the four scenarios under the lubricants generated category were also calculated using data from the 2020 DOE report. Lubricants collected but not refined include processes such as burning the used lubricants on-site, processing used lubricants into fuel oil, blending the used lubricants into high sulfur fuel oil, or exporting the used lubricants for combustion. Lubricants accounted for at the source are defined as used lubricants burned by generators, recycled on-site, landfilled, or dumped. Total volume flows for base oil imports and exports were estimated based on 2021 data from the EIA. Imports and export volume flows for each individual base oil were calculated using the volume fraction data in Table 1 (EIA, 2021b,c).

Reconciliation for volume flows was done using the STAN software package. Standard error of each volume flow used in STAN was estimated through an uncertainty analysis following the methods of Liang et al. (2021)

and Kawecki et al. (2018). This method is based on five different data quality indicator scores (DQIS): geographical, temporal, material representativeness, completeness, and source reliability (Liang et al., 2021; Kawecki et al., 2018).

Overall, 9.219 Mm^3 of base oils were produced in the U.S. in 2021. 2.099 Mm^3 of base oils were imported, and 5.470 Mm^3 were exported. This resulted in a net flow of base oils to the U.S. lubricant market of 5.849 Mm^3. Once the blending of additives into base oils was accounted for (i.e., finished lubricants), total U.S. consumption of lubricants in 2021 was 6.836 Mm^3. Approximately 3.008 Mm^3 of lubricants (44%) were lost-in-use, while the remaining 3.828 Mm^3 was considered generated lubricant waste, which includes re-refining to base oils, re-refining to non-base oils, lubricants collected but not re-refined, and lubricants accounted for at the source.

Volume flows of feedstocks, end-use, and end-of-life scenarios for U.S. lubricants are shown in Fig. 3. The two general end-use markets for lubricants are in automotive and industrial applications, which consumed approximately 3.588 and 3.249 Mm^3, respectively. The largest end-use category for automotive applications, as well as the entire U.S. lubricant market overall, was multigrade engine oils at 2.886 Mm^3, which represents 42.21% of the U.S. lubricant market. Multigrade engine oils are defined as oils that can meet a range of viscosity requirements, which is advantageous for applications that require a wide range of temperatures within which the oils will be used. The next highest end-use category was process oils at 1.373 Mm^3 (20.01% of market volume), which was the highest overall volume for any industrial application (Ramsey, 2001). The next highest applications were general industrial oils and metalworking fluids (both industrial applications), with volume flows of 0.716 and 0.682 Mm^3, respectively. These were followed by transmission and hydraulic fluids for automotive applications (0.546 Mm^3), industrial engine oils (0.477 Mm^3), and lubricants used in other automotive applications not including engine, transmission, or hydraulic fluids (0.156 Mm^3).

As all end-use applications received the same volume fraction estimate for both lost-in-use (44%) and lubricant waste generated (56%), the flows for both categories follow the same hierarchy as was previously described for end-use-volume flows. For generated lubricant waste, the largest volume flow was for lubricants collected not refined, at 1.275 Mm^3. This was followed by lubricants accounted for at the source (1.275 Mm^3), lubricants re-refined to base oils (0.817 Mm^3), and lubricants re-refined into non-base oils (0.545 Mm^3) (Fig. 4).

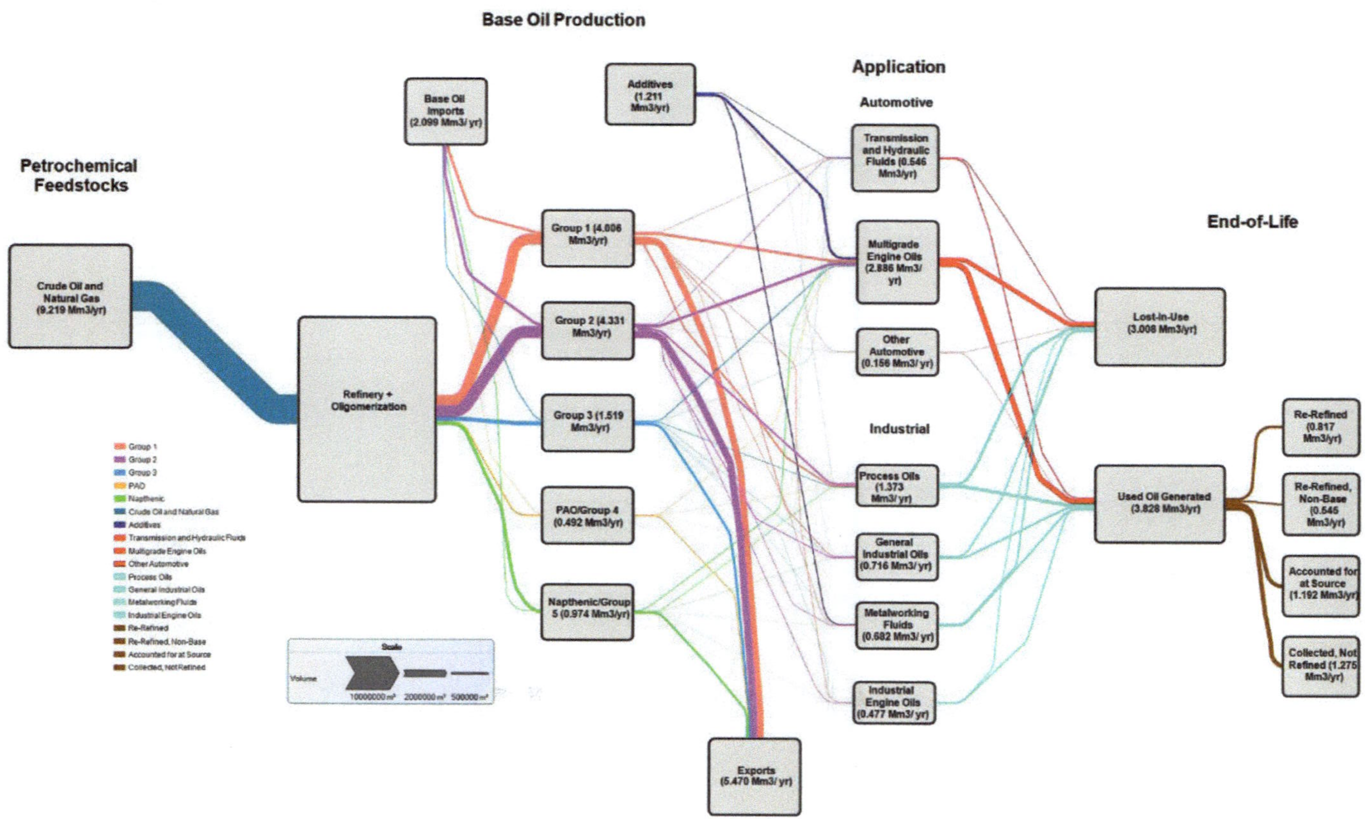

Fig. 3 Initial material flows of U.S. lubricants in 2021.

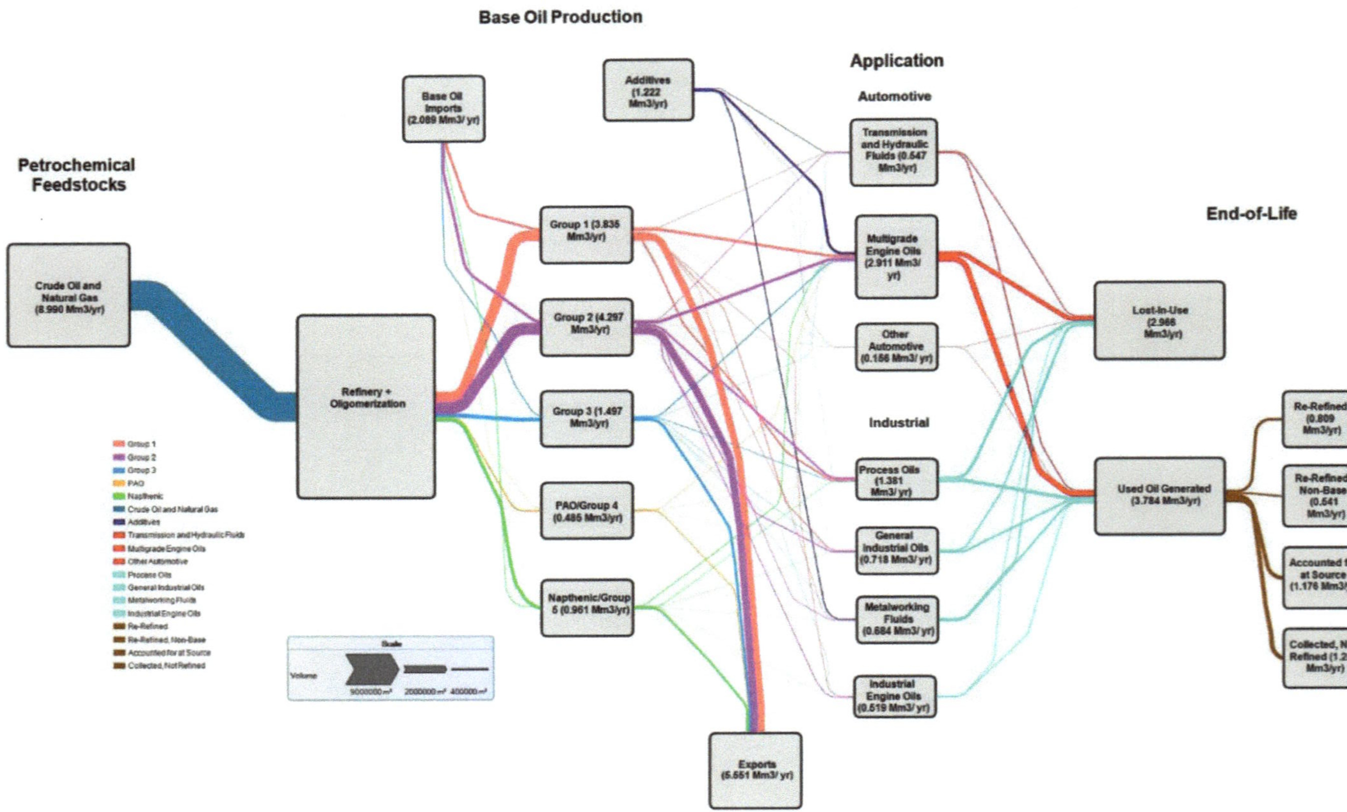

Fig. 4 Reconciled material flows of U.S. lubricants in 2021.

4. An integrated approach for supporting transitions to a circular economy

There is a recent trend for integrating LCA and MFA methodologies to evaluate the degree of circularity and assess the effects of circularity at different application levels (Corona et al., 2019). Gracida-Alvarez et al. (2022b) presented a circular economy (CE) sustainability framework that combined LCA and MFA to evaluate the environmental impacts and circularity metrics of different CE strategies for recycling waste plastic packaging. The CE strategies included increasing recycling rates and the integration of recycling technologies. This work used production of polyethylene terephthalate (PET) bottles to develop the framework due to PET's high recycling rate (around 29% of disposed PET bottles recycled) compared to other resins in the Unites States (NAPCOR and APR, 2018). However, only 21% of the recycled bottles are converted back into new bottles. Contamination and reduced mechanical properties of the recycled material limit their use to the production of low-value products (e.g., fibers, films, and strapping). Therefore, large amounts of virgin PET need to be combined with mechanically recycled PET to maintain the desired properties of the products. To overcome this issue, chemical recycling can be an alternative technology. Chemical recycling of PET includes technologies that convert waste bottles into new bottles (i.e., enzymatic hydrolysis) and technologies that convert fibers into bottles (i.e., methanolysis). The latter process represents an upcycling technology. For more details on this analysis please refer to Gracida-Alvarez et al. (2022b).

The comparison of GHG emissions (environmental impact), and virgin material use (circularity metric) of different scenarios: (1) current state, (2) mechanical maximized, (3) mechanical and chemical, and (4) mechanical, chemical, and upcycling is shown in Fig. 5. The "mechanical maximized" scenario showed 14% lower GHG emissions compared to the "current state," while the "mechanical and chemical" and "mechanical, chemical and upcycling" scenarios showed similar GHG emissions than those of the "current state." This is because in the "mechanical maximized" scenario, all waste PET bottles are recycled through mechanical recycling, which has lower GHG emissions compared to chemical recycling and upcycling (Gracida-Alvarez et al., 2022b). Chemical recycling is more energy and resource intensive. The comparison of virgin material use indicated higher reductions in the scenarios incorporating chemical recycling and upcycling

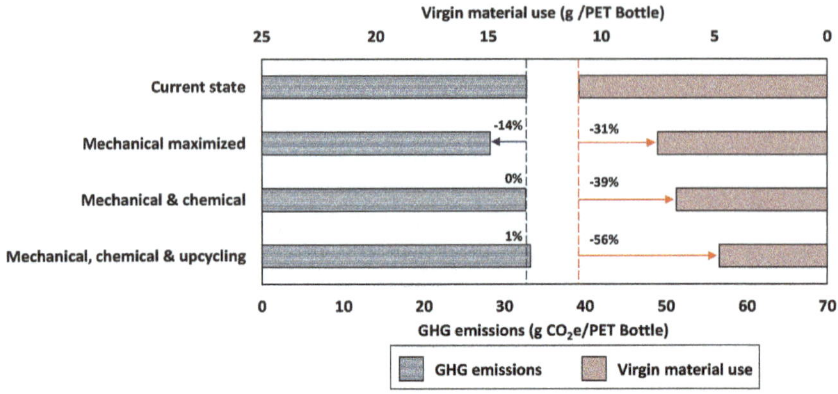

Fig. 5 Comparison of GHG emissions and virgin material use from implementing different circular economy strategies. The "Current state" considered the U.S. supply chain of PET bottles in 2017, where 4% of discarded PET bottles are converted into new bottles. In "Mechanical maximized," due to increase of the recycling rate, 38% of waste PET bottles were converted to new bottles. In "Mechanical and chemical," the increased number of bottles converted into new bottles were processed 50% through mechanical recycling and 50% through chemical recycling. The "Mechanical, chemical, and upcycling" incorporated bottle production from waste fibers diverted from landfill storage by increasing their recycling rate from 10% to 20%.

(39% and 56%, for "mechanical and chemical" and "mechanical, chemical, and upcycling," respectively) compared to the "current state" scenario. In the "mechanical and chemical" scenario, the incorporation of chemical recycling at a higher rate from the current state allowed the increased use of waste bottles not suitable for mechanical recycling, while in the "mechanical, chemical, and upcycling" scenario, fibers usually disposed in landfills were assumed to be diverted to new bottle production, which helped to improve the virgin material use metric. Fig. 5 also highlights the tradeoff between the environmental impacts and circularity metrics. For instance, the scenarios with increasing circularity (lower virgin material use) did not show GHG emissions reduction compared to the "current state," while the scenario with lower GHG emissions compared to the "current state" did not show the lowest virgin material use. This implies that reduced GHG emissions could be achieved under scenarios using less bottles from chemical recycling, but will lead to higher virgin PET usage in the supply chain. However, the chemical recycling processes evaluated in this analysis are still in the early stage of development and there is room for improvements (e.g., increase conversion yields, reduce energy requirements) that can increase process efficiency.

The outcomes of this combined LCA and MFA analysis can provide a guiding framework for a consistent and transparent evaluation of increased plastics recycling and identification of an appropriate plastics recycling approach with large-scale implications across value and supply chains. The optimal recycling technology selection depends on the polymer type, and in some cases, multiple recycling technologies need to be applied for different plastic waste streams (Schwarz et al., 2021). Linking outcomes of LCA and MFA with a techno-economic and market analysis is essential to develop effective policy intensives and increase recycling rates, enabling the circular plastic waste value chain. As discussed in previous chapters, process modeling, especially new pyrolysis process, is necessary to lay the groundwork for the systematic assessment, scale-up, and implementation of new plastic upcycling technologies. Current process models are not sufficient to simulate the performance of new plastics upcycling technologies. There is a need for the development of process simulation models for newly designed reactors with two (or three) moving phases to improve our understanding of reaction kinetics and transport phenomena. New technologies' process modeling also helps bridge the data gap between lab-scale experimental results and field-scale applications. Furthermore, the incorporation of techno-economic analysis (TEA) during the development stage of new technology is beneficial in determining the crucial performance parameters and cost drivers. The outcomes of this analysis can be used to evaluate economic feasibility, guide research and development, and quantify uncertainty and risk associated with new technologies.

5. Conclusions

Multiple driving forces will accelerate the circular economy around plastic materials, including but not limited to cost, environmental benefits, availability of collection/logistics systems and chemical recycling technologies, market conditions, policies and regulations, and social changes and adoption. Recently, there has been a growing interest in integrating different assessment tools to model the circularity transition and its impacts. A system-based multidisciplinary approach enables well-informed and meaningful decision-making that challenges current norms, values and practices of linear economy (Walzberg et al., 2021). Unfortunately, a lack of common sustainability metrics for measuring the performance and impacts of recycling technologies still makes it hard to compare results. A recently published report evaluated tools, models, and frameworks that have been funded by the U.S. Department of Energy Office of Energy Efficiency and Renewable Energy (EERE) to determine the extent of these applications in assessment of the CE (Upasani et al., 2022). In some cases, there is a need to include tools from complex systems science and operations and social science fields to assess multi-driving (and in most cases conflicting) factors (Walzberg et al., 2021). The review and discussion on systemic approaches provided in this chapter constitutes a first step in chemical upcycling of plastics waste. Future studies need to be conducted to include the following:

- The systematic approach should be able to help assess level of circularity for upcycling of plastic waste streams, systematically map and rank the circular economy opportunities at multi-scales, develop scenario for near- and long-term implementations, identify barriers (e.g., waste collection efficiency, fluctuating market price, etc.), and provide a wider transparency to determine possible unintended consequences.
- There is a need for establishment of common sustainability metrics for measuring the chemical upcycling plastic technology performance and quantifying the impacts of plastic recycling technologies to compare the results and assess the scaling effects of circular economy strategies.
- Integrating a variety of modeling tools promotes an effective transition from a linear to a circular economy by identifying relevant opportunities, challenges, and research gaps under a systemic and dynamic framework. This assessment should be conducted on a broader scale to identify the conflicting priorities (e.g., competition in the market for the same plastic waste stream as feedstock, tradeoffs between plastics circularity and environmental indicators, etc.) and unintended consequences from the circular economy strategy implementation.
- Various new plastics upcycling technologies currently at either embryonic stage or pilot-scale will be scaled up to large-scale applications in the near future. Implementation of such technologies should not only be viewed in terms of performance and economic feasibility at the specific scale, but also in terms of lifecycle and material flow analysis. The incorporation of systemic approaches during the development stage of new technology is beneficial in determining the impact of multiple driving forces, hence accelerating the circular economy around plastic materials.

Acknowledgments

The research was supported by the U.S. Department of Energy, Office of Energy Efficiency and Renewable Energy, Advanced Manufacturing Office and Bioenergy Technologies Office, under contract DE-AC02-06CH11357. The views expressed in the article do not necessarily represent the views of the U.S. Department of Energy or the United States Government or any agency or of any commercial entity. Neither the United States Government or any agency thereof, nor any of their employees or employees of contributing companies, makes any warranty, expressed or implied, or assumes any legal liability or responsibility for the accuracy, completeness, or usefulness of any information, apparatus, product, or process disclosed, or represents that its use would not infringe privately owned rights.

References

Antelava A, Damilos S, Hafeex S, et al.: Plastic solid waste (PSW) in the context of life cycle assessment (LCA) and sustainable management, *Environ Manag* 2019(64):230–244, 2019. https://doi.org/10.1007/s00267-019-01178-3.

Antonopoulos I, Faraca G, Tonini D: Recycling of post-consumer plastic packaging waste in the EU: recovery rates, material flows, and barriers, *Waste Manag* 126:694–705, 2021.

Argonne (Argonne National Laboratory): *GREET: Greenhouse gases, regulated emissions, and energy use in transportation*, Lemont, IL, 2021, Argonne National Laboratory. https://greet.es.anl.gov.

Benavides PT, Sun P, Han J, Dunn J, Wang M: Life-cycle analysis of fuels from post-use non-recycled plastics, *Fuel* 203:11–22, 2017. https://doi.org/10.1016/j.fuel.2017.04.070.

Benavides PT, Gracida-Alvarez UR, Lee U, Wang M: *Life-cycle analysis of conversion of post-use plastic via pyrolysis with the GREET model*, Lemont, IL, 2022, Argonne Nation Laboratory. Unpublished information.

Corona B, Shen L, Reike D, Rosales-Carreon J, Worrel E: Towards sustainable development through the circular economy—a review and critical assessment on current circularity metrics, *Resour Conserv Recycl* 151, 2019, 104498. https://doi.org/10.1016/j.resconrec.2019.104498.

EIA: *Refinery utilization and capacity*, 2021a. https://www.eia.gov/dnav/pet/pet_pnp_unc_dcu_nus_m.htm. Accessed January 2022.

EIA: *Petroleum and other liquids-imports*, 2021b. https://www.eia.gov/dnav/pet/pet_move_imp_dc_NUS-Z00_mbbl_a.htm. Accessed January 2022.

EIA: *Petroleum and other liquids-exports*, 2021c. https://www.eia.gov/dnav/pet/pet_move_exp_dc_NUS-Z00_mbbl_a.htm. Accessed January 2022.

Ellen MacArthur foundation: *Circular economy indicator methodology- an approach to measuring circularity*, 2019 https://emf.thirdlight.com/link/3jtevhlkbukz-9of4s4/@/preview/1?o.

Gracida-Alvarez U.R., Benavides P.T., Muhuri P., Lee U., Wang M. (2022a) Life-cycle analysis of plastic-to-plastic pathways via pyrolysis. In preparation.

Gracida-Alvarez UR, Xu H, Benavides PT, Wang M, Hawkins TR: *A circular economy sustainability analysis framework for plastics: Application for polyethylene terephthalate (PET)*, 2022b. Under review.

Jambeck JR, Geyer R, Wilcox C, et al.: Plastic waste inputs from land into the ocean, *Science* 347:768–771, 2015.

Jeswani H, Kruger C, Russ M, et al.: Life cycle environmental impacts of chemical recycling via pyrolysis of mixed plastic waste in comparison with mechanical recycling and energy recovery, *Sci Total Environ* 769, 2021, 144483. https://doi.org/10.1016/j.scitotenv.2020.144483.

Kawecki D, Scheeder PRW, Nowack B: Probabilistic material flow analysis of seven commodity plastics in Europe, *Environ Sci Technol* 52(17):9874–9888, 2018. Available at https://pubs.acs.org/doi/10.1021/acs.est.8b01513.

Khot M, Basavarajappa S: Plastic waste into fuel using pyrolysis process, *Int Res J Eng Technol* 04(09):1–4, 2017. Available at https://www.irjet.net/archives/V4/i10/IRJET-V4I1060.pdf.

Kunwar B, Cheng H, Chandrashekaran S, Sharma BK: Plastics to fuel: a review, *Renew Sustain Energy Rev* 54:421–428, 2016. Available at: https://doi.org/10.1016/j.rser.2015.10.015.

Law KL, Moret-Ferguson S, Maximenko N, et al.: Plastic accumulation in the North Atlantic subtropical gyre, *Science* 329:1185, 2010.

Lebreton L, Slat B, Ferrari F, et al.: Evidence that the great Pacific garbage patch is rapidly accumulating plastic, *Sci Rep* 8:1–15, 2018. https://doi.org/10.1038/s41598-018-22939-w.

Liang C, Gracida-Alvarez UR, Gallant ET, et al.: Material flows of polyurethane in the United States, *Environ Sci Technol* 55(20):14215–14224, 2021. Available at: https://pubs.acs.org/doi/full/10.1021/acs.est.1c03654.

LNG Publishing Company, Inc: *Lubes'N'Greases Factbook 2021–2022*, 13th edn, 2021. Available at https://www.lubesngreases.com/product/lubesngreases-factbook/.

NAPCOR (National Association for PET Container Resources) & APR (Association of Plastic Recyclers) (2018) Report on Postconsumer PET Container Recycling Activity in 2017. Available at http://www.plasticsmarkets.org/jsfcode/srvyfiles/wd_151/napcor_2017ratereport_final_1.pdf.

Nicholson SR, Rorrer JE, Singh A, et al.: The critical role of process analysis in chemical recycling and upcycling of waste plastics, *Annu Rev Chem Biomol Eng* 13, 2022. 14.1–14.24 https://doi.org/10.1146/annurev-chembioeng-100521-085846.

Phanisankar BSS, Vasudeva RN, Manikanta JE: Conversion of waste plastic to fuel products, *Mater Today Proc* 33:5190–5195, 2020. https://doi.org/10.1016/j.matpr.2020.02.880.

Ramsey P: *Multigrade oil—to use or not to use*, 2001, Available at https://www.machinerylubrication.com/Read/204/multigrade-oil.

Schwarz AE, Ligthart TN, Bizarro DG, De Wild P, Vreugdenhil B, van Harmelen T: Plastic recycling in a circular economy; determining environmental performance through an LCA matrix model approach, *Waste Manag* 121(2021):331–342, 2021.

Upasani S, Walzberg J, Ravikumar R, et al.: *Mapping the opportunity space to model the circular economy using tools funded by the DOE office of energy efficiency and renewable energy*, 2022, Available at https://www.nrel.gov/docs/fy22osti/81693.pdf.

Walzberg J, Lonca G, Hanes RJ, Eberle AL, Carpenter A, Heath GA: Do we need a new sustainability assessment method for the circular economy? A critical literature review, *Front Sustain* 1(620047):1–20, 2021. https://doi.org/10.3389/frsus.2020.620047.

Further reading

EPA: *U.S. National recycling goal*, 2020, Available at https://www.epa.gov/recyclingstrategy/us-national-recycling-goal.

Index

Note: Page numbers followed by "*f*" indicate figures and "*t*" indicate tables.

R

S

T

FSC
www.fsc.org
MIX
Papier | Fördert gute Waldnutzung
FSC® C083411

Kolibri 360 GmbH
Ferdinand-Jühlke-Straße 7
99095 Erfurt, Deutschland
produktsicherheit@kolibri360.de